# Methods and Materials in Microelectronic Technology

## THE IBM RESEARCH SYMPOSIA SERIES

COMPUTATIONAL METHODS IN BAND THEORY
Editors: P. M. Marcus, J. F. Janak, and A. R. Williams

COMPUTATIONAL SOLID STATE PHYSICS
Editors: F. Herman, N. W. Dalton, and T. R. Koehler

SPARSE MATRICES AND THEIR APPLICATIONS
Editors: D. J. Rose and R. A. Willoughby

COMPLEXITY OF COMPUTER COMPUTATIONS
Editors: R. E. Miller and J. W. Thatcher

COMPUTATIONAL METHODS FOR LARGE MOLECULES AND LOCALIZED STATES IN SOLIDS
Editors: F. Herman, A. D. McLean, R. K. Nesbet

ION IMPLANTATION IN SEMICONDUCTORS AND OTHER MATERIALS
Editor: Billy L. Crowder

STIFF DIFFERENTIAL SYSTEMS
Editor: Ralph A. Willoughby

OPTIMAL ESTIMATION IN APPROXIMATION THEORY
Editors: Charles A. Micchelli and Theodore J. Rivlin

TOPICS IN SURFACE CHEMISTRY
Editors: Eric Kay and Paul S. Bagus

ADVANCES IN DIGITAL IMAGE PROCESSING: Theory, Application, Implementation
Editor: Peter Stucki

THE PHYSICS AND CHEMISTRY OF LIQUID CRYSTAL DEVICES
Editor: Gerald J. Sprokel

COMPUTATIONAL METHODS IN CHEMISTRY
Editor: Joachim Bargon

DETERMINATION OF SURFACE STRUCTURE BY LEED
Editors: P. M. Marcus and F. Jona

METHODS AND MATERIALS IN MICROELECTRONIC TECHNOLOGY
Editor: Joachim Bargon

# Methods and Materials in Microelectronic Technology

Edited by
Joachim Bargon
*IBM Research Laboratory*
*San Jose, California*

PLENUM PRESS • NEW YORK AND LONDON

Library of Congress Cataloging in Publication Data

International Symposium on Methods and Materials in Microelectronic Technology (1982: Bad Neuenahr-Ahrweiler, Germany)
Methods and materials in microelectronic technology.

(The IBM research symposia series)
"Proceedings of the International Symposium on Methods and Materials in Microelectronic Technology, held September 29–October 1, 1982, in Bad Neuenahr, Federal Republic of Germany"—T.p. verso.
Includes bibliographies and index.
1. Integrated circuits—Very large scale integration—Design and construction—Congresses. 2. Photolithography—Congresses. 3. Ion beam lithography—Congresses. 4. X-ray lithography—Congresses. I. Bargon, Joachim. II. Title. III. Series.
TK7874.I5924 1982 621.381′7 84-13352
ISBN 0-306-41803-7



Proceedings of the International Symposium on Methods and Materials in Microelectronic Technology, held September 29–October 1, 1982, in Bad Neuenahr, Federal Republic of Germany


A Division of Plenum Publishing Corporation
233 Spring Street, New York, N.Y. 10013



Printed in the United States of America

# PREFACE

The papers collected in this volume were presented at the International Symposium on Methods and Materials in Microelectronic Technology. This symposium was sponsored by IBM Germany, and it was held September 29 - October 1, 1982, in Bad Neuenahr, West Germany.

The progress of semiconductor and microelectronic technology has become so rapid and the field so sophisticated that it is imperative to exchange the latest insight gained as frequently as it can be accomplished. In addition, it is peculiar for this field that the bulk of the investigations are carried out at industrial research and development laboratories, which makes some of the results less readily accessible. Because of these circumstances, the academic community, which among other things, is supposed to communicate the progress in this field to students of different disciplines, finds it rather difficult to stay properly informed.

It was the intent of this IBM sponsored symposium to bring together key scientists from academic institutions, primarily from Europe, with principal investigators of the industrial scene. Accordingly, this symposium exposed technologists to scientists and vice versa. Scientific advances often lead directly to technological innovations. In turn, new technologies are often arrived at empirically and, because of that, are initially poorly understood. Scientific inquiry then attempts to probe these processes and phenomena in order to achieve a better understanding. Thus science and technology are intricately interconnected, and it is important that technical exchange between technologists and scientists is facilitated, since the problems are typically interdisciplinary in nature.

Bad Neuenahr, a spa in West Germany, initially seemed to have nothing in common with microelectronics. The conference attendees soon found out, however, that this town is quite familiar with 'bits' and 'chips', whereby the former stands for a particular type of German beer, and the latter referred to the gambling casino next door to the conference hall.

We would like to express our appreciation for the support obtained by IBM Germany and to Plenum for commissioning this book. We want to thank Mrs. Erika Röhm, from Stuttgart, Germany, for her professional support as conference secretary in Bad Neuenahr. J. B. expresses his gratitude for the support received while at the IBM Research Laboratory in San Jose, California, in particular from Mrs. Pauline Hodnick, his secretary, who helped with the editing and the revisions of the manuscripts. He also wishes to thank his daughters Susanne and Gesine for generating the subject index.

Joachim Bargon
IBM Research Laboratory
San Jose, California
Symposium Chairman

Paul Schweitzer
IBM Germany
Sindelfingen
Symposium Manager

CONTENTS

# IMPACT OF MICROELECTRONICS - TECHNICAL, ECONOMICAL AND SOCIAL ASPECTS

Dieter Seitzer

Universität Erlangen-Nürnberg
Lehrstuhl für Technische Elektronik
D-8520 Erlangen

## INTRODUCTION

Scientists make research, politicians carry responsibility? This is a provocative question. The purpose of this contribution is to improve the necessary information exchange between those who develop microelectronics technology and those who are affected by it. By communicating its various implications in industry, economy and society, the objective of an improved sharing of the function of technological development and responsibility for the consequences can be approached. Aspects of technology itself, especially those of physical limitations that are the subject of companion papers, are excluded from consideration. Extensive use is made of source material presented in "Microelectronics and Society - For Better or For Worse"[1], of a series of public papers edited under the title "Microelectronics, Information, Society"[2] and an earlier paper.[3]

## APPLICATIONS OF MICROELECTRONICS

Disregarding the subjects of companion papers such as the significant areas of materials, structures, devices and circuits there are basic issues on the systems level that are sometimes put under the notion of the "VLSI dilemma":

1. The requirement for large volumes (Fig. 1)
2. The development cost or the "Moore Law" (Fig. 2)

3. The packaging, access or pin problem (Fig. 3)
4. The interface between IC manufacturer and user, or between IC components and systems houses.

All 4 issues are indications of the level of complexity that has been reached in VLSI today. Large volumes in the order of 1 million chips per year are needed per VLSI circuit, because the increasing initial investment has to be shared by large volumes of circuit production in order to arrive at sufficiently low cost per circuit for the final user. According to G.E. Moore[4] the design cost in person months doubled in every 2-2/3 years over the last 2 decades and continue to do so. Including inflation, the cost doubles every 2 years. In order to avoid losing the increased density achievable on the chip level, by bonding, interconnection and packaging, the number of circuits accessible per pin has to be increased sufficiently. The solution leads to communication-like structures employing multiplexing, switching that share the package interconnections efficiently. The interface to the user represents a manyfold problem: It is not only a question of specifications but also of know-how exchange or protection and, last but not least, a question of rear and forward integration as pointed out later. The best answer to all the 4 problems in terms of a circuit is the memory circuit: It is flexible in application such that large volumes are needed, its design profits from a modular structure, the number of pins shows a logarithmic increase with the size of the memory matrix, and it is easy enough for each user to be sold ready from the shelf. However, the market place is extremely competitive leading to dramatic rates of change and price drops.

## Possible areas of VLSI application

The potential of VLSI technology is tremendous and includes further cost decrease, if the requirements mentioned before can be met. However, it is by far not yet used to its full extent. The world market in IC's was around 17 • $10^9$ DM in 1980 and is estimated to double until 1985. It is only a fraction of about 5% of electronic equipment production. So there is a high multiplication factor of added value for the IC market. In a study, it is estimated that 25000 applications of electronic circuits in data processing, telecommunications, consumer and control are possible out of which only about 5% have been developed until 1980. So it seems justified to look briefly on those possible application areas. Fig. 4 shows a survey of examples of electronic equipment and system applications structured into 8 sections.[5] The classical areas of processional applications will continue to develop and grow through permanent cost/performance improvement of electronic circuits. In telecommunications, new services such as videotext will be

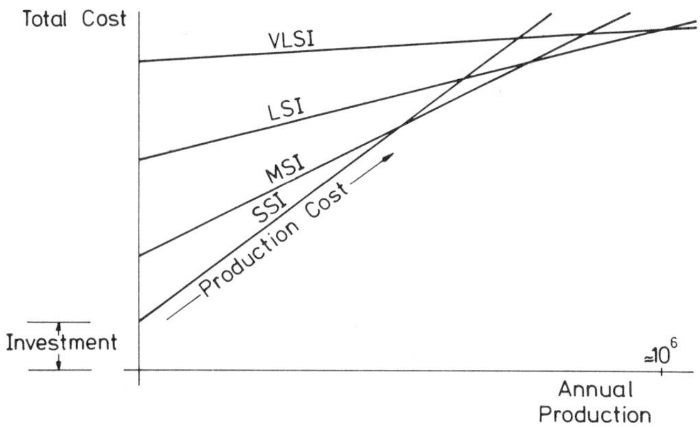


Fig. 1: Cost versus annual production volume of integrated circuits for various technologies

VLSI: Very large scale integration
LSI: Large scale integration
MSI: Medium scale integration
SSI: Small scale integration

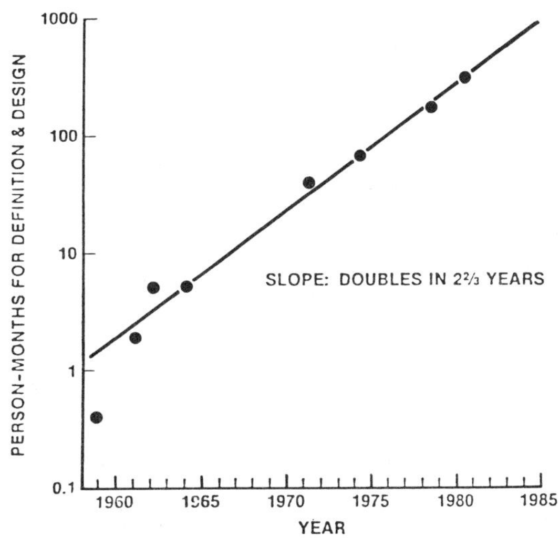


Fig. 2: Increase of integrated circuit development time versus time.[4]

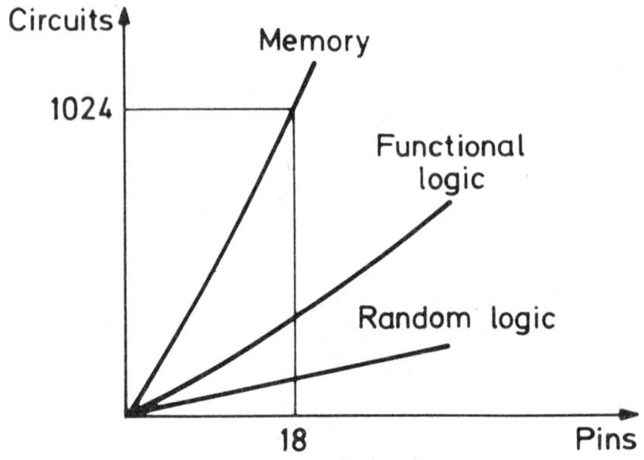


Fig. 3: Number of circuits that can be accessed by package pins for various types of circuits.

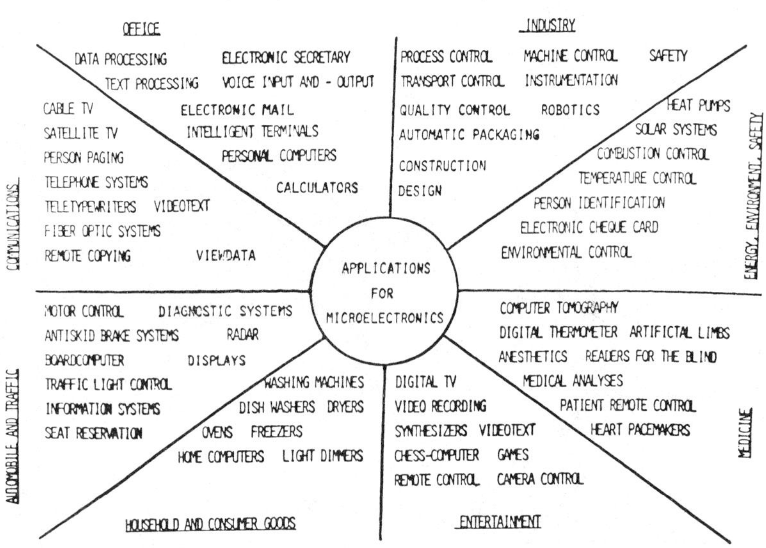


Fig. 4: Survey of possible application areas for integrated circuits.[5]

offered and new media such as optical fibres for communications as well as broadcast satellites, will increase the market also of IC's. Speech synthesis that has achieved reasonable quality levels, can add new features to information services. The merging of telecommunications and information processing will help to distribute intelligence and control.

Industrial applications will see the substitutes of mechanical by electronic apparatus, in the first place. After substitution, it is observed that new functions are added that were previously too difficult or too expensive to build. As a third step, new concepts of equipment occur that are based on information technology. New requirements such as saving of energy and protection of environment but also of strategic objects, such as power plants, surveillance of water and air pollution, will require new telemetry and related equipment. Medical applications will range from electronic aids to powerful tools for the doctors diagnosis and treatment. Last but not least, and here the multiplication factor is most obvious, each individual is a potential and manifold user for electronics. It started with electronic watches and pocket calculators and continues with electronic games, personal computers, kitchen and home appliances and the automobile as an electronic system.

## The need for standardisation

At first glance, the list of applications looks very impressive. Further inspection, however, reveals the fact that only few of them are feasible economically under the 4 constraints of the VLSI dilemma, i.e., meet the volume requirement to justify a custom design. Most of them are viable only if the circuits can be shared by other applications. Thus, standardisation becomes a key to the application of microelectronics. It can be broken down into 4 functional components:

1. Standardisation of hardware
2. Standardisation of signals, or data representation
3. Standardisation of functional description, or programming
4. Standardisation of operator functions

This separation is based on the hypothesis that all information handling equipment can be represented in the form of a piece of hardware (electronic set up) that accepts an input signal upon which it reacts by an output signal according to a function that is determined by its structure. The function eventually can be modified by an operator or upon external environmental conditions (Fig. 5a).

The ideal case for the hypothesis is, nearly trivial to say, a computer itself, but also the whole field of information processing is a model for it. In the case of a microcomputer, the hardware is standardised in terms of CPU, memory, input/output blocks. It processes data between input and output under program control that can be changed by the user. The success of information processing is based similarly only on its modularity and universality: It has managed to make its equipment, in the form of the computer, universal and widely independent of its application, the information is structured in the universal form of numbers and letters that can be combined to groups according to standard rules given as programming languages, and the external control is operated via standard equipment (keyboards, display).

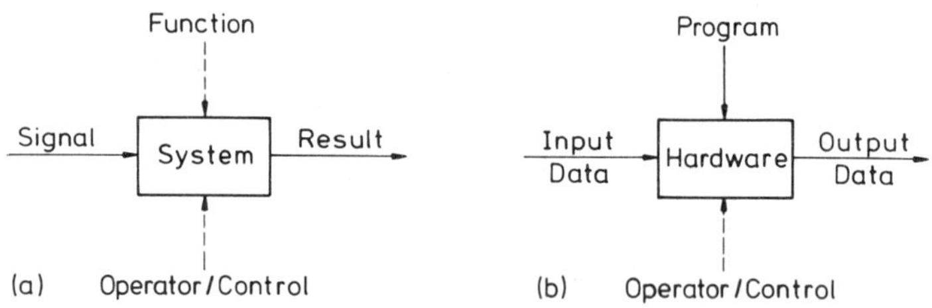


Fig. 5: Structure of information handling equipment
a) General piece of electronic equipment
b) Computer as an example

It is essential for all other applications to adopt the data processing structure as a model to organise their required functions. Communications and process control have achieved it by application of pulse code modulation to their analog signals. Continuous signals such as sounds, images and physical quantities can be replaced by samples taken in a distance in time specified by the signal bandwidth. After modulation they are sequences of ONES and ZEROS and can be processed in the same manner as data by digital hardware. The "personalisation" or "customisation" of the function is done by the program, i.e., by software. Therefore, software is the remaining major problem, and it is the progress of software as a technology that becomes the key to many sophisticated applications. With increased volume of application of standard hardware, a customisation on each level can be observed, i.e., one chip processors, special purpose processors (digital filters), controllers, etc. Automtion will also become the remedy to solve the second major problem of VLSI design, layout and development cost of IC technology itself. The design of VLSI circuits itself has to make better use of the capabilities of information technology standardising its tools to the extent that potential users can do their design themselves. The utmost efficiency in using silicon area has to be traded against the efficiency and correctness of designing the circuit automatically in order to minimize overall cost. The purpose is to achieve faster turnaround and economical feasibility also for lower volume applications. The thesis is that the leading edge applications can develop only if the base for microelectronics use is widened correspondingly. The idea of structured design by simplifying design rules and layout, by using standard cells, gate arrays or programmed logic arrays, is nothing else than standardisation on various levels.

The need for standardisation is the entrance fee to the economic proliferation of microelectronics. It means to restrict freedom and design cleverness and to sacrifice optimisation in favour of independent modules of design and fabrication that can be automated, individually tested, cascaded, and universally applied to justify automation and to manage complexity. This process leads to "integration by integration,"[6] i.e., the integration of circuits leads to an integration of industries, such as mechanical and electrical, telecommunications and data processing etc., but also of services such as business and public administration.

## IMPACT ON INDUSTRY

Microelectronics will affect the industry in two ways:

1. Structure and operation of enterprises
2. Workers, workplace and work itself.

The classical structure of an enterprise is characterised by the way that a product takes through development, production, marketing, and administration. Microelectronics have introduced significant changes in each of these functional areas; its major impact, however, is on production and organisation. The development of VLSI circuitry made it possible to build an increasing number of originally separate functions into one tiny component, the integrated circuit chip. The value added in manufacture has decreased from 70% to about 50% when going from electromechanical to digital electronic switching equipment, the rest to 100% being 13% for components and 17% of material previously and

42% for components, 7% for material today. Other examples are: 1000 assembly steps for a mechanical watch, only 5 components now. The production of an electromechanical teleprinter takes 75 hours, an electronic one needs only 17.7 hours. In this case, a microprocessor replaces 936 mechanical parts. (Fig. 6)

A mechanical calculator needs 9 hours of assembly, an electronic one 1 hour. Within a sewing machine a microprocessor replaces 350 mechanical parts, one CNC machine two skilled workers, one robot up to 4 workers. Consequently, there are heavy cuts in the manpower needed: The production personnel fell from 45% to 25-30% between 1970 and 1980, a major U.S. telecommunications manufacturer reduced its labour force from 39000 in 1970 to 17000 in 1980, a similar UK manufacturer from 88000 in 1974 to 65000 in 1979 and a case from Sweden gives figures of 15300 in 1975 to 10300 in 1978.[7] As an action of defense, the big-product manufacturers apply a "rear integration" strategy into the components field. On the other hand, microcomponent manufacturers, through "forward integration," penetrate into the equipment and system manufacturers market (Fig. 7).

As pointed out in the section on applications, the structure of the hardware equipment has to be standardised to a large extent; they become "black boxes," whereas specialisation or customisation is achieved by software. Due to the accelerated innovation rate of microelectronics, the average lifetime of products went down from about 10 years for mechanical equipment to 5 years for computers or even to 2 years for minicomputers.

The phenomenon of black boxes, on the other hand, creates new flexibility in production. Traditional assembly lines disappear to make room for more integrated ones. This, in turn, establishes associated manpower structures of assembly units of 20 ... 30 persons that are responsible for a complete product including testing and repair. On a larger scale this leads to new opportunities for small companies that are flexible enough to adapt themselves to short product cycles and rapid innovation.

These circumstances lead to radical changes also in relative costs: More for development and planning (from 1-2% to 6-10%), more toward service oriented rather than manufacturing companies. The composition of the labour force changes (Fig. 8); in the case of a telex machine manufacturer the amount of qualified workers reduced from 82% to 35%, whereas engineers and technicians increased from 5 to 30%.[7]

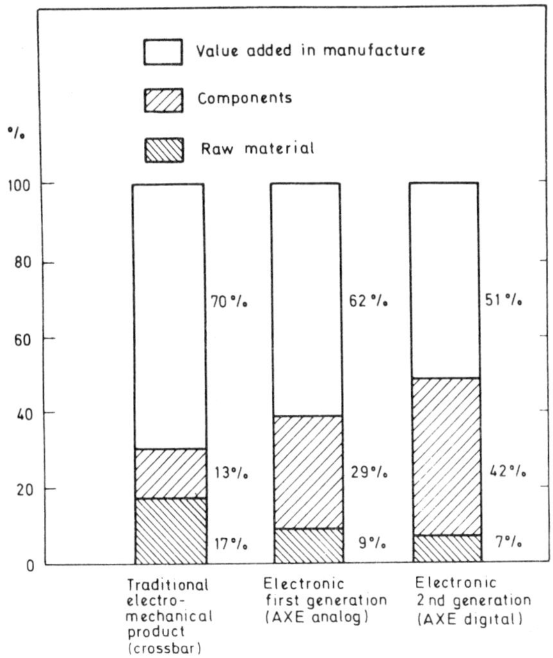


Fig. 6: Structure of manufacturing cost when going from mechanical to electronic equipment.[7]

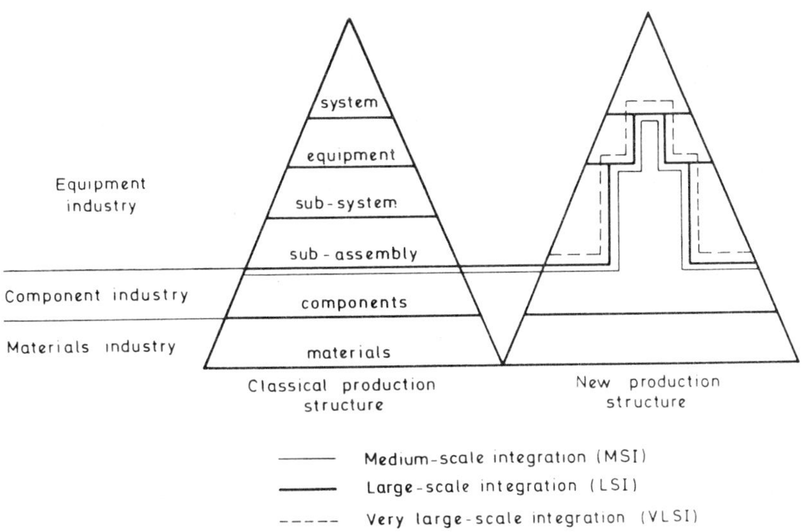


Fig. 7: Phenomenon of forward integration due to circuit integration.[7]

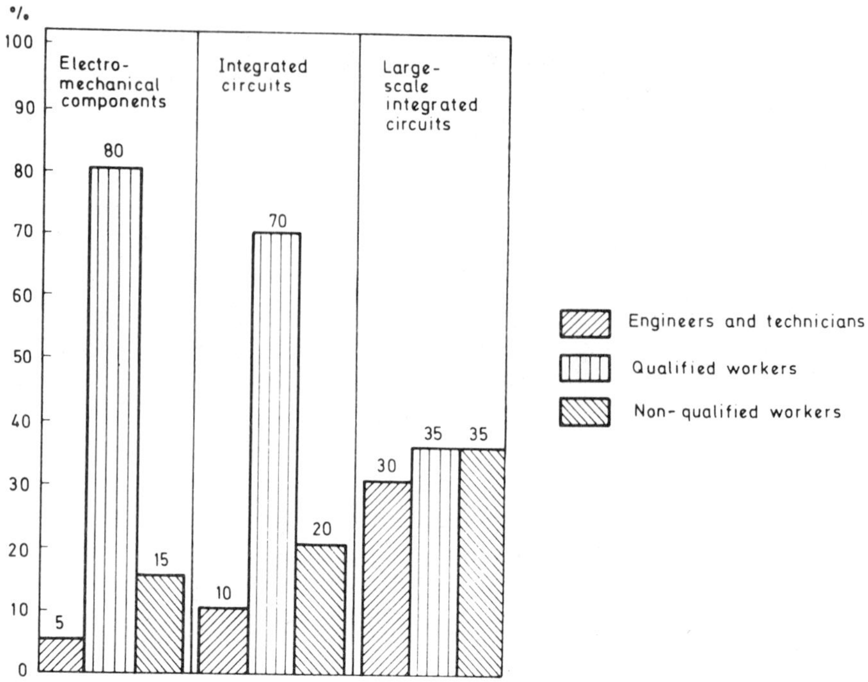


Fig. 8: Change in the composition of the labour force when going from mechanical to integrated electronic equipment.[8]

Besides the effect of reducing the numbers of jobs, microelectronics has effects on the quality of the work to be done. The job content, its meaningfulness, the working environment, (health, safety, supervision) the learning involved, social contacts etc., are changed. A special case of concern are visual displays based on CRTs, with keyboards. Their use can have ergonomic effects such as eye-strain, stress, fatigue, headaches, and even increased isolation as well as the feeling of being supervised in performance by counting the number of input/output actions, error corrections etc.

## IMPACT ON ECONOMY

Microelectronics is considered a new key technology, comparable in its effect to other technologies introduced previously such as the steam engine, the railway, electricity, telecommunication etc., to quote some examples. A key technology has a number of features that are met by microelectronics:

1. The potential applications are very widespread such that all or many areas of the total economy are affected. Microelectronics, in this context, can be said to have one major impact: Automation in production and administration. It will make production so cheap and flexible that automation will proliferate also into small size companies. At the same time, office work will be affected so heavily, that public and administrative services, that are themselves the infrastructure of many organisations, will be modified tremendously.

2. The decrease of cost associated with microelectronics will drop substantially for still some time to come and will put heavy economical pressure in favour of its introduction.

3. The first innovators will have a significant competitive advantage. The relative share of markets might change significantly on a national and even international scale.

As an example, which will give this section a national touch, the case of machine industry in the Federal Republic of Germany will be considered:[9] The four major industrial groups in Germany are machine, electrical, motor vehicles and chemical industry. They represent about 40% of the revenue of the manufacturing industry. However, their export share is much higher than that proportion. Among them the machine industry has a leading position in that it holds a share of between 20% and 25% of all western countries about equal with the US and twice as much as Japan; together they have more than 2/3 of the world market.

A key section of machine industry is that of computer numerically controlled (CNC) machine tools. It is here where mechanics and electronics interface and overlap most, therefore the special term "mechatronics" has been created. The progress in this field depends heavily on the amount of microelectronics employed. For CNC machines, the production in 1981 in Germany was $1.7 \times 10^9$ DM, in Japan $3.0 \times 10^9$ DM, whereas 2 years before it was $1.0 \times 10^9$ DM in both countries. Another source quotes 22000 CNC-machines for 1980 in Japan, and 4700 in Germany.[10] These figures show the rapid growth in Japan with strategic consequences due to the penetration of CNC machines into other industrial areas.

A study carried out in the German machine industry[11] shows that 7% of the enterprises did not consider microelectronics as yet. 56% of them are in the phase of planning and information collection. 25% of them are in the phase of development and testing. 12% of them produce and sell products with microelectronics. Due to the long duration of development and construction, the danger of being too late becomes obvious. More pessimistic statements call the "machine industry" the "watch industry of the 80's."[10]

Another innovation due to microelectronics is the development of robots that can be used instead of human labour in assembly, joining and handling. A case examined again in Germany was the introduction of 10 robots in five companies for welding, paint spraying and handling of pieces: A total of 46 people in non-qualified jobs were affected; 7 become redundant, 28 moved out of the sector, 10 changed jobs within the sector, one moved into it. Another case study of 40 applications of robots in a major vehicle manufacturer showed that 1 robot replaced 4 workers and created one new job.

A further area affected by information handling equipment is office work that can be automated by text processing: Among 2.7 million office jobs considered, 42% can be formalised and 25% to 30% can be automated. In public administration the savings potential is even higher: 75% can be formalised and between 25% and 38% could be automated.[12]

Some of the figures and facts might be of interest in this international context, i.e., to compare the gross national product with production and application of integrated electronics:[5]

| | USA | Japan | Western Europe |
|---|---|---|---|
| Gross national product in fraction of all western countries | 1/3 | 1/8 | 1/4 |
| Integrated circuit production | 2/3 | 1/4 | 1/10 |
| Application of integrated circuits | 1/2 | 1/4 | 1/4 |
| Ratio application/gross nat. prod. | 1.5 | 2 | 1 |

Germany uses about 1/3 of Western Europe, about the same as France and England together. It is further interesting to compare the electronic equipment production by application fields for the 2 years 1978/1985,[5] (Fig. 9).

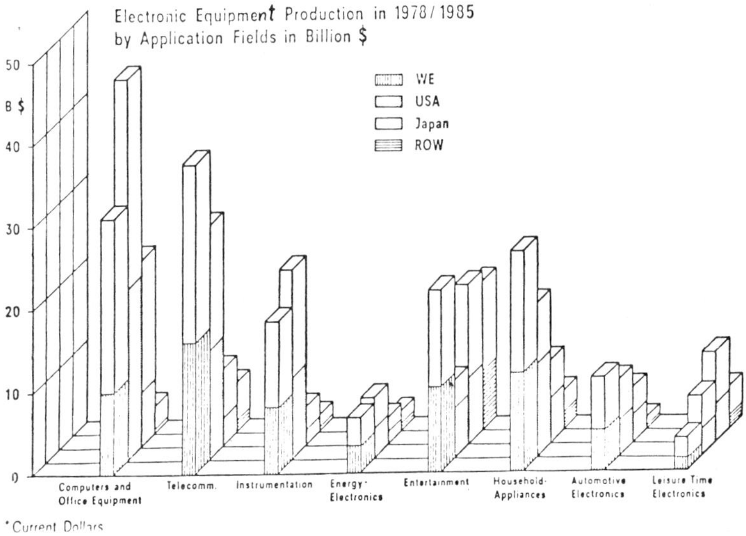


Fig. 9: Electronic equipment production in 1978/1985 by application fields and geographic distribution (1978: base of column, 1985: top).[5]

It shows the leading role of the US in computers, of Western Europe in telecommunications and of Japan in leisure time electronics and its growing share in entertainment electronics. The figure indicates also a severe problem: The gap to the rest of the world which will grow rather than reduce in the future.

## IMPACT ON SOCIETY

The impact of microelectronics on society will be operated through information technology. Information will become as important as material and energy. Major characteristics are the bridging of time by means of storage, bridging of space by means of telecommunication, automation and mechanisation of all functions that can be formalised and quantised. It is in this area that much speculation is done that can turn out to be right or wrong; it is in this area also, that the ambivalence of information usage will become obvious. However, some statements can be made that show tendencies. It is not to claim that things will or must happen in that way, but to strengthen our imagination in order to be able to act and react more thoughtfully.

Information systems will make more information available in a shorter amount of time to many users. Data banks will be available to the individual as well as to all types of organisations. Decisions are aided and public affairs can become more transparent. On the other hand, information, useful in the hand of a doctor, can be detrimental in other hands. Criminal information systems can help in protecting the individual; a national data center as well can be abused to supervise the individuals of a society to the extent of absolute control.

The quality of administrative services can be improved. On the other hand, direct contact between human individuals will be reduced leading to unresponsiveness of public administrations to the wants and needs of their clients. One salient case of complaint became publicly known: The former German Bundeskanzler Helmut Schmidt expressed his concern, because he could no longer read and understand his electricity bill!

This indicates two other aspects: The power balance among public and private organisations may change as well as the one between the bureaucracy and the individual. As well, the balance between rational thinking and non-rational creative thinking will be affected in favour of the first one. The computer is used as a projective screen, where simulation of issues taken from the real world takes place. This implies that this formalised information is removed from its pragmatic context while being processed. When the output is fed back, the results may distort the physical reality. At least a loss of flexibility as compared to human information handling will occur. So, vital decisions affecting survival should not be left to machines alone, they continue to require human intervention. Information technology does not itself create social effects, but it enlarges the weaponry of social forces. For instance, the desire for equality and more orderly arrangements is a powerful stimulus towards an ever more perfect administration. But we do not know how much degree of disorder we need to stay innovative and creative. Order, as we know from information theory, means loss of freedom for action and to make decisions.

The question is: What will happen to the individual human being? Will mankind become enriched by information or become isolated by watching screens? Steinbuch[13] introduces the term "informational deficiency": It means that man is unable to absorb the tremendous amount of information available to him due to limited mental processing and storage capabilities. It is a small computer that is permanently overloaded. In spite of this, man has survived and by being clever, has produced progress and useful results. By taking over what earlier generations had learned, he could continue and learn new things; by trusting other men's knowledge, abilities and skills, he multiplies his own power. So, the stronger artificial intelligence grows, man has to also develop tradition as a means to convey knowledge and human understanding and has to trust in other people as a means of defense of human behaviour.

One final observation has become obvious: A perfect administration relying on the availability of information, always and everywhere becomes vulnerable. While protection from abuse enforces centralization of storage, its vulnerability requires distribution.

What remains to be said: The pace of change is dictated less by the available technology than by cultural factors and social context. Therefore, it will differ widely from country to country and between the highly industrialised countries and the rest of the world. We should understand this circumstance as a challenge and a fair opportunity to master information technology!

## CONCLUSIONS

It has been attempted to assess the impact of microelectronics by devoting equivalent depth or superficial consideration to technical, economical and social aspects. It is obvious that microelectronics, as a base of information technology, carries an enormous potential to influence our lives in a similar way as material and energy.

There are some prerequisites to make full use of its potential: The large investments needed for sophisticated equipment and design labour have to be shared by large volumes of circuit fabrication. Therefore, a lot of effort has to be spent for opening sufficiently many applications and sufficiently fast to maintain the balance between investment and its return: The information flow between technologists, designers, users and educators has to be organised to avoid "application," design," "investment" or "education" gaps.

The possible consequences are manifold: The computer as a powerful instrument will serve as a model for structuring microelectronics technology itself, but also industry and the infrastructure of society: Standardisation, modularity, automation and formalisation will be the key features of this integration procedure.

The ambivalence of information technology will turn for better or worse in its social context: Automation will make free of physical labour, meaning the loss of many jobs as the reverse side of the coin. However, if microelectronics is a key base technology, it will create more jobs than it destroys as any other new technology did before (Fig. 10). So, the social context should be made to favour innovation by microelectronics rather than prevent its application. This is a tedious task because it is much

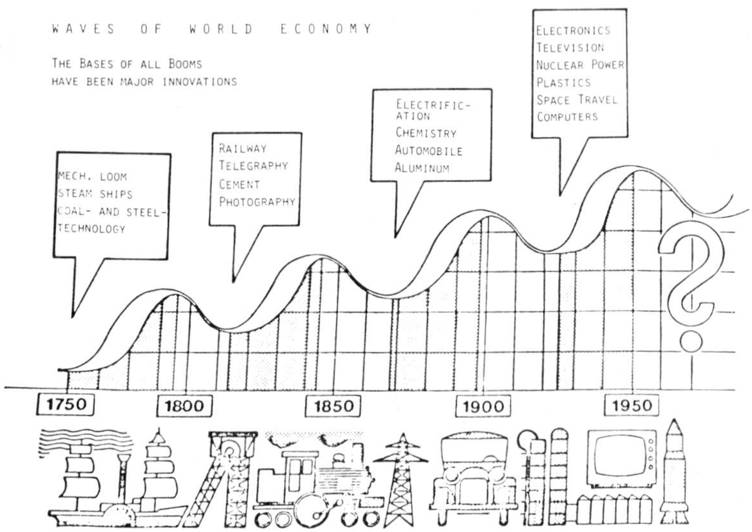


Fig. 10: Waves of world economy introduced by various technologies.

easier to predict how many jobs will get lost than to estimate how many innovations will occur.

Although information technology will favour formalised thinking and objective numerical assessment parameters, man should neither have fear nor compete in that aspect with the computer. Instead, we should concentrate ourselves to use the free time becoming available to develop creative imaginative thinking and to learn enough to keep pace with the rapid changes and movement of information technology.

## REFERENCES


1. G. Friedrichs, A. Schaff, (editors), "Microelectronics and Society," For Better or For Worse, Pergamon Press, Oxford 1982; deutsche Fassung: "Auf Gedeih und Verderb," Europaverlag, 1982.

2. H. Niemann, W. Schüssler, D. Seitzer, (Herausgeber), "Mikroelektronik, Information, Gesellschaft," Springer-Verlag, Heidelberg (to be published).

3. D. Seitzer, "Ein Beitrag zur Standortbestimmung der Mikroelektronik," Archiv der Elektrischen Übertragung Band 35 (1981), S.27-33.

4. G. E. Moore, "VLSI: Some Fundamental Challenges," IEEE Spectrum, April 1979.

5. E. Hofmeister, "Wirtschaftsfaktor Mikroelektronik-Nationale und Internationale Aspekte," Mikroelektronik, Information Gesellschaft, Springer-Verlag, Heidelberg.

6. N. C. de Troye, "Integration durch Integration," Elektrotechnik und Maschinenbau 95 (1977), S.287-292.

7. B. Lamborghini, "The Impact on the Enterprise," For Better or For Worse, Pergamon Press, Oxford 1982.

8. J. Evans, "The Worker and the Workplace," For Better or For Worse, Pergamon Press, Oxford 1982.

9. U. Thomas, "Technologie, Politik und Innovation," Mikroelektronik, Information, Gesellschaft, Springer-Verlag, Heidelberg.

10. H. Hinz, "Technischer Fortschritt im Zwielicht-Zur Technologie und Innovationspolitik der Gewerkschafter," Mikroelektronik, Information, Gesellschaft, Springer-Verlag, Heidelberg.

11. H.-J. Warnecke, "Entwicklungen in der Fertigungstechnik durch den Einsatz von Mikroprozessoren," Mikroelektronik, Information, Gesellschaft, Springer-Verlag, Heidelberg.

12. G. Friedrich, "Microelectronics and Macroeconomics," For Better or For Worse, Pergamon Press, Oxford 1982.

13. K. Steinbuch, "Die Informatisierung der Gesellschaft," Mikroelektronik, Information, Gesellschaft, Springer-Verlag, Heidelberg.

# MILESTONES IN SILICON SEMICONDUCTOR TECHNOLOGY

Erhard Sirtl

Heliotronic GMBH

Burghausen, Germany

## INTRODUCTION

Shortly after metallurgical-grade silicon (96-98 %) had been commercially available via the arc furnace process, selected samples of this polycrystalline product were the basis of a very early electronic device - the crystal detector. During the many empirical tests in his laboratory (by 1920 they literally exceeded the number of 30 000 different substances) Pickard[1] discovered the superiority of silicon compared to other substances in 1906 and shared the priority with two other independent investigators[2,3]. The strong evolution of vacuum tubes in the later decades seemed to restrict the use of such detectors to radio amateurs. Thirty years later, however, the difficulties in the high radio frequency area brought this simple rectifier system and an element called silicon into focus again.

## IMPROVED POINT CONTACT DEVICE TECHNIQUES

A survey of the stages of development between 1940 and 1960 is given in Fig. 1. Although reactions were known already in the early days of this crystal detector renaissance that would have provided enormously higher purity levels (CVD of $SiH_xCl_{4-x}$), rather simple metallurgical upgrading (leaching and remelting of MG-Si) had been the main choice.

Even this kind of treatment enabled the metallurgist already to be fairly reproducible in the average crystal quality after controlled solidification of the remelted material. Typical ingots, as they were obtained at Bell Labs (Fig.2), show a distinct separation

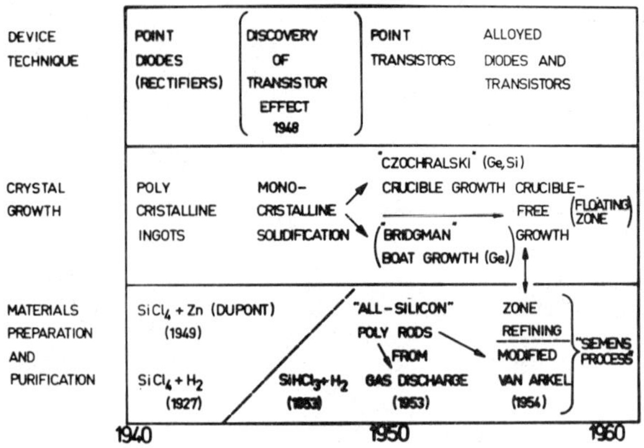


Fig. 1 History of high-purity silicon technology (1940-1960)

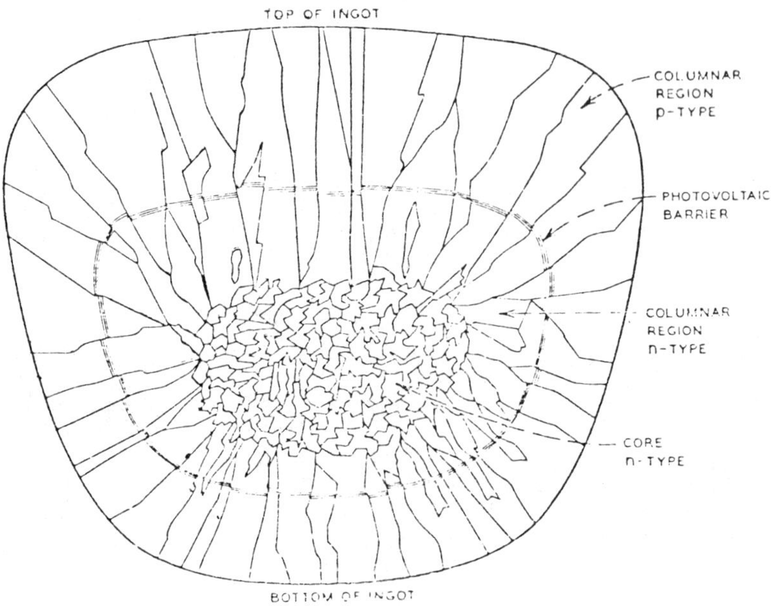


Fig. 2 Schematic of early silicon ingot (ref. 4)

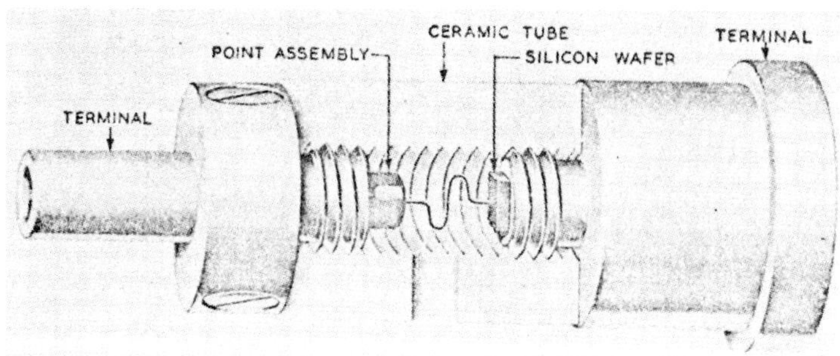


1N21

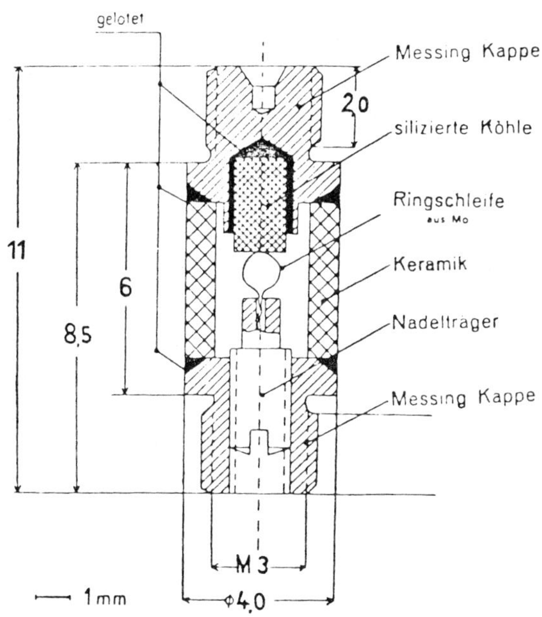


ED 705

Fig. 3 Typical airborne US (ref. 4) and German (ref. 5) rectifiers from World War II

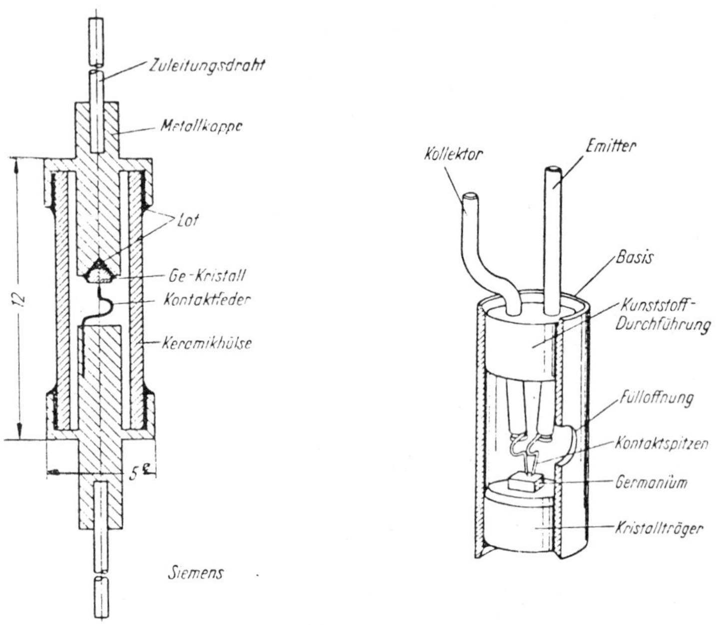


Fig. 4 Improved point contact diode (Siemens) and transistor (Bell), both on germanium basis (see ref. 11)

line between center and periphery after etching. Silicon platelets prepared from each zone showed inverse rectification behavior. Already in 1941 they were named n-type and p-type areas and identified (in the majority) as phosphorus- or boron-doped, respectively. Later on, ideas about the general rôle of III-V elements had been corroborated by intensional doping with Al, Ga, As and Sb[4].

There was fierce competition during World War II between Anglo-American and German technologists in this most important microwave battlefield. Surprisingly the concepts and set-ups had been quite similar on both sides which is shown by the comparison of a Bell and a Telefunken rectifier (Fig. 3). Further improvements were made after 1945, when silicon quality had been improved by the so-called Dupont process which was based on a vapor-phase reaction between $SiCl_4$ and Zn[6]. The discovery of the transistor effect[7] and progress in the understanding of p-n junctions soon established a new situation on the background of monocrystalline specimens. They had been grown first from germanium, but soon after the first small silicon rods were prepared by the Czochralski technique from quartz crucibles[8] or by crucible-free float-zoning[9,10]. Improved diodes based on this material and early types of transistors maintained the point contact system for some time as it is shown in Fig. 4. Increased knowledge was collected in those years on the nature of the material in close vicinity to the point contact interface.

## FIRST CONCEPTS BASED ON PLANAR P-N JUNCTIONS

As important as they have been for gaining fundamental knowledge, melt grown p-n or p-n-p (n-p-n) structures had no basis for economic device production (Fig. 5a). The alloyed-junction transistor has been a longer lasting system (Fig. 5b) although of limited range of applicability. The mesa technique (Fig. 5c) finally had brought the active part of silicon platelets to one side and initiated diffusion and epitaxy techniques.

Alloying techniques played a major part in power device production, whereas the mesa technique has truly been a precursor of the tremenduously successful "planar technique" based on the special properties of the native oxide layer of silicon and its photolithographic behavior.

Increasing demands in purity etc. initiated efforts in preparing a hyperpure silicon quality. This, in turn, has been a very good basis for fundamental investigations on the nature of many elements in the silicon lattice and their kinetics. This type of research work held on until it became overruled by the urgency of device-oriented problems due to the dramatic evolution in micro- and power electronics. Pertinent text- and handbooks still collect most of their data from work that has been published during this very fruitful scientific period.

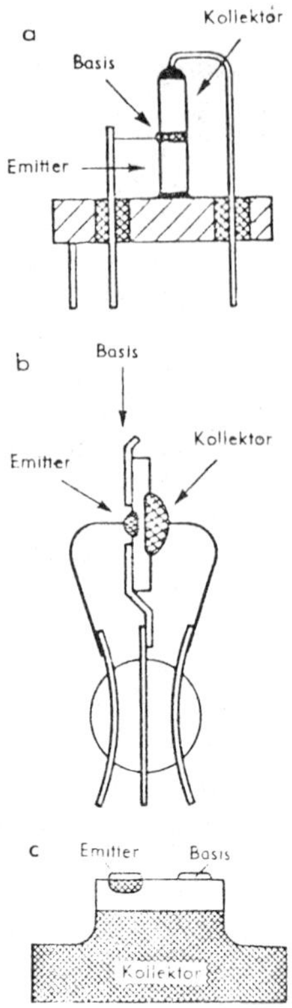


Fig. 5 Schematic of an as-pulled, alloyed-junction, and mesa transistor (ref. 13)

The early planar junction technology already had the benefits of operating with hyperpure CVD material[14] from the so-called Siemens process (see Fig. 1). Very soon it was accepted worlwide as a reliable starting material for further preparation of monocrystals either by the floating-zone or the crucible-pulling process. Figs. 6 and 7 show the basic operational steps of this concept - a technology that has not changed its general feature since in "conventional" electronics.

The increasing developmental efforts in single crystal growth from the melt were accompanied by defect characterization which concentrated in these years on grain and twin boundaries, edge dislocation (Fig. 8) and doping gradients. Nearly all of it was done on the basis of metallographic methods. The family of defects to be considered became more diversified by including epitaxial film deposition: Partial and screw dislocations, stacking faults, localized twins (Fig. 9) - all of them generated under experimental conditions that we had been calling "monocrystalline overgrowth" in Germany, before the word "epitaxy" (or in the later years: homoepitaxy) became the final technical term for it.

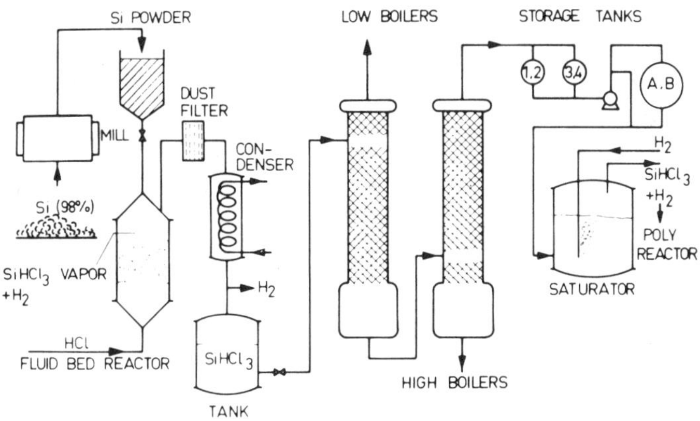


Fig. 6 Preparation and refining of trichlorosilane (flow chart), after ref. 12

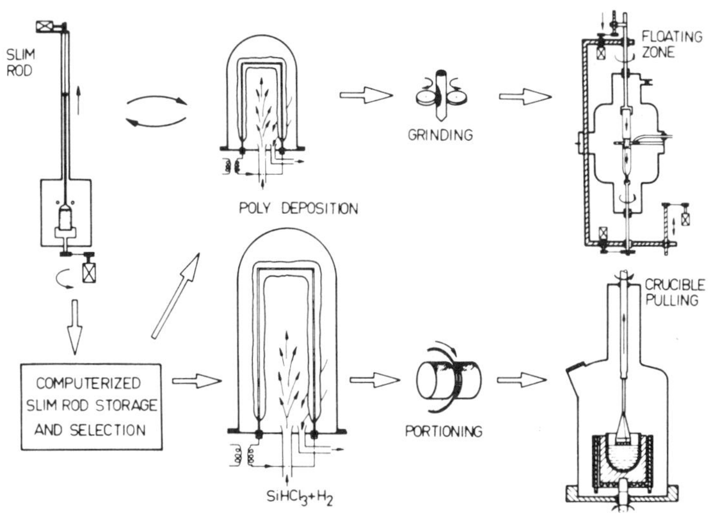


Fig. 7 Schematic of production of poly- and monocrystalline silicon (flow chart), after ref. 12

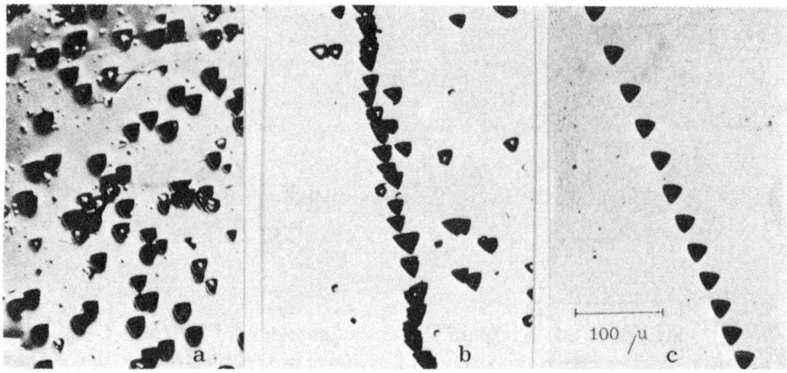


Fig. 8 Micrographs of dislocation patterns (ref. 12)
a) homogeneous distribution
b) low-angle grain boundary (lineage)
c) glide plane (slippage)

Preferentially etched, (111) plane

Fig. 9 Secondary growth twin configuration and stacking faults on (111)

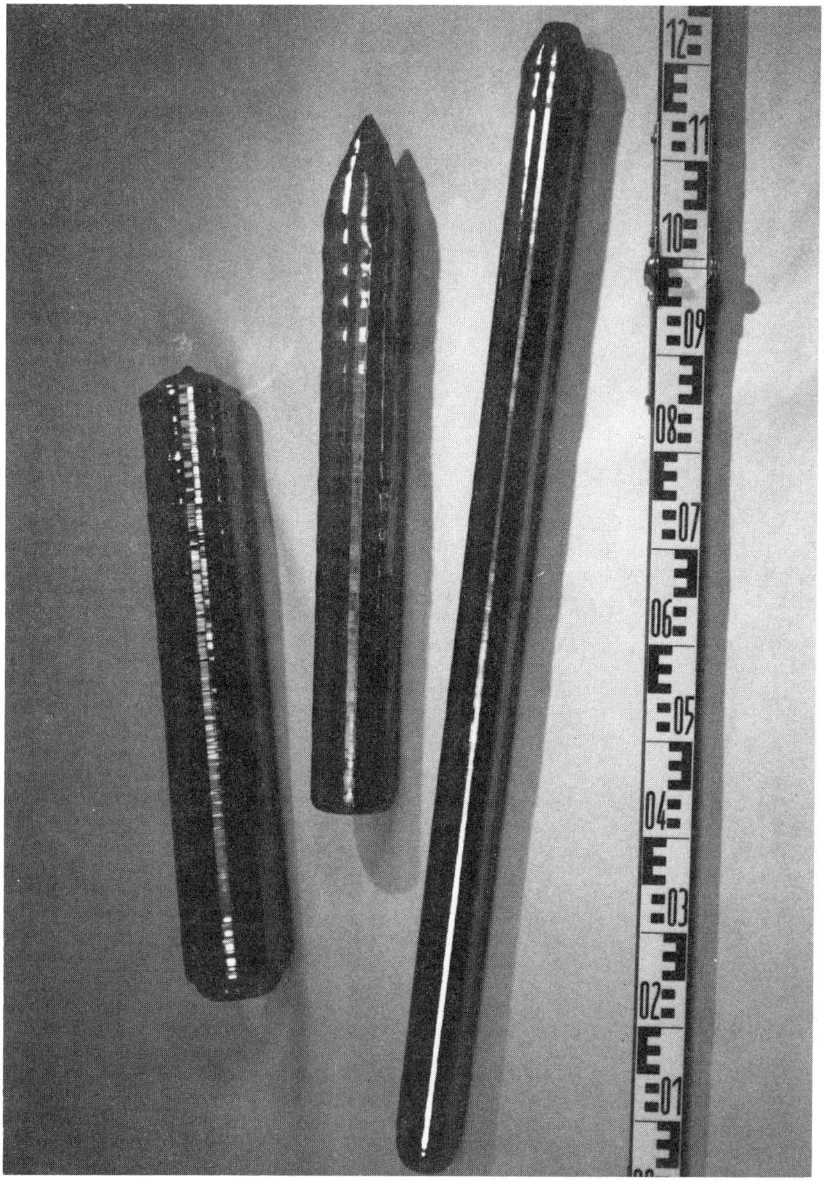


Fig. 10 Records of Wacker dislocation-free crystals in 1975. 3" and 4" dia: float-zoned; 5" Czochralski-pulled 10 units of calibration $\hat{=}$ 1 meter

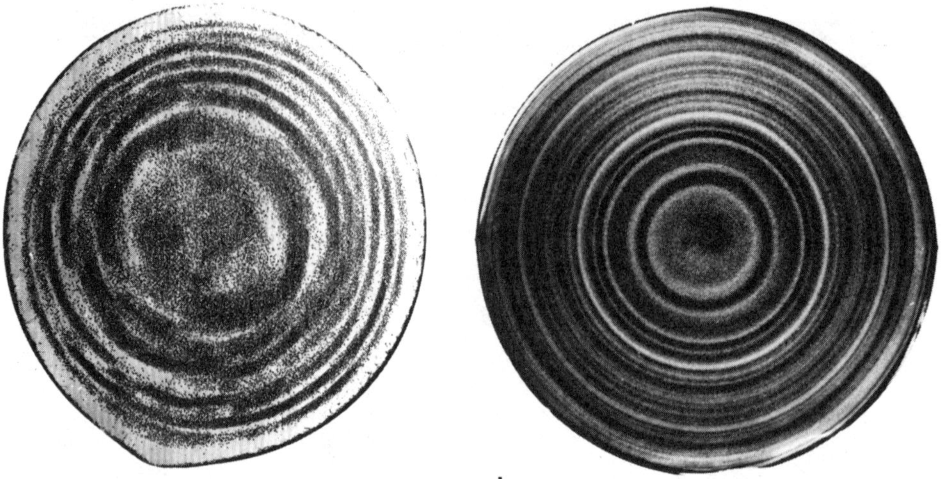

a b

Fig. 11 Swirl patterns in DF single crystals (cross-section)
a) float-zoned (FZ), X-ray topogram from ref. 12
b) Czochralski-grown (CZ), after preferential etching
(Courtesy of W. Zulehner, Wacker-Chemitronic)

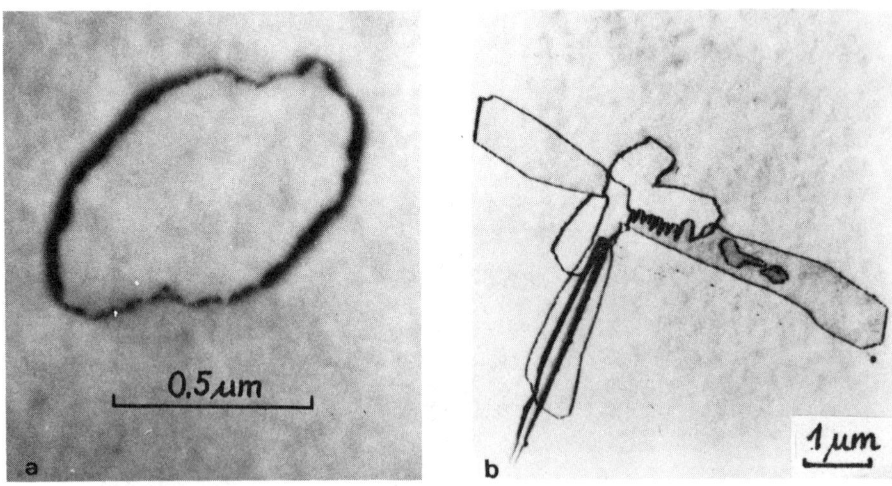


Fig. 12 a) Simple and b) complex version of dislocation loops
(swirl defects) in float-zoned crystals
(Courtesy of B. Kolbesen, Siemens A.G.)

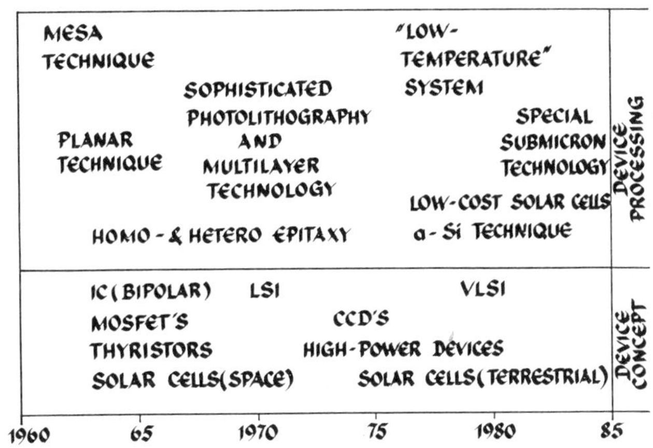


Fig. 13 Important fields of device research between 1960 and 1985

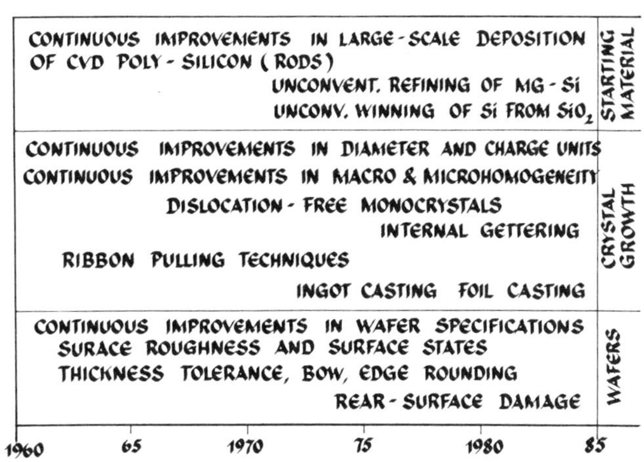


Fig. 14 Important subjects in material development between 1960 and 1985

MICROELECTRONICS AND HIGH-POWER DEVICES BASED ON SOPHISTICATED WAFER TECHNOLOGY

With the introduction of photolithography and the improvements in diffusion and thin film techniques different new routes in silicon technology became a necessity:

The silicon platelet had to be turned into a real wafer with gradually increasing dimensions. This has been possible only by switching from dislocated to dislocation-free specimens due to the basic phenomenon of an increasing agglomeration of edge dislocations to form low-angle grain boundaries when exceeding rod diameters of about 1". In addition, wafer (surface) quality conditions had to be improved tremendously owing to the trend of microminiaturization. Moreover, economical considerations including standardization (where possible) not only did support the call for large-diameter wafers, but also stimulated the growth of factories and equipments, which had been constructed more and more under the aspects of cost savings through industrial large-scale production.

This may be illustrated in the field of crystal growth by the dimensions of dislocation-free crystals depicted in Fig. 10. The maximum experimental diameters meanwhile have exceeded the dimensions depicted here by an inch or more and it remains to be seen whether this will lead to significant changes in future device processing technology. The big milestone in crystal growing is certainly the method of "necking" a given seed crystal prior to the formation of a bulk crystal of given diameter. The fundamental experiment of Dash[15] has been conveyed here into a standard technique where many kilograms of silicon are held by such a seed - sometimes as small as a few millimeters only - in a top-to-bottom (crucible pulling) or a bottom-to-top (floating zone) position.

In our eyes the highest challenge for both sides - the material supplier and the device processing engineer - has been the mastering of the entirely different defect situation ("swirl" and "haze") in dislocation-free (DF) silicon crystals. This applies to the tremendous sophisticated efforts in diagnostics and the different ways to overcome problems by preparative means as well. Typical macroscopic patterns on cross-sections of DF crystals are shown in Fig. 11. The swirl-like arrangement of many localized defects was responsible for name-giving here, but in many Czochralski-type crystals their appearance in form of concentric zones still is called a "swirl-defect" pattern. Whereas de Kock[16] has initiated a thorough investigation on the kinetics of their formation, the real nature of some categories of such localized defects had been revealed first by Föll et al.[17]. Some characteristic micrographs are shown in Fig. 12. We still have to deal with phenomena based on localized defects, particularly below the $\mu$m range (dislocation loops, bulk stacking faults, precipitates). Some of them appear under special

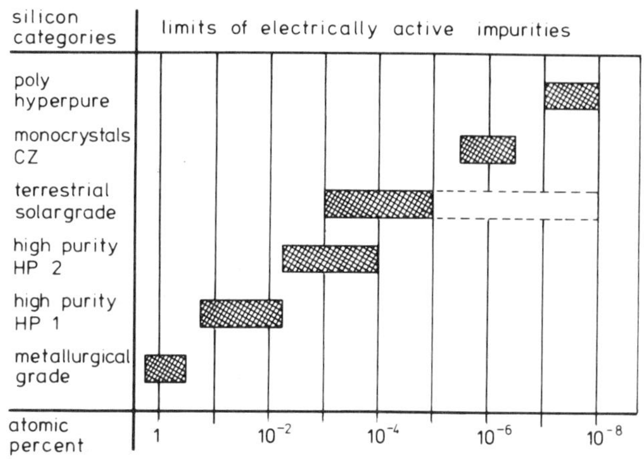


Fig. 15 Quality ranges for different silicon products HP 1 high purity, basic stage of refining; HP2 high purity, advanced stage of refining; TSG terrestrial solar grade, final stage of refining

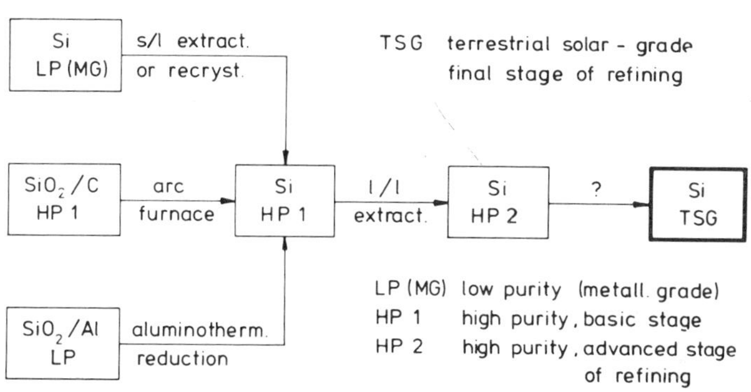


Fig. 16 Different approaches to large-scale production of Si for terrestrial solar cells (symbols see Fig. 13)

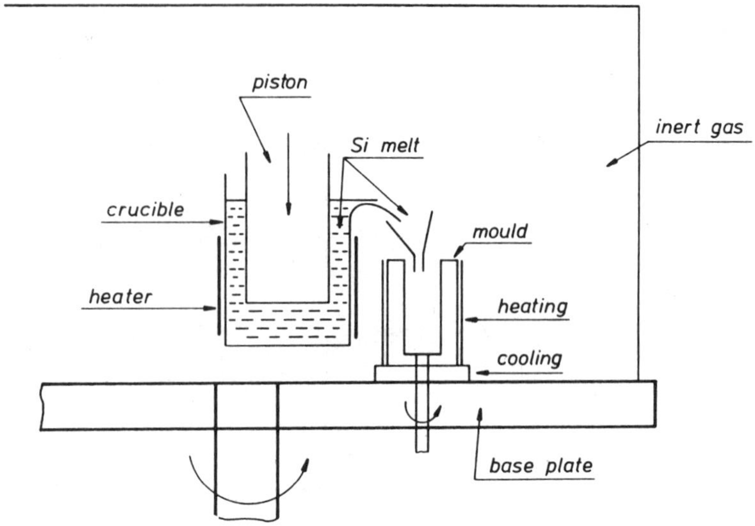


Fig. 17 Schematic view of an equipment for semicontinuous production of multigrain silicon ingots by mold casting

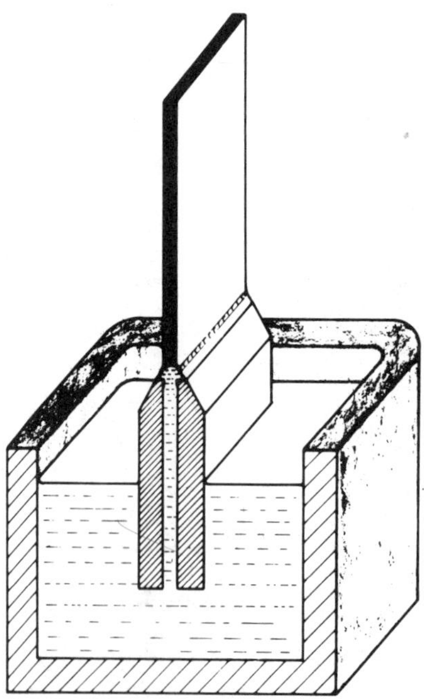

Fig. 18

Schematic graph of edge-defined film-fed growth (EFG) resp. capillary action shaping technique (CAST)

Fig. 19 Polycrystalline silicon slice
100 x 100 mm

conditions without intension, others had been deliberately created (internal gettering).

The emphasis in most of these studies either has been put on minimizing already such defects by appropriate modifications during material preparation or to annihilate (optimize) them through sophistication of the existing device processes. As the majority of defects today are developed during certain steps in device processing, the kinetics of their formation became a particularly interesting object for research scientists. This has led in the meantime to very promising investigations on the nature and concentration of intrinsic point defects and their interaction with the most common dopants and impurities.

Such knowledge may, in the long run, enable us to create the standard material for a future submicron device technology - an extremely homogeneous and reproducible silicon crystal. Other efforts, namely those trying to shift device processing conditions more and more from $T > \frac{2}{3} T_m$ (m melting point) toward $T < \frac{2}{3} T_m$ are considered to support this. Figs. 13 and 14 finally try to illustrate the milestones in device conceptualization between the years 1960 and 1985 in coordination with important material developments. It includes wafer processing technology which in part had to be modified later on to overcome the problems of localized defects.

## SOLAR CELL TECHNOLOGY (TERRESTRIAL)

In the charts of Fig. 13 and 14 typical keywords are representing a new technology package that is entirely different compared to practically every other applicational field. The reason for it cannot be explained only by the nature of an extremely large-sized discrete device - the solar cell. It is much more the necessity to develop completely new sub-technologies in every discipline for the "simple" purpose of drastic cost reduction. The target to create systems which are planned to compete with conventional power sources in the long run has become the hallmark of this new area of exploration.
In many of the technical approaches we have learned of meanwhile there is a strong resemblance with the silicon technology of the very early history of silicon as a semiconductor. For economical reasons alone some of the crude processes of the past have been reinvestigated. Besides promising purification studies under new aspects the usefulness of polycrystalline silicon in different textures has been the real surprise here. In combination with adjusted device techniques it seems to be a very promising basic concept for developing low-cost systems in photovoltaics.

To characterize the difference between conventional and inconventional technologies, four different figures selected from a review article[18] will be shown. Fig. 15 indicates the purity level of solar-grade silicon which is considered to meet future requirements, whereas Fig. 16 outlines the promising low-cost (low-energy) approaches to reach solar-grade quality through pyrometallurgical procedures.

In Fig. 17 a future continuous block-casting unit is depicted and the schematic drawing of a ribbon-pulling system based on capillary forces (Fig. 18) is representative here for many different methods to directly fabricate silicon sheet material. Finally, an actual-size silicon slice (Fig. 19) illustrates the special nature of materials used in this new and highly evolutionary technical field. At least during investigation of the interaction of intrinsic and extrinsic point defects benefitial overlapping can be expected here with the results of basic studies in advanced electronic-grade material. It can be hoped, too, that interdisciplinary coordination in process engineering will lead to interesting concepts that may be able to bridge the present gap between unconventional and conventional techniques.

## REFERENCES

1) G.W. Pickard, US Patent 836,531 (1906)
2) H.H.C. Dunwoody, US Patent 837,616 (1906)
3) L.W. Austin, Phys.Rev. 24, 508 (1907)
4) see review article of J.H. Scaff, Metallurg.Trans. 1, 561 (1970)
5) K. Seiler in "Naturforschung und Medizin in Deutschland 1939--1946" (FIAT-Reports), Vol. 15/16, I-5.2, p.272 ff.
6) D.W. Lyon, C.M. Olson, and E.D. Lewis, J.Electrochem. Soc. 96, 359 (1949)
7) J.Bardeen and W. Brattain, Phys.Rev. 74, 232 (1948)
8) G.K. Teal and E. Buehler, Phys.Rev. 87, 190 (1952)
9) P.H. Keck and M.J.E. Golay, Phys.Rev. 89, 1297 (1953)
10) R. Emeis, Z.Naturforsch. 9a, 67 (1954)
11) W. Heywang and H. Henker, Z. Elektrochem. 58, 283 (1954)
12) see e.g. H. Herrmann, H. Herzer, and E. Sirtl in "Advances in Solid State Physics", Vol. 15, H.J. Queisser (Ed.), Vieweg, Braunschweig, 1975, p. 279
13) H. Teichmann, "Halbleiter" Bibliographisches Institut, Mannheim, 1962
14) F. Bischoff, German Patents DBP 1 102 117 and 1 140 549 (both 1954)
15) W.C. Dash, J. appl. Phys. 29, 736 (1958)
16) A.J.R. de Kock, Phylips Res. Rep. Suppl. 1 (1973)
17) H. Föll, U. Gösele and B.O. Kolbesen, J. Crystal Growth 40, 90 (1977)
18) J. Dietl, D. Helmreich, and E. Sirtl, "Solar Silicon" in Crystals: Growth, Properties and Applications, Vol. 5, J. Grabmaier (Ed.), Springer, Berlin-Heidelberg, 1981, p. 43

# ELECTRONIC TRANSPORT IN SEMICONDUCTOR MATERIALS

Hans Joachim Queisser

Max-Planck-Institut für Festkörperforschung

7000 Stuttgart 80, Heisenbergstr.1, FRG

Recent developments in research and applications of electronic transport in semiconductors are reviewed. The possibility of achieving low carrier densities, the purity and perfection of single crystals, low effective carrier masses, and high dielectric constants have established semiconductors as exceedingly versatile in contrast to metals for understanding and utilizing electronic transport. The individual electron scattering mechanisms, which determine mobilities, can be separated and studied in detail. Ionized-impurity scattering has become of renewed interest. Hot-carrier effects, being accessible in semiconductors, have recently found a new method of investigation by means of optical excitation. Carriers far from equilibrium, such as ballistic electrons, are studied in specially structured samples. High-resistivity semiconductors still present many problems, caused by compensation, by interplay of localized versus extended states, and by very long dielectric relaxation. Interfaces have captured much of the recent attention. Grain boundaries are significant for polycrystalline silicon, as used in integrated circuits or solar cells. The transport across such boundaries is governed by thermal activation over a potential produced by localized electronic states associated with the boundary. Quantum effects and quasi-two-dimensional transport give rise to novel features at semiconductor surfaces and interfaces. Quantization can be further enhanced by magnetic fields, this two-fold quantization led to a new resistance standard based only on fundamental constants.

INTRODUCTION

The transport of charge carriers in semiconductors is reviewed in this article. Much progress in the detailed understanding of electronic transport has recently been achieved especially for this important class of materials. Variability and technological control of semiconductors has resulted in increased specificity of electronic transport with more sharply defined energies and momenta of the carriers, time-resolved studies, and better understanding of phenomena far from equilibrium. This progress parallels that of microelectronics as a whole. Materials technology and geometry control have led to the advances in transport physics, which in turn are utilized for new device applications.

This article opens with a short review of the standard description of electronic transport, involving the concept of mobility. The specific advantages of semiconductor materials are discussed next. The second chapter treats the various physical processes which determine carrier mobilities, describes recent advances in obtaining particularly high values of the mobility and discusses non-equilibrium effects. Chapter 3 alerts to the more troublesome cases of high-resistivity phenomena. The last chapter covers the increasingly important points which arise for electron transport in structured materials, rather than in simple bulk samples. Geometrical confinement is essential, and interfaces play a special role here, with new quantum effects arising.

## Kinetic Concept of Conduction

Mobile charge carriers, negative electrons or positive holes, are characterized by effective masses, m, which in general differ from the mass $m_o$ of the free electron. The effective mass is a convenient way to assess the influence of the crystal forces acting upon the carriers.[1-3] Replacing the mass by an effective one therefore acknowledges the individuality of a given material and its specific electronic band structure E(k), where the energy E of an electron having a wave vector k is fixed. An electric field F changes the momentum $(mv_d)$ of the carriers according to

$$d(mv_d)/dt + mv_d/\tau_m = eF \tag{1}$$

where t is the time and $\tau_m$ is introduced as an average momentum relaxation time, describing the return to equilibrium of a perturbed momentum distribution. The second term in Eq.(1) contains through $\tau_m$ all physical processes of electron-crystal interaction which lead to momentum transfer of the carriers when drifting through the crystal. For steady state, the first term in (1) vanishes, leading to

$$v_d = (e/m)\tau_m F = \mu F \tag{2}$$

which defines the mobility $\mu$, a most important concept for the microscopic description of electronic transport.

The local current density j relates to the local field F by means of a conductivity $\sigma$, $j=\sigma F$, thus

$$\sigma = ne\mu = (ne^2/m)\tau_m \qquad (3)$$

Equation (3) establishes a link between the macroscopic quantity $\sigma$ and parameter $\tau_m$ which has to be explained by quantum-mechanical microscopic models for each specific scattering process. There is usually a competition between such processes; a simple approximation to account for this competition is to add the inverse relaxation times.

The Boltzmann equation is frequently used to evaluate relaxation times.[1-4] In special cases, a simplifying relation between $\tau_m$ and the carrier energy $\varepsilon$ can be used; introducing a constant $\tau_o$:

$$\tau_m = \tau_o(\varepsilon/k_B T)^r \qquad (4)$$

where $k_B$ is Boltzmann's constant and T is the lattice temperature. The exponent r varies between -0.5 to +1.5 for the different scattering processes.[1-3]

A major complication arises from the velocity distribution of the carrier ensemble. We are not dealing with one uniform velocity, rather have to consider a distribution function $f(v_x,v_y,v_z)$. The relaxation time is to be averaged as

$$<\tau_m> = \int_{-\infty}^{+\infty} \tau_m(v)v_z(-\partial f/\partial v_z)d^3v / \int_{-\infty}^{+\infty} f_o d^3v \qquad (5)$$

for current flow in the z-direction. $f_o$ is the unperturbed distribution function.

The literature on transport phenomena is huge, the reader is here referred to recent summaries.[1-3] We cannot go into further details here. Let us summarize the most important parameters which describe electronic transport and have to be explained for each case, which also, however, provide adjustable variables for a wealth of transport phenomena. First, there is the density n of participating carriers, electrons and holes; they can be provided by appropriate doping. Only for very high temperatures is the intrinsic carrier density $n_i$ of significance. Here we have already a major difference to metals, whose carrier density cannot be controlled nor adjusted. Secondly, the mobility $\mu$ is relevant; it varies widely with temperature, type of material, and perfection. Third, the distribution function, f, has to be considered; it depends on the field and on the density of states provided by host and doping. Further, however, we have to abandon the simplifying assumption of homogeneous bulk transport whenever we deal with complex, structured samples, such as typical semiconductor devices, especially for modern high-speed and highly integrated structures. Steady-state assumptions also no longer hold if we want to understand transient and very high frequency behavior. Despite the tremendous complexity and variability of this problem of transport, the use of semiconductor materials has led to significant advances; next we discuss their outstanding role.

## Semiconductors: Propitious Materials

Semiconductor materials, especially the elements silicon and germanium as well as the compounds, such as GaAs as a prototype for the $A^{III}$-$B^{V}$ semiconductors, can today be prepared with extreme purity and crystalline perfection. Such perfection is prerequisite for controlled doping. Therefore, very small densities of charge carriers (down to $10^{8}cm^{-3}$) as compared to metals ( $10^{20}cm^{-3}$) can be achieved, which in turn is necessary to accomplish extrinsic and local effects and phenomena far from equilibrium. Metals with approximately one electron per host atom are so densely flooded that, for example, no deviations from Ohm's law of strict proportionality between current and voltage have been observed.[1] Semiconductors, on the other hand, have between $10^{-14}$ electrons per atom (for semiinsulating GaAs) up to about $10^{-2}$ electrons per host atom for degenerately doped materials.

These small numbers of carriers facilitate experiments in which the carriers are accumulated in specific regions, such as charge packets in real space and in time or electron populations in particular regions of momentum space, seen in stressed semiconductors, where some band minima are lowered in energy, thus provide exclusive states of occupancy. Modern trends in semiconductor transport also aim towards achieving sharper energy distributions than usually encountered.

The small mass, the low density of carriers, the extreme purity and perfection and also the relatively large dielectric constants all facilitate attainment of hot-electron effects. The electric field feeds energy of a carrier with a rate $\mu e F^2$. Intercarrier-scattering usually establishes a carrier temperature $T_e$, which exceeds the lattice temperature T. The temperature differential generates a loss rate proportional to the temperature difference. Such hot-electron effects were expected and discovered early in semiconductor history and remain a lively topic peculiar to this family of solid state materials. An entirely new aspect has been added by the stringent requirements of modern microelectronics, where steep gradients of dopant concentrations and high electric fields are natural consequences of very small structural dimensions. A novel feature, called ballistic transport can be observed, where the carrier may be regarded as in collisionless motion for times and lengths which are resolvable. In the following chapters we discuss these aspects of semiconductor electronic transport and begin with an analysis of carrier mobility.

## CARRIER MOBILITY

Various mechanisms lead to scattering in a crystal. Some are inevitable, such as phonon scattering in a material at nonzero temperature, others - such as impurity scattering - are extrinsic, thus might be controlled. We first summarize the most important processes leading to relaxation at low fields, then mention methods to influence and improve mobilities and finally discuss non-equilibrium effects.

### Contributions to Mobility

Hall-effect measurements enable us to determine the mobility over wide ranges of temperature. The characteristic temperature-dependence leads to identification of the individual components. Table 1 furnishes a brief summary.

TABLE 1 Temperature Dependence for Various Scattering Processes in Semiconductors

| | |
|---|---|
| Acoustic phonons | $T^{-1.5}$ |
| Piezoelectric phonons | $T^{-0.5}$ |
| Polar optical phonons of energy $\hbar\omega$ | $T^{0.5}[\exp(\hbar\omega/kT)-1]$ |
| Neutral impurities | $T^{0}$ |
| Ionized impurities | $T^{+1.5}$ |

Hall measurements are relatively indirect; a more direct attack, especially on the distribution function would be desirable. Optical techniques are here available. The electron distribution in a conduction band may be probed by investigating the spectral line shape of a luminescence band originating from transitions of free electrons out of the conduction band into sharply defined levels, say of acceptor atoms. The high-energy tail of the luminescence contains information on carrier temperature; a shifted Maxwellian distribution can be used as an approximation. Luminescence has been measured as a function of (low) electric fields for GaAs, yielding a direct, strictly optical determination of carrier mobility.[5]

At the highest temperatures, phonon scattering dominates. Electrons in polar materials suffer strong scattering with optical phonons, the temperature dependence is here governed by the density of thermally excited phonons. Acoustic phonons usually rule the mobility near room temperature with their well-known $T^{-1.5}$ dependence. The mobility rises with falling temperature until ionized impurities begin to scatter. Then the mobility decreases again with decreasing temperature, because colder carriers are slower, therefore suffer stronger scattering at a Coulombic point defect.[4] Contributions from lattice defects, such as dislocations or stacking faults are expected, but have not yet been clearly demonstrated.

An example of a mobility curve rich in features is presented in Fig.1:

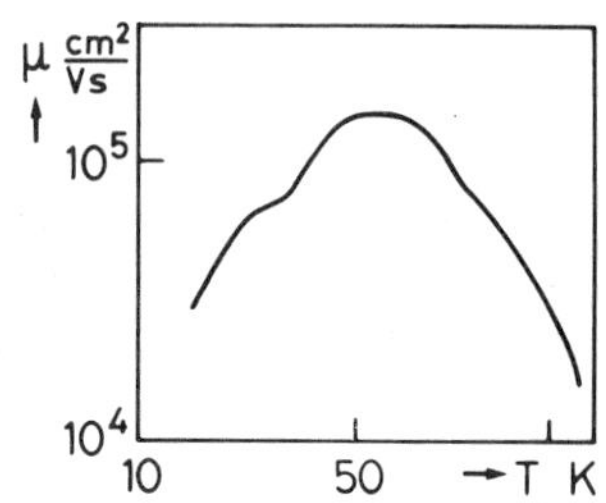


Fig.1. Schematic for a mobility-temperature curve, doubly logarithmic plot, typical for GaAs

Advances in Mobility Improvement

Impurity scattering can be reduced if only the necessary dopant impurities are incorporated into the lattice and compensation by charged minority dopants is avoided. The mobility in the ionized-impurity-regime can, however, be enhanced considerably by illumination with photons exceeding the band gap energy.[5-7] Such light creates excess carriers which screen the Coulomb potential of the impurity and thus reduce the scattering; quantitative explanation has been achieved.[6,7] Mobility measurements are highly sensitive to stray light in a Hall-effect apparatus!

An ingenious way to enhance mobilities without sacrifying dopant atoms is indicated in Fig.2. The method is called "modulation doping".[8] The parent dopant atoms are separated from the donated carriers by a potential wall. Such sharp spatial separations can today be achieved, for example, by molecular-beam-epitaxy.[9] The lower-lying conducting region receives the free carriers while being of high purity without dopant scattering centers. Enhanced mobilities have been observed at reduced temperatures, where ionized impurities would normally dominate the mobility.[8]

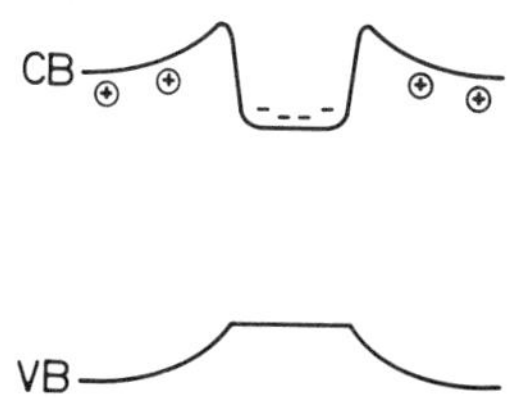


Fig.2. Principle of modulation-doping. The donors ⊕ are spatially separated from the electrons -. CB is the edge of the conduction band, VB valence band.

Non-Equilibrium Effects

Hot-electron effects can easily be observed in semiconductors.[10] A variety of methods, including microwave and optical techniques, are available today.[11,12] Much use is made for the analysis by simulating the scattering events in Monte-Carlo calculations, since it is in general very difficult to describe these non-equilibria in closed analytical form. Laser illumination can conveniently generate ensembles of very hot carriers, far above the minimal energy of the band edges.[13] Recent advances in understanding the processes leading to cooling of such distributions have relied heavily on picosecond spectroscopy.[14]

Examples of the definiteness and specificity of electronic transport, realizable only in semiconductors, are electron transfer both in momentum space and real space.[15] Consider first momentum transfer. Assume a semiconductor having subsidiary minima above the lowest minimum with smaller mobility - because of larger effective masses - in the upper minima. A critical field exists, beyond which the electrons have been heated so much, that their energy is sufficient to reach the higher-lying minima. The reduced average mobility can lead to a negative differential resistance, which is a desirable physical feature for achieving high-frequency oscillations.[16] This phenomenon is useful for high-frequency generation and competes with impact ionization and transit effects for device applications. Figure 3 gives an example of a velocity versus field relation in GaAs, Ge and Si.

Transfer of electrons into regions of differing transport conditions in real space have become particularly interesting with the advent of technology to fabricate well-defined regions of exceedingly small dimensions, such as feasible through molecular-

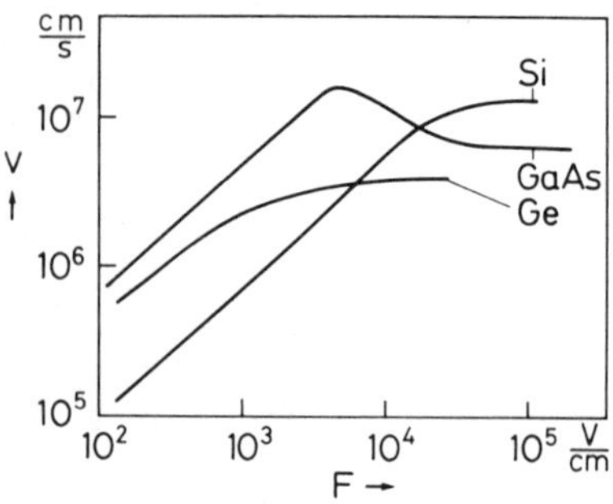


Fig.3. Velocity versus field relation for electrons in various semiconductors

beam-epitaxy or other present techniques used for large-scale integration. Electron emission from silicon to $SiO_2$ is one of the variations and plays a role in MOS-devices.[17] Modulation-doping was used for experiments involving real-space transfer in $Al_xGa_{1-x}As$; new features in the Ohmic range were observed and especially in the high-field-regime, when the average electron energy becomes comparable to the band edge differential of these heterostructure materials, thus can overcome the potential barrier and enter the region of differing conductance.[15] The switching times thus depend strongly on the materials parameters, for example exponentially on the compositional parameter x and can be made to vary from picoseconds to very long times. These novel effects are thus a typical result of utilizing highly structured semiconductor samples.

Hot-electron relaxation effects have now assumed an important role as a research topic for device application. Velocity overshoot is a typical example.[18] The average drift velocity - and not just the high-velocity tail of a distribution! - can be made to exceed the quasi-static v(F) value. Highly correlated motion of the electrons in either space or time is necessary for this purpose, which may be achieved with sufficiently large and rapid transients or by means of sufficiently large spatial non-uniformity. A rapidly varying electric field, i.e. over a distance comparable with a mean-free-path, can accelerate a carrier to the velocity

$$v_o = (\varepsilon_s/m)^{1/2} \quad , \qquad (6)$$

where $\varepsilon_s$ is the energy difference between upper satellite valley and lowest valley (0.3 eV for GaAs, 0.53 eV for InP). The resulting velocities for compound semiconductors are in the range of $10^8$cm/sec; carriers attain this velocity without suffering any energy-transferring collisions; ballistic motion is obtained under these conditions. Such velocity is, of course, desirable for fast devices because transit times limit the high-frequency operation.

Field-effect transistor structures have been used to prove these concepts and to study phenomena such as velocity overshoot in detail. Direct measurements of overshoot have been achieved by submillimeter-wave conductivity in inversion layers of silicon MOS-field-effect transistors, which have been claimed to be observed at sufficiently low frequencies in the frequency domain, corresponding to a time domain in the picosecond range.[19] The availability of electric pulses in the picosecond time domain, as derived from optical pulses generated by mode-locked lasers, promises direct studies of these effects.[20] The advances in submicron lithography will greatly stimulate this field; buried metal base fingers in submicron dimensions in compound semiconductor structures have been proposed for cut-off frequencies near 1000 GHz![21]

## PROBLEMS OF HIGH-RESISTIVITY MATERIALS

Opposite to the materials with rapid motion and large conductance are the many varieties of semiconducting materials with high resistivity. Here we still face many difficulties in obtaining definitive materials conditions, in making reliable measurements with trustworthy interpretations, and in providing universal theories. The advancement from the well-defined semiconductor regime into the domain of insulators proves difficult.

### Compensation

Doping with just one type of dopant is desirable for definiteness. Inadvertent compensation by the opposite type of dopant is, however, often almost inevitable. A thermodynamic tendency favors such compensation, say of an n-type semiconductor by minority acceptors, since the solubility of acceptors increases with increasing electron concentration. Group-IV atoms in the $A^{III}$-$B^{V}$ compound semiconductors, are often amphoteric, acting as donors on one sublattice, as acceptors on the other sublattice.[22] Semiconductors of type $A^{II}$-$B^{VI}$ have an even stronger tendency to lead to compensation, presumably by charged native defects, such as vacancies. These materials continue to suffer from this technological deficiency, they can in general not be doped both n- and p-type, thus p-n junctions are not feasible.[23]

This situation of predominant compensation is bothersome. The degree of compensation is difficult to measure. Attempts have been made to determine the ratio of majority to minority dopants from the mobility in the ionized-impurity regime. The total number of charged impurities can become large, while the excess of the majority is small, thus the carrier concentration is also small. Therefore the mobility should be depressed by compensation. However, the model of ascribing all mobility-depressing effects solely to compensation[24] may be misleading, as has been shown by studies which utilize a detailed analysis of optical transitions.[25]

Compensation, on the other hand, is a way to achieve insulating regions, which are needed for substrates or other portions of semiconductor devices. The semi-insulating GaAs, doped with chromium or oxygen, is the best known example, although still incompletely understood and subject of complete specialists' conferences.[26]

Compensation leads to the formation of localized electron states inside the forbidden energy gap of the semiconductors, these are "tail states", whose density decreases exponentially when going away from the band extrema. Silicon-doped GaAs with deeply extending tail states is used in efficient infrared emitters.[27] Such material can, however, also be used to perform definitive studies of "dispersive electron transport" with time-resolution in the picosecond range.[28] Dispersive transport arises when carriers are released and retrapped, thus intermittently occupy low-lying, spatially localized immobile states and higher-lying, delocalized current-carrying states. Basic features of this model could be proven, using the well-defined and doping-adjustable semiconductor material.[28]

The interplay of localized and delocalized states is a central problem of solid state physics with consequences for electronic transport.[29] Amorphous semiconductors have revitalized this old field.[30] Amorphous materials typically show a conductance dominated by hopping between localized sites and a conductivity governed by

$$\sigma = \sigma_o(T) \exp (AT^{-1/4}) \tag{7}$$

A "mobility edge" separates the localized from the delocalized states. Charge carrier localization now plays an important role in metallic conduction, where very interesting effects of minimal conductance and phase transitions are being discussed.[31] This subject is beyond the scope of this article, however, it appears that semiconductors provide superior specimens for a definite study of these effects of current interest.[32]

The Relaxation Case

A novel regime of transport arises when the dielectric relaxation time $\tau_r$ exceeds the minority carrier lifetime $\tau_\ell$.[33] In metals or doped semiconductors, the dielectric relaxation time $\tau_r=\kappa/\sigma$ (where $\kappa$ is the dielectric constant) is usually unmeasurably short, therefore any perturbations of the local potential are quickly readjusted by rearrangement of the free carriers. The neutrality condition therefore holds; a significant simplification in transport theory!

For such general description of transport, a set of differential equations has to be solved. First, Poisson's equation linking electric fields with charge densities. Secondly, a continuity equation relating the rates of change of charge densities with current sources and sinks and with recombination and generation of electrons and holes. Thirdly, the current is defined as being caused by local fields and diffusion, and finally one needs an empirical function to describe the recombination, which is usually set proportional to the deviation $(np-n_i^2)$ from equilibrium. The treatment of these general equations plus necessary boundary conditions for the specific case becomes difficult if $\tau_r>\tau_\ell$. Then space charge effects dominate, and a number of unusual phenomena arise.[34] A sublinear current-voltage relation holds for p-n diodes.[35] Injection of minority carriers can lead to an increase in resistance, quite contrary to what we are used to in standard semiconductor physics!

A series of studies[34,35] and computer simulations[36] has shown that it is extremely dangerous to cling to the familiar concepts of measuring electrical properties and to interpret them microscopically for these high-resistance materials! The determination of a simple conductivity becomes already questionable, since resistance no longer scales with sample length, and Hall-effect data may be completely misleading.[37] High-resistivity materials are thus not only difficult to assess for chemical purity, crystallinity and perfection, but usually also defy a definitive electrical characterization. These pitfalls are, however, not always realized in the literature, because it seems often regarded as trivial that the traditional concepts and techniques remain valid.

A completely new concept for charge transport is invoked to explain the conductivity of conducting polymers, such as polyacetylen. Solitons are thought to move through the crystal, these are irregularities in the regular alternating sequence of single and double C-C bonds. Such a model would explain the observed absence of magnetic susceptibility which one would expect if free carriers were present.[38]

GEOMETRICAL CONFINEMENT

Current and voltage are integral quantities, current density and electric field are the meaningful local quantities for current transport. The very nature of semiconductors is their ability to sustain appreciable gradients of conductivity and carrier densities, thus local changes of these and other quantities can and must be measured. Notice that Shockley in his famous and now almost historical first book on semiconductors showed microprobing and contacting as being the most important novel experimental techniques in photographs of the frontispiece.[39] The idea of guiding, controlling and influencing the flow of charge in highly structured semiconductor specimens remains a central theme of modern semiconductor research and technology.

Contacts, being essentially metal/semiconductor interfaces, are a technologically vital part, yet even today incompletely understood composites. Although Schottky's basic ideas are still valid, there is still often more magic and mystery in proper preparation, especially for cases, where contact must be made to lowly doped regions. Laser-annealing of such interfaces is a recent addition to the metallurgy of contacts.[40]

Grain Boundaries

Very early semiconductor history started out with irregular polycrystalline samples. The victorious development of single-crystal technology was an essential part in achieving the present state of the art in circuit integration. However, polycrystalline silicon has recaptured a good deal of activity and interest recently. One reason is that polycrystalline silicon might have a chance to become an economical material for solar cells with efficiencies exceeding 10% even for large-area devices under mass production conditions.[41] Secondly, silicon-gate technology for field-effect transistors uses polycrystalline material with a need for as high a conductance as possible to properly furnish the current to the gates even at the necessary reduction of conductor cross-section. Therefore an understanding of physical processes in polycrystalline materials is necessary, which amounts to an understanding of the grain boundaries.[42]

Grain boundaries are regions of high defect concentration. The simple model of a grain boundary as just being a regularly spaced array of dislocations is probably too naive. Impurities segregate preferentially at grain boundaries, because usually these are the last portions of the material to solidify. Oxygen seems to play a particularly important role in silicon grain boundaries.[43] The electrical effect is surprisingly small for many cases: at room temperature many grain boundaries apparently contribute very little excess resistance. Twin boundaries, being of slightly impaired perfection, are only active if substantial amounts of extra impurities are present, which then tend to precipitate there.[44]

The grain boundary constitutes an area of localized defects which may be electrically charged. The charge then leads to a potential barrier. Carriers can overcome such barriers via thermal activation. The current density is therefore given as a thermionic emission[42]

$$j = A^{*}T^{2} \exp(-e(\emptyset_{B}+\zeta)/kT)\left\{1-e^{-eV/kT}\right\} \qquad (8)$$

where A* is a (modified) Richardson constant, $\emptyset_B$ is the barrier height. $\zeta$ gives the Fermi level position in the grain, V is the applied voltage.

Many measurements of the conductance and the space-char layers (via capacitance data) have been done, in particular with Si polycrystalline samples or at individual grain boundaries in large-grain poly-Si or specifically grown bicrystals. The general idea of thermionic transport above the boundary has been verified.[42] Photoconductance and photocapacitance data at judiciously low applied voltages has led to a determination of the defect density of states within the forbidden gap for Si grain boundaries.[45] Grain boundaries show a photosensitive behavior because the charge dominates the magnitude of the potential wall, which in turn governs exponentially the flow of current, see Eq.(8). Selenium, for a long time a widely used but little understood semiconductor, seems to be ruled in its transport behavior by such internal potential barriers.

## Space-Charge Complications

We have already learned that the formation of space-charge is an effect typical for semiconductors, when compared to metals. Charge may be trapped at fixed sites in a semiconductor, especially at the so-called "deep" impurities[46] with their comparatively large thermal ionization energies. Surface states are the other prominent example for such localized charge. Space-charge implies potential variations of the electrical potential, which in turn may strongly influence and restrict the spatial regions where a current will flow. Electrical measurement of surface conductance is thus actually an important tool to study surface charges. Neglect of these effects becomes especially dangerous when the conducting region is small in at least one dimension, for example in a thin epitaxial layer above an insulating substrate.

Persistent photoconductivity in thin semiconductor layers is a particularly fascinating example. The photo-induced excess conductance persists even after illumination, its magnitude often follows a proportionality with the log of photon dose. Two models have been proposed.[47] One assumes microscopic potential barriers at deep impurities: the impurity relaxes and creates a potential wall which hinders recombination and lets the excess carrier remain free to contribute to conductance. The second model assumes macroscopic barriers instead, for example at interfaces or at dopant fluctuations. The electron-hole pair is separated - similar to solar cell action - one carrier diffuses to the other side, where it might get trapped, the other remains free. The charge trapping, however, acts like a charged gate electrode and modulates the conductance channel by altering the width of the space charge layer. This latter model has been experimentally proven for epitaxial GaAs samples;[47] the conductance behavior therefore depends very strongly on trap concentrations not in the conductance path itself but rather in relatively distant adjacent portions.[48] Needless to say, completely erroneous results of mobilities or carrier densities would be obtained if such effects were ignored![49]

## Quantum Effects at Interfaces

Potentials with rapid spatial variation lead to quantization effects, which in turn modify electron transport. Two-dimensional systems have here brought a wealth of new efforts and applications. The most important system consists of inversion and accumulation layers at semiconductor-insulator interfaces. The metal-silicon dioxide-silicon system was particularly useful. Variations of the metal-gate potential can easily be utilized to vary the two-dimensional charge carrier density in the silicon at the $Si/SiO_2$-interface. In general, heterostructures display such quantum effects, provided the interfaces are sharply defined to within at least one nanometer. Semiconductor heterojunctions, especially for the system $GaAs/Ga_xAl_{1-x}As$, and superlattices fulfill this condition. Once again, it has become evident that materials control to within atomic dimensions regarding geometry and chemical composition leads to entirely new and useful physical phenomena. A recent review exhaustively covers electronic properties of two-dimensional systems.[50]

Two-dimensional transport requires a special description. One usuallyobtains a sheet conductivity $\sigma_{(2)}$

$$\sigma_{(2)} = eN_s\mu \, , \tag{9}$$

where $N_s$ is an areal charge density. $\sigma_{(2)}$ may also be expressed by a transconductance $g_d$ between source and drain of field-effect transistor by

$$\sigma_{(2)} = g_d L/W \tag{10}$$

where L and W are length and width of the MOSFET channel. The mobility definition is also different and presents complications. Often, an effective mobility is defined:

$$\mu_{eff} = \sigma_{(2)}/C_{ox}(V_g-V_t) \qquad (11)$$

where $C_{ox}$ is the oxide capacitance per unit area, $V_g$ is the gate voltage, and $V_t$ the threshold gate voltage. This definition already indicates what types of measurements are used to define transport parameters. The fraction f of nontrapped versus total charge may vary widely, indicative of another problem to determine and characterize transport. A detailed critical discussion of methods and results is contained in the recent review by Ando, Fowler and Stern.[50]

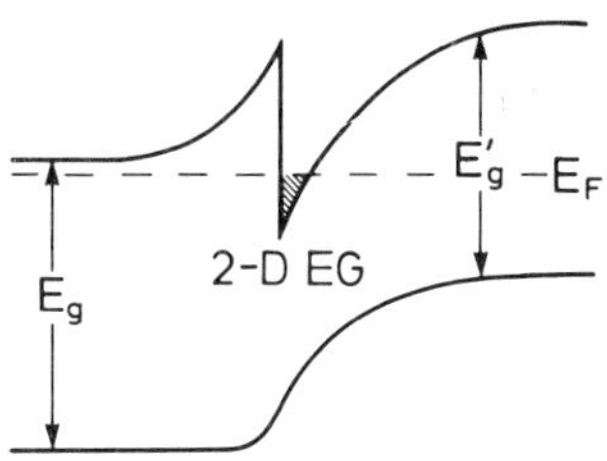


Fig.4. Heterojunction band scheme leading to a two-dimensional electron gas

Figure 4 shows the arrangement to obtain a two-dimensional electron gas at a heterojunction interface between two semiconductor materials. The band bending at the interface leads to a sharply localized triangular potential well containing the electron gas. Subbands arise from the quantization, which can be analyzed by means of optical spectroscopy, such as Raman scattering, and transport measurements including magneto-transport; Shubnikov-de Haas experiments are here particularly useful. The details of this very rapidly moving field are beyond the scope of this general survey.

## Quantized Resistance Standard

We conclude this article with a brief report on a quantum transport effect which has provided a new standard of the Ohm and led to a surprising new way to determine the Sommerfeld fine structure constant. The electron gas at an interface is subjected to a second quantization by means of a magnetic field. The density of states distribution now consists of a sequence of sharp Landau peaks. The number $n_L$ of states per Landau level is

$$n_L = eB/h \qquad (12)$$

where B is the field, and h is Planck's constant. There are no available states between Landau levels. If the Fermi level is placed in between two such levels, then there can be no scattering of the electrons at sufficiently low temperatures. As a result, the Hall conductivity

$$\sigma_{xy} = e^2/h = 2\pi c \cdot \alpha \qquad (13)$$

is now only a function of elementary constants and does not depend on material parameters nor on geometry. The fine structure constant $\alpha$ can thus be obtained.

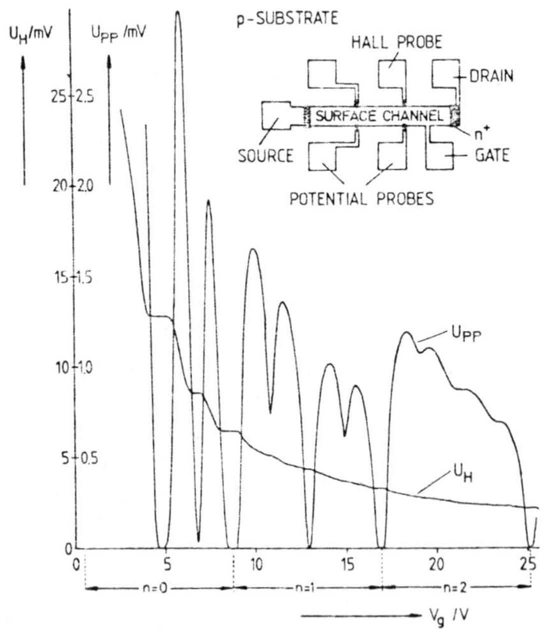


Fig.5. Hall voltage $U_H$ and potential drop $U_{pp}$ as a function of gate voltage $V_g$ for a silicon MOS field effect transistor at 18 T and 1.5 K (after Ref.51)

Figure 5 shows data and a typical structure used. Recent new results[52] suggest more detailed effects also when the Landau levels are only filled to 1/3, which has been tentatively explained by ordering phenomena such as a Wigner triangular lattice or charge-density-wave formation.

The quantity $h/e^2$ represents a resistance. Therefore the von-Klitzing-effect of quantized Hall resistance finds an application in establishing a new standard for the unit "Ohm", which is defined without references to specific materials or geometry. Many of the national bureaus of standards are therefore working on this novel effect of quantized electronic transport at semiconductor interfaces.

## CONCLUSIONS

Electronic transport in semiconductors has become a highly active field again, stimulated especially by the advances in semiconductor materials perfection and geometry control. Finely structured composite semiconductor samples with chemical composition and doping control to almost atomic dimensions offer challenging opportunities to ever more exact guidance of the motion of electrons and usage for electronic devices of improved performance, especially for very high frequencies.

## REFERENCES


1. K.H. Seeger, Semiconductor Physics, Springer Series in:"Solid State Sciences", Vol.40, Springer-Verlag Heidelberg (1982).
2. B.R. Nag,"Theory of Electrical Transport in Semiconductors", Pergamon, Oxford (1972).
3. Handbook of Semiconductors, Vol.I, "Band Theory and Transport Properties", W. Paul, editor, North Holland, Amsterdam (1982).
4. D. Chattopadhyay and H.J. Queisser, Rev. Mod. Phys. 53:745 (1981).
5. W. Bludau, E. Wagner, and H.J. Queisser, Solid State Commun.18: 861 (1976); for a review on photoexcitation, see R.G. Ulbrich, Solid-State Electronics 21:51 (1978).
6. M.A. Paesler and H.J. Queisser, Phys. Rev. B17:2625 (1978).
7. D.E. Theodorou and H.J. Queisser, Phys. Rev. B19:2092 (1979).
8. H.L. Störmer, J. Phys. Soc. Jpn. 49, Suppl.A:1013 (1980).
9. K. Ploog, Annu.Rev.Mater.Sci. 11:171 (1981).
10. E.M. Conwell, High Field Transport in Semiconductors, Solid State Phys. Suppl.9, Academic Press, New York (1967).
11. Proceedings of the Hot-Electron-Conference 1977, Solid State Electronic 21, pp. 1 to 323 (1978).
12. Proceedings of Third Internat. Conference on Hot Carriers 1981 in: J. de Physique 42, supp. 10 (1981).
13. C. Weisbuch, Ref.11, p.179; R.C.C. Leite, Ref.11, p.177.
14. J. Shaw, Ref.12, p. C7-445.
15. K. Hess, Ref.12, p. C7-3.
16. A historical review by C. Hilsum, Ref. 11, p.5.
17. T. Ning, Solid-St.Electron. 21, 273 (1978).
18. H. Kroemer, Ref. 11, p.61.
19. S.J. Allen, Jr., D.C. Tsui, F. DeRosa, K.K. Thornber, and B.A. Wilson, Ref. 12, p.C7-369.

20. P.R. Smith, D.H. Auston, and W.M. Augustyniak, Appl. Phys. Lett. 39:739 (1981).
21. L. Eastman in "Festkörperprobleme", Vol. XXII, P. Grosse, editor; Vieweg, Wiesbaden (1982).
22. J.M. Whelan, J.D. Struthers, and J.A. Ditzenberger, Proc. Int. Conf. Semicond. Prague, p. 966 (1961).
23. See, for example, P.J. Dean, Czech. J. Physics 30:272 (1980).
24. D.L. Rode, in "Semiconductors and Semimetals", ed. by R.K. Willardson and A.C. Beer, Academic Press, New York, Vol. 10, chapter 1 (1975).
25. T. Kamiya and E. Wagner, J. Appl. Phys. 48:1928 (1977).
26. Proc. Int. Conf. Semi-Insulating Compounds, Evian (1982) to be published.
27. H.J. Queisser, J. Appl. Phys. 37:2909 (1966).
28. E.O. Göbel and W. Graudszus, Phys. Rev. Lett. 48:1277 (1982).
29. N.F. Mott, "Metal-Insulator-Transition", Taylor & Francis, London (1974).
30. Electronic Transport in Amorphous Semiconductors is reviewed by P. Nagels in: Amorphous Semiconductors, ed. by M.H. Brodsky, "Topics in Applied Physics", Vol.36, Springer, Heidelberg (1979), p.113.
31. D.J. Thouless, Phys. Rep. 13C:93 (1974). A recent summary in "Anderson Localization", ed. by Y. Nagaoka and H. Fukuyama, Springer Ser. in: Solid State Sciences, Vol.40; Springer, Heidelberg (1982).
32. For a recent contribution concerning localization in semiconductors, see G.A. Thomas, Proc. 16th Int. Conf. Physics of Semic., M. Averous, ed., North Holland, Amsterdam (1982.
33. H.J. Queisser in: "Solid State Devices",1972, P.N. Robson, editor, Inst. of Physics Series 15, London and Bristol, p.145 (1973).
34. W. Van Roosbroeck and H.C. Casey, Jr., Phys. Rev. B5:2154 (1972).
35. H.J. Queisser, H.C. Casey, Jr., and W. van Roosbroeck, Phys. Rev. Lett. 26:551 (1971).
36. J.-C. Manifacier and H.K. Henisch, J. Appl. Phys. 52:5195 (1981).
37. M. Ilegems and H.J. Queisser, Phys. Rev. B12:1443 (1975)
38. S. Roth and M. Peo, Colloid Polym. Sci. 259:279 (1981)
39. W. Shockley, "Electrons and Holes in Semiconductors", D. van Nostrand, Princeton (1950).
40. H. Tews and C. An, J. Appl. Phys. 53:5339 (1982).
41. B. Authier; H. Fischer,in two articles in "Festkörperprobleme", Vol. XVIII; J. Treusch, ed., Vieweg, Braunschweig (1978) describe polycrystalline Si with columnar structure and solar cells from this material, pp.1 and 19; also see ref.42.
42. "Grain Boundaries in Semiconductors", ed. by H.J. Leamy, G.E. Pike, and C.H. Seager, North-Holland, New York (1982).
43. H.J. Queisser, J. Phys. Soc. Jpn. 18, suppl. III, 142 (1963); D. Redfield, Appl. Phys. Lett. 40:163 (1982).

44. H.J. Queisser, J. Electrochem. Soc. 110:52 (1963)
45. J. Werner, W. Jantsch, and H.J. Queisser, Solid State Commun. 42:415 (1982).
46. H.J. Queisser "Festkörperprobleme", Vol.XI; O. Madelung, editor, Vieweg, Braunschweig (1917), p.45; S. Pantelides, Rev. Mod. Phys. 50:797 (1978).
47. H.J. Queisser and D.E. Theodorou, Phys. Rev. Lett. 43:401 (1979). A review by M.K. Sheinkman and A. Ya. Shik, Fiz. Tekh. Poluprovodn. 10:206 (1976) [Sov. Phys. Semicond. 10:128 (1976)]
48. D.E. Theodorou, H.J. Queisser, and E. Bauser, Appl. Phys. Lett. 41: (1982).
49. D.M. Collins, D.E. Mars, B. Fischer and C. Kocot, J. Appl. Phys. 53: (1982).
50. T. Ando, A.B. Fowler, and F. Stern, Rev. Mod. Phys. 54:437 (1982)
51. K. v. Klitzing, G. Dorda and M. Pepper, Phys. Rev. Lett. 45:494 (1981).
52. D.C. Tsui, H.L. Stormer, and A.C. Gossard, Phys. Rev. Lett.48: 1559 (1982).

# RECENT ADVANCES IN THE THEORY OF IMPURITIES AND DEFECTS IN SEMICONDUCTORS

Sokrates T. Pantelides

IBM Thomas J. Watson Research Center

Yorktown Heights, New York 10598

## INTRODUCTION

Scientific advances often lead directly to technological innovations. On other occasions, new technologies emerge from experimentation and are based on processes and phenomena that are poorly understood. Scientific inquiry then attempts to probe these processes and phenomena in order to achieve better understanding and, in turn, point to new directions for technological advances. In this way, science and technology are intricately interconnected. Nevertheless, the connections are often hard to see as most of us usually focus on either scientific or technological research. This symposium has brought together technologists and scientists, and one of our aims is to understand each other and get a feeling for the work pursued in areas outside our own expertise. I have the honor of being one of the few scientists who have been asked to address this audience which consists mostly of highly talented technologists. In my talk, I shall review a number of recent advances in my field of expertise, with an attempt to place them in a perspective relating to technology in general and microelectronics in particular. Since time is limited, I will not attempt a broad review. Instead, I will focus primarily on theoretical work carried out by myself and a group of valuable collaborators in the last few years. Work by other groups will also be reviewed where appropriate in order to give a fairly comprehensive picture of the current state of the art.

My general theme will be impurities and defects in semiconductors. As we all know very well, pure semiconductors would be poor conductors and also poor insulators, fairly useless for practical applications. The utility of semiconductors derives from our ability to alter their transport properties over wide ranges by doping them with various impurities either uniformly or inhomogeneously. For example, we can dope Si with P or other group-V elements and obtain varying degrees of n-type conductivity. Similarly, we can dope Si with B or other group-III elements and obtain varying degrees of p-type conductivity. By inhomogeneous doping, we can produce pn junctions which are the building blocks of electronic devices.

Impurities that are used as dopants to make a semiconductor either n-type or p-type are known as *shallow impurities* for reasons that will become transparent later on. Shallow impurities are only a small class of impurities, however. Most of the rest, known as *deep*, play a drastically different role in determining the properties of a given semiconductor. Deep impurities, or, in general, deep defects are usually undesirable in devices, since they act mostly as recombination centers for electron-hole pairs, thus reducing the number of carriers available for conduction. Nevertheless, in some applications, one wants to kill excess carriers very quickly, and deep impurities come in very handy. They are also useful for light-emitting diodes, if they introduce an efficient radiative recombination mode.

The science part of the impurity story concerns our understanding of how and why each impurity species alters the properties of the host. Once more, we have an interplay between experiment and theory, each taking the lead at different times, probing deeper and producing more detailed understanding. For shallow impurities, fairly detailed understanding was achieved almost thirty years ago. You are all probably familiar with its essential aspects. Take a phosphorus atom, for example, substituting for a Si atom. The substitution corresponds to adding a positive charge on a Si nucleus and an extra electron. We, therefore, have the analog of a hydrogen atom except for two important modifications: (a) the positive charge is embedded in a dielectric medium, and (b) the electron, if separated from the positive charge, is not a free electron, but a Bloch electron. It can, however, be viewed as free with an effective mass $m^*$. The resulting hydrogenic states are scaled by a very large factor, making the ionization energy of Si:P only 45 meV. It is because this ionization energy is small compared with the band gap of 1.2 eV that these impurities are known as *shallow*. A similar picture applies for group-III impurities in Si which introduce shallow states near the valence-band edge. At room temperature, when only a few electrons can be thermally excited across the band gap, virtually all shallow impurities are thermally ionized making carriers available for conduction.

The theoretical framework for describing shallow impurities is known as effective-mass theory. This theory has been quite successful. Detailed quantitative understanding of all the properties of shallow impurities is still incomplete, but recent developments are rather technical and inappropriate for review here. One aspect of shallow impurities, however, generated considerable interest in the last decade, spurred largely by technological relevance. Miniaturization of devices and large-scale integration has led to the use of *heavily doped* semiconductors, especially Si. At high doping concentrations, the impurity levels broaden into a band which eventually merges with the host conduction or valence band so that the net result is a shrinkage of the forbidden band gap. Since the value of the gap is an important parameter in transistor design, experimental and theoretical investigations have been undertaken to elucidate the effect. The first part of my talk will deal with this problem.

In the remainder of my talk I will discuss deep centers, *i.e.,* deep impurities and defects. Unlike shallow impurities, deep centers constitute a strong perturbation and generally introduce bound states in the midgap region, far from the band edges (hence the designation "deep"). Theoretical descriptions of the properties of deep centers have been evolving very slowly in the last thirty years. Only in the last few years theoretical methods were developed and implemented providing descriptions of deep centers at the same level of sophistication and accuracy as currently feasible for perfect crystals. I shall not go into the mathematical and other technical aspects of these developments. Instead, I shall review the highlights of the accomplishments achieved by these calculations with emphasis on the interrelations between theory, experiment, and technological applications.

## HEAVILY-DOPED SILICON

As we already saw in the Introduction, a single shallow impurity, *e.g.,* P in Si, introduces a hydrogenic bound state below the conduction band edge, as shown in Fig. 1a (the "excited" Rydberg states lying even closer to the band edge are not shown). As the concentration of such impurities increases, it is natural to expect the impurity level to broaden into a band (Fig. 1b). When a certain critical concentration is exceeded, the impurity band merges with the host band, as shown schematically in Fig. 1c. The net result is a narrowing of the forbidden energy gap.

Band-gap shrinkage induced by heavy doping was first detected experimentally by optical absorption in 1967.[1] The effect acquired significance a year later when Kaufmann and Bergh[2] recognized its role in limiting the efficiency of transistors. Conventional theory assumed a constant band gap and predicted that increasing the doping of the emitter would increase its efficiency as an injector of minority carriers into the base. Experience confirmed this prediction for moderate levels of doping. For high doping levels (higher than about $10^{19}$ $cm^{-3}$ in Si), however, it was found that

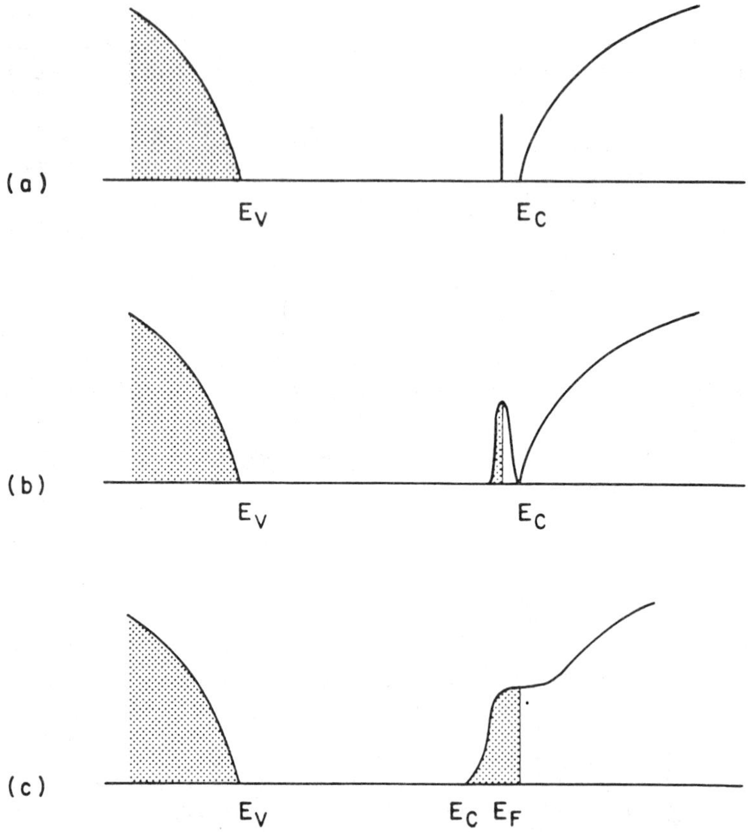


**Figure 1.** Schematic illustration of the origins of band-gap shrinkage. (a) Valence- and conduction-band densities of states and the bound state of a single shallow impurity; (b) Impurity band formed by a high concentration of shallow impurities; (c) Merging of the impurity band with the host conduction band at even higher concentrations, resulting in a net gap reduction.

the injection efficiency no longer increased. Kaufmann and Bergh[2] were the first to attribute this phenomenon to a shrinkage of the band gap due to heavy doping. In the 1970's, with the advent of miniaturization, heavy doping became more common because it allows for thinner depletion regions that are necessary to accommodate the potential difference between the n and p regions of a junction. In 1976, Slotboom and de Graaf[3] carried out a series of measurements on bipolar transistors from which they extracted values of the gap shrinkage. The striking result was that these values were significantly different from those extracted earlier from optical-absorption experiments.[1] These discrepancies remained a puzzle for many years. The difficulties were reviewed in 1977 by Keyes[4] and confirmed by new optical experiments by Schmid[5] in 1981. At about the same time, Schmid, Thewalt and Dumke[6] reported photoluminescence measurements which yielded a third set of gap shrinkages, incompatible with either of the earlier two! We have collected representative values of the three sets of gap shrinkages in Fig. 2. The discrepancies are indeed substantial.

Theoretical understanding of the gap shrinkage induced by heavy doping evolved in a piecemeal fashion. In the 1960's and early 1970's the effect was viewed as described schematically in Fig. 1, *i.e.,* in terms of an impurity band merging with the continuum. Quantitative calculations based on this notion were difficult, however. In 1976, Inkson[8] laid the foundations of a rigorous theoretical description and, in 1980, Mahan[9] gave a comprehensive treatment. More extensive calculations were reported in 1981 by Berggren and Sernelius.[10] These papers treated the problem from a different point of view. One starts out with a high concentration of impurities. If the effect of the positive ions is at first ignored, the extra electrons occupy the lowest conduction band up to a Fermi energy $E_F$. Electron-electron interactions then cause a fairly rigid lowering of the conduction bands and raising of the valence bands. When interactions with the positive ions are included, additional rigid shifts are obtained (Fig. 3).

Theoretical values of the gap shrinkage were difficult to compare with experimental values because of the serious discrepancies among the values extracted from different experiments. The state of the art is best illustrated by Fig. 7 of Ref. 10, which reveals that no meaningful comparison of different theoretical results could be made with experimental values. In recent work by Selloni and Pantelides,[11] this difficulty was removed by focussing attention to the quantities that are actually measured by experiment. Calculations were carried out for the photoluminescence spectra, and direct comparison with the measured spectra of Ref. 6 revealed very good agreement (Fig. 4). The calculations followed along the lines of Refs. 9-10 assuming an ordered lattice of impurities. One significant difference was the inclusion of electron-ion interactions via a full calculation instead of second-order perturbation theory

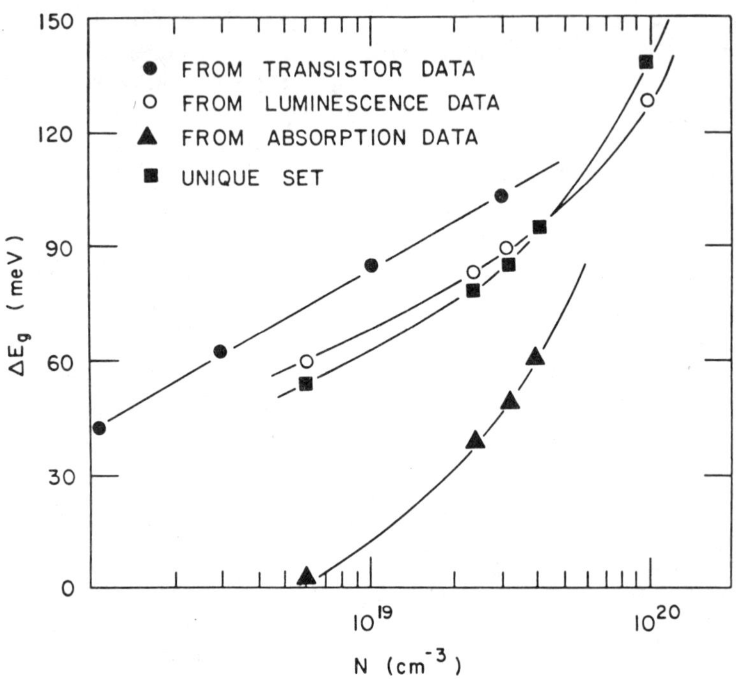


**Figure 2.** Gap shrinkages as a function of impurity concentration in Si as extracted from optical-absorption (Ref. 5), photoluminescence (Ref. 6), and transistor data (Ref. 3). The "unique set" is obtained from luminescence data using the theory of Ref. 11 and is consistent with both optical-absorption and transistor data (from Ref. 12).

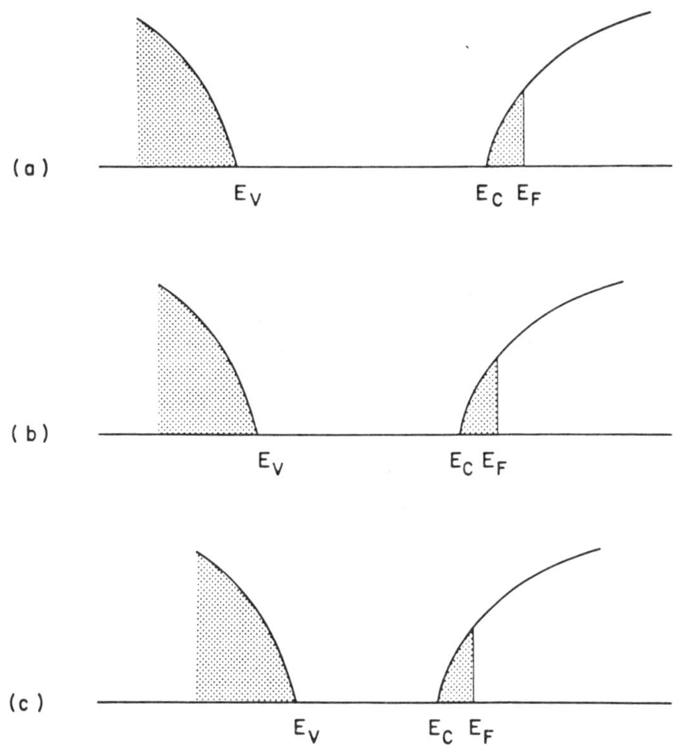


**Figure 3.** Schematic illustration of a different point of view regarding the origins of gap shrinkage. (a) Ignoring the effect of the positive ions, all the extra electrons occupy the host conduction bands up to a Fermi energy $E_F$; (b) Electron-electron interactions cause rigid shifts of both the conduction and valence bands; (c) Electron-ion interactions cause additional rigid shifts.

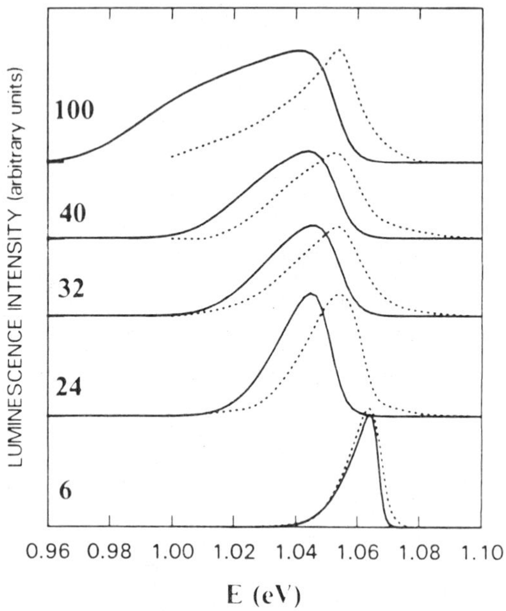


**Figure 4.** Theoretical (solid curves, Ref. 11) and experimental (dotted curves, Ref. 6) photoluminescence spectra of heavily-doped Si for five different impurity concentrations. The numbers labeling the curves are the concentrations in units of $10^{18}$ $cm^{-3}$. The spectra shown are phonon replicas at 58 meV below the zero-phonon spectra.

and the inclusion of multivalley interactions. This effect, which arises from the existence of six equivalent minima (valleys) in the lowest Si conduction band, was neglected in all previous theoretical calculations. It was found to be significant, as demonstrated in Fig. 5. In this figure, we plot the high-energy edge of the photoluminescence spectra which corresponds to $E_g+E_F$, where $E_g$ is the actual band gap at a given impurity concentration. If intervalley interactions are left out, agreement with experiment is quite poor.

In more recent, still unpublished work, Pantelides and Selloni[12] used the results of Ref. 11 as a guide and probed the analysis of the experimental data that led to the gap-shrinkage values shown in Fig. 2. The origins of the discrepancies were identified. It was concluded that the actual data are fully consistent with one another, and the discrepancies in the gap-shrinkage values were consequences of different poor assumptions made in extracting these values from the data. Here, we will review briefly the main aspects of this work.

First, it is straightforward to show that the photoluminescence and optical-absorption data are mutually consistent. We note that optical absorption ought to begin at precisely the values $E_g+E_F$ extracted from the photoluminescence spectra. In Fig. 6 we show the original optical-absorption spectra of Schmid[5] and mark the values of $E_g+E_F$ with short vertical lines. We see that each vertical line lies slightly to the left of the dip in the absorption curve, as it should. (The decreasing absorption below $E_g+E_F$ is due to free carriers. The superposition of a decaying and a rising curve leads to a dip slightly to the right of the absorption onset, as shown in the insert of Fig. 6.) It is clear that, if one wishes to extract values of $E_g+E_F$ directly from the optical-absorption spectra, one would need a detailed fit of the data *in the region of the dip*, which requires an accurate theory of free-carrier absorption. Schmid actually fit the data *at higher energies*, treating the gap as a free parameter, and his procedures led to incorrect values for the gap. For example, he concluded that the gap remains unchanged (zero shrinkage) for $N=6\cdot10^{18}$ $cm^{-3}$ on the basis of the high-energy fits. If this were true, the onset of across-the-gap absorption ought to be at an energy *to the right* of the onset of absorption in pure material. Schmid's own data (Fig. 6) show that the opposite is in fact true, demonstrating the inadequacy of his fitting procedures.

The above discussion suggests that the luminescence and absorption data ought to yield the same gap shrinkages, but the process of extracting such values is much simpler in the case of the luminescence data. Indeed, once the values of $E_g+E_F$ are extracted from luminescence data, the theoretical values of Ref. 11 can be used to obtain values of $E_g$ and hence values of the gap

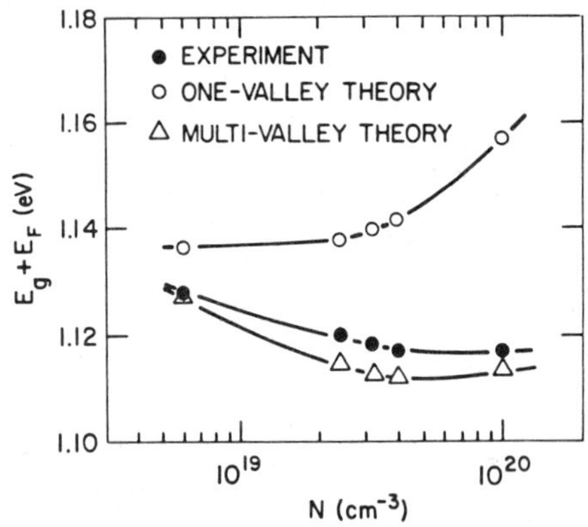


**Figure 5.** Experimental and theoretical positions of the high-energy edges of the zero-phonon photoluminescence spectra of heavily doped Si (from Ref. 11).

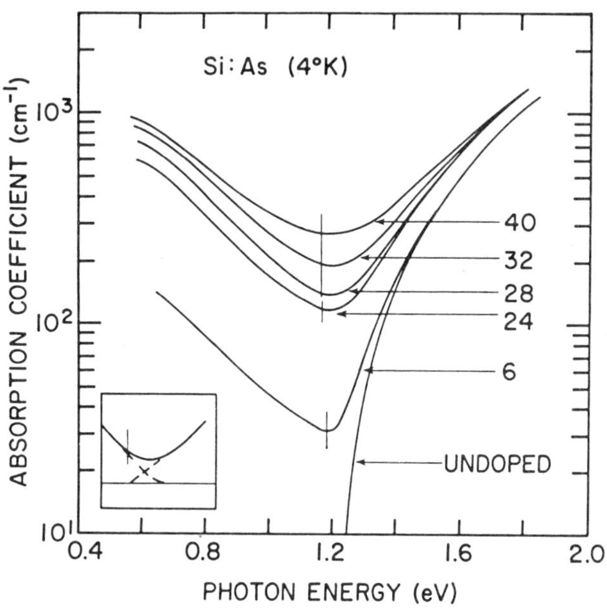


**Figure 6.** Optical-absorption spectra of heavily-doped Si (Ref. 5) at 4°K. The short vertical lines indicate the position of $E_g+E_F$ as extracted from the photoluminescence spectra of Ref. 6 and mark the onset of absorption across the gap. In the inset we show how the decaying free-carrier absorption and the rising across-the-gap absorption are superposed so that the dip is slightly to the right of $E_g+E_F$ (from Ref. 12).

shrinkage. These values of the gap shrinkage are also shown in Fig. 2. They are consistent with optical-absorption data as discussed above.

The reconciliation of the absorption and luminescence data with the transistor data of Slotboom and de Graaf[3] is considerably more involved and is described in detail elsewhere.[12] The main point is that the quantity those authors call the "band gap" is not the true band gap. They use the expression

$$pn = A(T)\exp(-E_g/kT) \tag{1}$$

for the pn product in the heavily-doped bipolar-transistor base. This expression is carried over from the case of lightly-doped semiconductors but is not valid in the case at hand. Many authors, including those of Ref. 3 (see also Refs. 4, 5 and 7) recognized the fact that $E_g$ in Eq. (1) is some "effective" gap, but differences from the true gap were not deemed significant. A careful theory[12] reveals that the actual expression for the pn product can be put in the form of Eq. (1), but the quantity $E_g$ is then given by

$$E_g = E_{g,actual}(T,N) - S(T,N) + D(T,N)$$

where $E_{g,actual}$ is the actual band gap which can legitimately be compared with values extracted from luminescence and absorption data as discussed above, S is a quantity that arises from the use of appropriate degenerate statistics, and D is a quantity that arises from properly averaging over configurations in order to take into account the disorder in the impurity distribution. S can be calculated exactly as a function of temperature T and concentration N. Its values are not negligible and, in fact, make the discrepancies of Fig. 2 even worse. D can be calculated only approximately by assuming a model for disorder. Using a relatively simple model, we find values of D which are substantial, opposite in sign from S, yielding band-gap values in the range of values extracted from luminescence and absorption spectra as discussed above. The uncertainties in disorder theories, however, suggest that the transistor data are not suitable for the extraction of accurate gap values.

The overall conclusion is that accurate values for the gap shrinkage are conveniently extracted only from luminescence spectra. It can then be shown that the optical-absorption and transistor data are consistent with these values.

## DEEP IMPURITIES AND DEFECTS

As we already noted in the Introduction, deep centers affect the properties of the host material very differently from shallow impurities. The most important difference is that deep centers introduce bound states in the midgap region, which act as "stepping stones" to enhance electron-hole recombination across the band gap. Many experimental techniques have been developed over the years for the study of such deep centers. In most cases, one uses external probes to cause transitions of electrons from the bands to the localized state or *vice versa* while monitoring the process directly or indirectly. With the help of appropriate theory, one can then extract information about ionization energies, capture and emission cross sections, and other useful properties. One of the most difficult problems, however, is to establish a reliable identification of the defect that gives rise to the observed energy levels. For years, deep levels resulting from the incorporation of a given impurity X in a host were casually attributed to X occupying a substitutional site. Recently, however, evidence has been mounting that the majority of deep centers are more likely to be complexes.[13] After all, shallow impurities are very similar to host atoms so that substitution causes a weak perturbation and hence a shallow localized state. In contrast, deep levels arise from strong perturbations caused by impurities which are significantly different from host atoms. It would, therefore, be reasonable to expect such impurities to prefer some bonding arrangement which is different from that preferred by host atoms. Take Si, for example, as a host and consider P and Au impurities. A P atom at a Si site can be stable since four of its five valence electrons can easily play the role of the four valence electrons of the Si atom that used to be there. An Au atom, on the other hand, has only one valence electron and cannot reform the four bonds that the Si atom used to have with the four nearest neighbors. Lattice relaxations, pairing with other impurities, rebonding, or a combination of the above are then likely to occur for a net gain of energy. Identification is, therefore, one of the most important tasks that must be accomplished for any deep center. Identification is, however, one of the most difficult tasks as well, because it is inevitably a process of elimination. Reliable theoretical predictions about the properties of different candidates can be essential in achieving identification and establishing understanding of the role of each deep center.

The theory of deep centers evolved very slowly over the last thirty years. It is not appropriate to review here the various methods and their shortcomings. A fairly detailed review may be found in Ref. 13. In the last few years, however, significant progress has been achieved, and theoretical descriptions of deep centers became possible at the same level of sophistication and accuracy as was possible for perfect crystals.[14] The breakthrough was accomplished by implementing practical versions of methods using Green's functions. These methods exploit the underlying periodicity of the host *and* the localization of the defect potential. I will not describe any aspects of the mathematical apparatus employed in these calculations. Instead, I shall use selected examples of results to illustrate the kind of progress that has been achieved through these calculations.[15] In particular, I shall use examples to illustrate how the recent calculations made valuable contributions in four different areas: (1) the determination of the electronic structure of deep centers and checking the validity of existing semiempirical models; (2) the identification of defects; (3) the prediction of new effects; and (4) the development of new models to study defect complexes.

Electronic Structure

One of the primary aims of the theory of deep centers is to determine the defect-induced changes in the electronic structure of the host. In particular, one wishes to predict the number, energy positions, wavefunctions, *etc.*, of the localized states associated with a given center. Let me take the single vacancy in tetrahedral semiconductors as an example. The electronic structure of these materials can be understood quite simply in terms of bonds formed between $sp^3$ hybrid orbitals on adjacent atoms. When an atom is removed, there are four "dangling bonds" on the four nearest neighbors pointing in the direction of the vacant site. Thus, one expects the vacancy to have localized states resulting from these dangling bonds. Simple symmetry considerations suggest that the four dangling bonds would give rise to an s-like singlet state (fully symmetric combination of the four dangling bonds) and a p-like triplet state (a combination with alternate positive and negative signs). In the absence of detailed calculations, however, one cannot reliably tell the energy positions or even the ordering of these states. It has been assumed, however, that, in the absence of lattice relaxation, the dangling-bond-like states have $sp^3$ character and that departure from $sp^3$ character is a sign of relaxation of the surrounding lattice.[16] It has also been argued that the ordering of the s-like and p-like dangling-bond states in the III-V compounds may be opposite of that occurring in Si.[17]

The recent calculations obtained accurate descriptions of vacancies in Si and some III-V compounds and provided a check of the above qualitative pictures. It was indeed found that, in the absence of lattice relaxation, the vacancy has an s-like and a p-like localized state built primarily from the dangling bonds. It was found, however, that the p-like level lies above the s-like level in both Si and III-V compound semiconductors, a result whose consequences turn out to be important (see Section III.2 below). Furthermore, whereas the wavefunctions of these states have $sp^3$ character in Si, the same is not true in the compound semiconductors (Fig. 8). For example, in Fig. 7, we show the charge density of the p-like state of the Si vacancy[14,18] illustrating its $sp^3$ hybrid character. In Fig. 8, we show the charge density of the p-like state of the Ga vacancy in GaP[19] and see that it has almost pure p character on the nearest neighbors.

In addition to the dangling-bond-like localized states, vacancies were found to have a number of other localized states which are degenerate with the

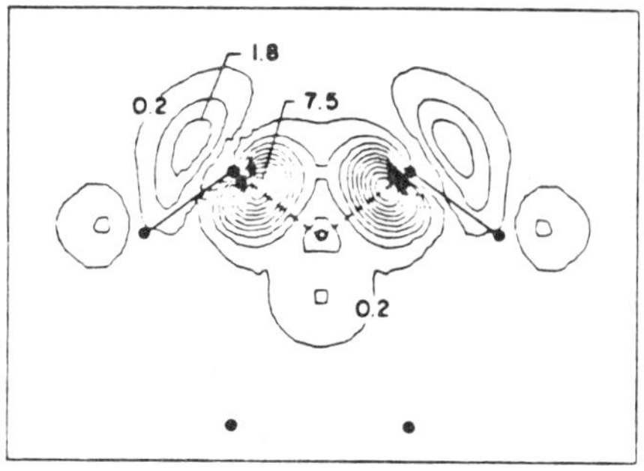


**Figure 7.** The charge density of the p-like state of the Si vacancy. Note the $sp^3$-like character at the nearest neighbors (from Ref. 14).

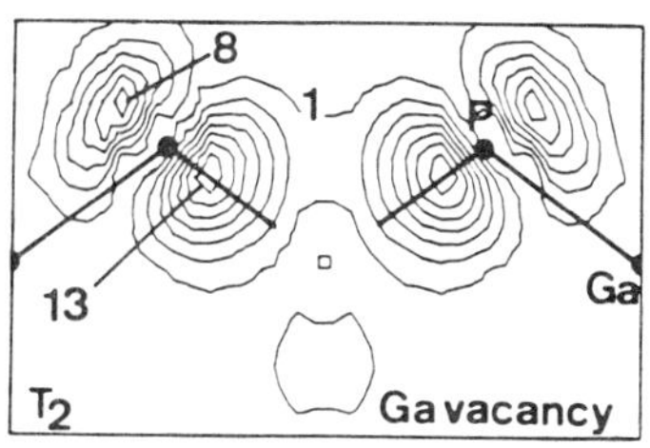


**Figure 8.** The charge density of the p-like state of the Ga vacancy in GaP. Note the almost pure p character at the nearest neighbors (from Ref. 18).

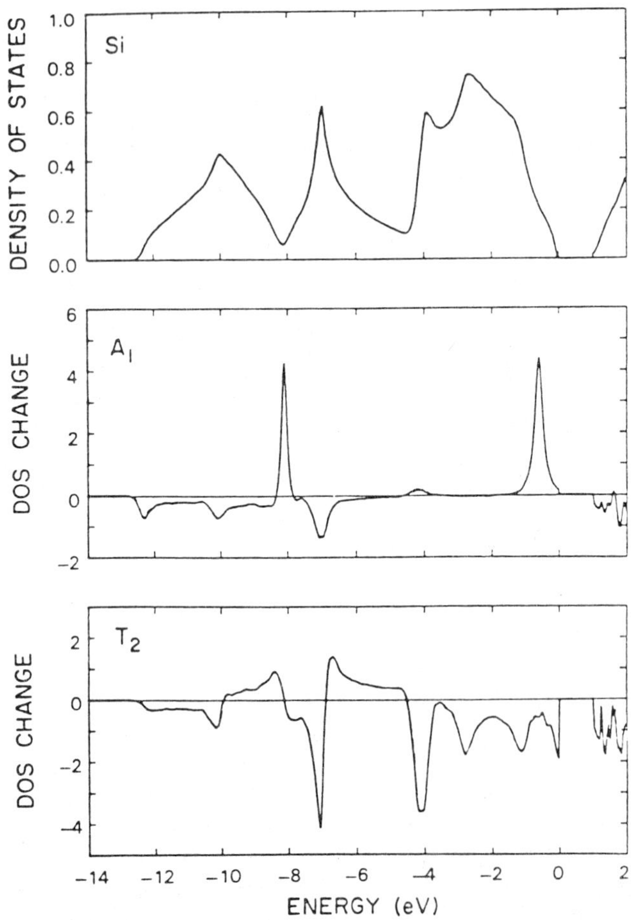


**Figure 9.** The density of states of a perfect Si crystal (top panel) and the changes in the density of s-like ($A_1$) and p-like ($T_2$) states induced by a vacancy (from Ref. 14, 18).

perfect-crystal band continua. In Fig. 9 we show the detailed results for the vacancy in Si. The origins and wavefunction character of these states are discussed in Ref. 18. Finally, we should remark that similar detailed calculations and analysis of the resulting localized states has been carried out for a variety of impurities in Si and the III-V compounds.[15,20,21] We will not attempt a review of those results here.

## Identification of Defects

We have already mentioned that the identification of defects is a very difficult task. In Si, electron paramagnetic resonance (EPR) proved to be a very valuable tool in identifying irradiation-induced defects such as the vacancy, vacancy-impurity complexes, *etc.*[22] In compound semiconductors, however, broad lines and other difficulties made EPR impractical. In 1976, Kaufmann *et al.*[23] overcame these difficulties and successfully used EPR to identify the antisite defect in GaP. Soon after, Kennedy and Wilsey[17] reported the identification of the Ga vacancy in GaP. One of the assumptions in that work concerned the ordering of the dangling-bond-like s-like and p-like states. In Si, it had been known that the p-like state lies above the s-like state and is only partially occupied, triggering a distortion of the surrounding lattice.[24] The observed EPR signal in GaP, however, revealed the center had the full tetrahedral symmetry of an atomic site. Thus, in attributing the observed signal to the Ga vacancy, Kennedy and Wilsey[17] suggested that the ordering of the levels is reversed in GaP, namely, the s-like level lies above the p-like level. Detailed calculations,[19] however, revealed that the p-like level actually lies above the s-like level by about 1 eV for all charge states and for either inward or outward symmetric relaxation of the neighbors (Fig. 10). This result led the authors of Ref. 19 to suggest that the observed EPR signal is not due to Ga vacancies. A reexamination of the experimental data of Ref. 17 revealed that carbon impurities at Ga sites would be a suitable candidate. Kennedy and Wilsey had ruled out carbon on the grounds that it is likely to be shallow, and thus have a very different EPR spectrum. Calculations,[19] however, demonstrated that carbon is, in fact, deep and would be consistent with all the available data. Since the publication of Ref. 19, a number of experimental efforts have been undertaken to test the two conflicting assignments of the observed EPR signal, but no new definitive conclusions have been reached.

## Predictions of New Effects

We have already mentioned that the Si vacancy is unstable against symmetry-breaking (Jahn-Teller) distortions because it has a partially occupied bound state in the gap. Several charge states of the vacancy were identified from EPR experiments, but a number of observations remained unexplained.[24]

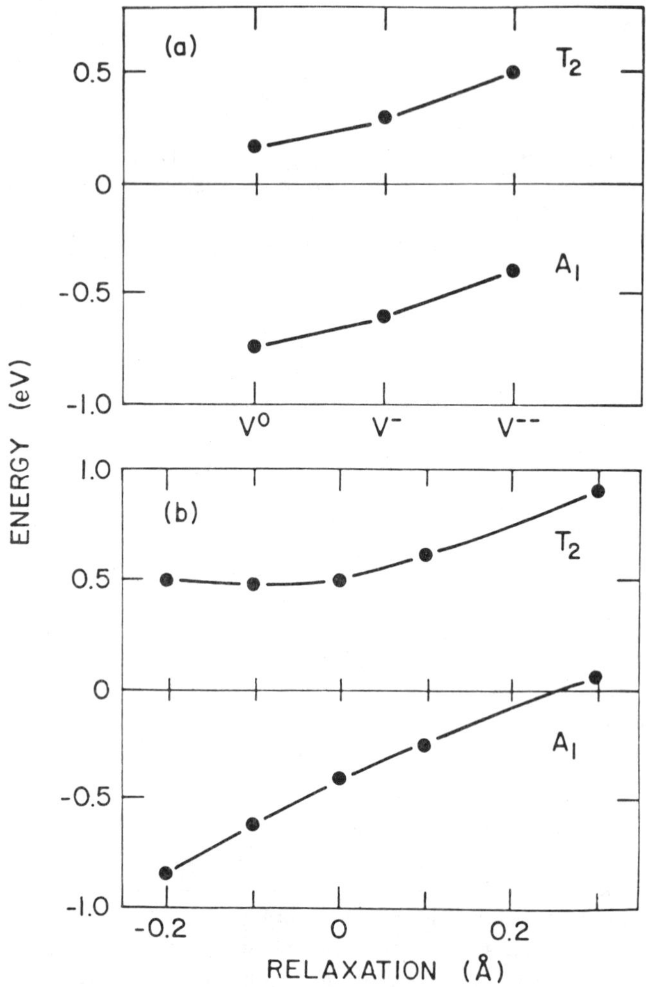


**Figure 10.** The energy positions of the p-like ($T_2$) and s-like ($A_1$) states of the Ga vacancy in GaP as functions of charge state and symmetric inward and outward relaxation of the neighbors (from Ref. 19).

Detailed calculations of the effect of distortions on the energy levels of the vacancy were carried out[25,26] and supplemented by a phenomenological model of the restoring forces that counter the Jahn-Teller-induced distortions.[25] The results indicated that the vacancy is likely to be a negative-U center. The existence of such centers was postulated by Anderson[27] as a possible explanation of the absence of free spins in chalcogenide glasses. Normally, deep centers have a positive U, which means that if it costs an energy $\epsilon$ to take an electron from the valence bands and put it in an empty bound state in the gap, it would cost more energy, $\epsilon + U$, to put a second electron in the same level. The extra energy U is the result of electron-electron repulsions. A negative-U center can occur if lattice distortions are a strong function of the charge state of the center. It is then possible for lattice distortions to provide the energy needed to overcome the electron-electron repulsion so that the effective U is negative. It was predicted in Ref. 25 that the vacancy in Si is such a center and can exist only in a ++ or neutral state. The prediction was consistent with the observation that the + state of the vacancy could be detected only after photoexcitation.[24] Since the publication of Ref. 25, additional experiments have confirmed that the vacancy in Si is indeed a negative-U center.[28]

## Complex Defects

The parameter-free Green's-function calculations that resulted in accurate descriptions of many point defects become very time-consuming and can strain the storage capacity of modern computers if used to describe complex defects such as multivacancies, vacancy-impurity complexes, *etc.* In a recent study,[29] it was demonstrated that the accurate results obtained for the single vacancy can be used to develop a simple model and obtain a systematic description of multivacancies. Calculations for multivacancies were carried out using a semiempirical Hamiltonian which was defined by requiring that it yields a satisfactory description of both the perfect-crystal band structure *and* the single-vacancy localized states. Once such a Hamiltonian is constructed, calculations for multivacancies are quite straightforward using the Green's-function technique. It was noted that the resulting bound states in the gap region could be understood as arising primarily from the dangling bonds as we saw to be true for the single vacancy. In fact, it was concluded that these localized states could be described quantitatively by simply expanding their wavefunctions in terms of the dangling bonds and fitting the necessary matrix elements to the Green's-function results for the single vacancy. Thus, the dangling-bond-like localized states of any multivacancy can be obtained quantitatively by simply diagonalizing a matrix whose size is equal to the number of dangling bonds. This simple scheme can be formally justified by recognizing that the basis functions need not be viewed as simple dangling bonds, but as generalized basis functions having the symmetry properties of

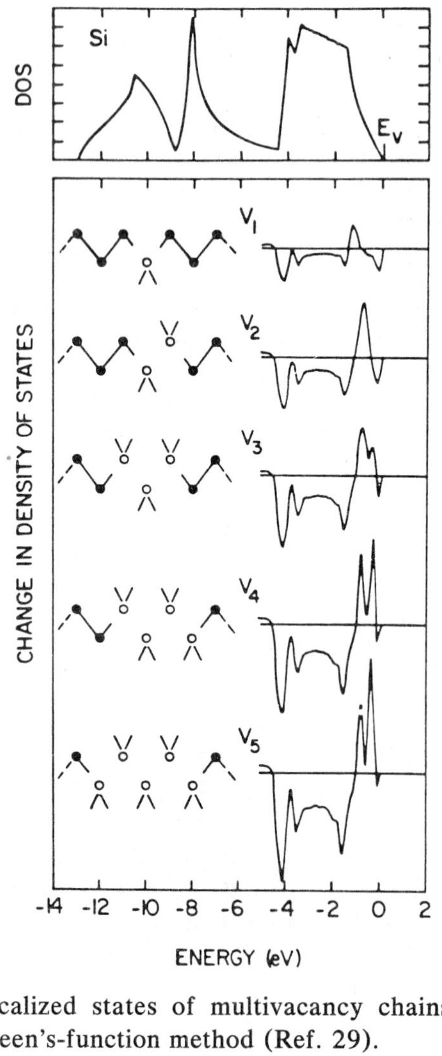


**Figure 11.** The localized states of multivacancy chains in the gap region calculated by the Green's-function method (Ref. 29).

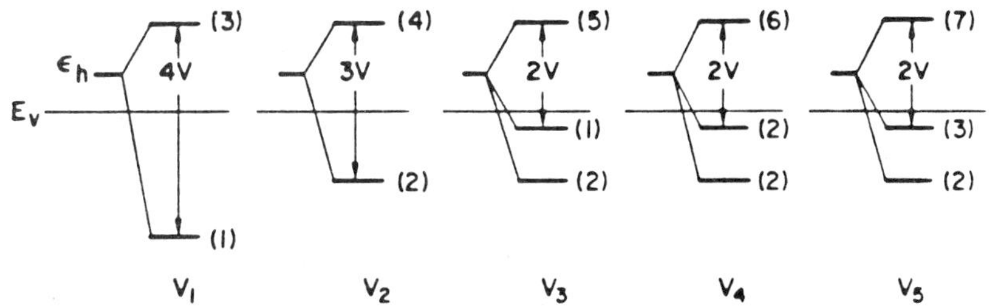


**Figure 12.** The localized states of multivacancy chains in the gap region calculated by the "dangling-bond model" (Ref. 29).

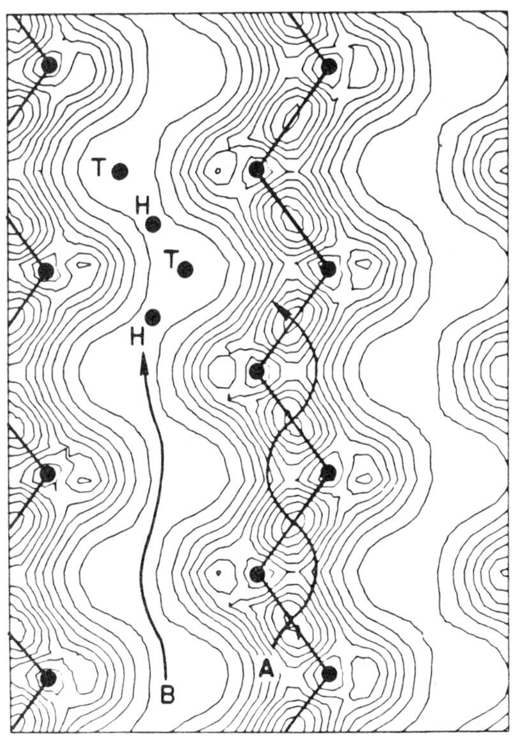


**Figure 13.** The charge density of a perfect Si crystal in the (110) plane. T and H denote the high-symmetry tetrahedral and hexagonal interstitial sites. Path A was proposed in Ref. 31 as the low-temperature migration path for the self-interstitial. Path B was proposed as the more likely migration path in Ref. 29.

dangling bonds and whatever backbond structure is necessary to produce exact solutions in the case of the single vacancy. A fuller discussion may be found in Ref. 29. The success of the simple model is illustrated in Figs. 11 and 12. Figure 11 shows the Green's-function results for the multivacancy chains along the (110) direction. Figure 12 shows the corresponding results of the "dangling-bond model."

The self-interstitial in Si has provided another example of using the accurate, parameter-free calculations as benchmarks to develop models for more complicated systems. Parameter-free Green's-function calculations were carried out for a Si atom at the high-symmetry tetrahedral site (Fig. 13).[30] A semiempirical Hamiltonian was then constructed, fitting the parameters so as to reproduce those results. The semiempirical Hamiltonian was then used to determine the properties of the self-interstitial at various points along a path in the interstitial region.[29] These studies led to a reexamination of the accepted interpretation of experimental data according to which the self-interstitial migrates efficiently along path A in Fig. 13.[31] It was concluded that path B (Fig. 13) is more likely to be preferred. A fuller discussion of this work is given in Ref. 29.

## ACKNOWLEDGMENTS

I am greatly indebted to J. Bernholc, I. Ivanov, N. O. Lipari, M. Scheffler, A. Selloni, and J. P. Vigneron for their fruitful collaboration in various aspects of the work described in this review. This work has been supported in part by the Office of Naval Research under Contract No. N-00014-80-C-0679.

## REFERENCES

1. A. A. Vol'fson and V. K. Subaschiev, Fiz. Tekh. Polupravodn. **1**, 397 (1967) *Sov. Phys. Semicond.* **1**, 327 (1967); M. Balkanski, A. Aziza, and E. Amzallag, *Phys. Status Solidi* **31**, 323 (1969).

2. W. L. Kaufmann and A. A. Bergh, *IEEE Trans. Electron Devices* **ED-15**, 732 (1968).

3. J. W. Slotboom and H. C. de Graaf, *Solid State Electron.* **19**, 857 (1976).

4. R. W. Keyes, *Comments Solid State Phys.* **7**, 149 (1977).

5. P. E. Schmid, *Phys. Rev.* **B23**, 5531 (1981).

6. P. E. Schmid, M. L. W. Thewalt, and W. P. Dumke, *Solid State Commun.* **38**, 1091 (1981).

7. For a review of early theoretical work, see R. A. Abram, A. J. Rees, and B. L. H. Wilson, *Adv. Phys.* **27**, 799 (1978).

8. J. C. Inkson, *J. Phys. C* **9**, 1177 (1976); P. A. Sterne and J. C. Inkson, *J. Appl. Phys.* **52**, 6432 (1982).

9. G. D. Mahan, *J. Appl. Phys.* **51**, 2634 (1980).

10. K. F. Berggren and B. E. Sernelius, *Phys. Rev.* **B24**, 1971 (1981).

11. A. Selloni and S. T. Pantelides, *Phys. Rev. Lett.* **49**, 586 (1982).

12. S. T. Pantelides and A. Selloni, to be published.

13. See, for example, discussion of this issue in the review article by S. T. Pantelides, *Rev. Mod. Phys.* **50**, 797 (1978).

14. The first such results were reported simultaneously by two different groups: J. Bernholc, N. O. Lipari, and S. T. Pantelides, *Phys. Rev. Lett.* **41**, 895 (1978); G. A. Baraff and M. Schlüter, *Phys. Rev. Lett.* **41**, 892 (1978).

15. For other recent reviews, see M. Scheffler, Festkörperprobleme **22**, 115 (1982) and M. Jaros, *Deep Levels in Semiconductors* (Adam Hilger, Bristol, 1982).

16. See, *e.g.*, U. Kaufmann and J. Schneider, Festkörperprobleme **20**, 87 (1980).

17. T. A. Kennedy and N. D. Wilsey, *Phys. Rev. Lett.* **41**, 977 (1978).

18. J. Bernholc, N. O. Lipari, and S. T. Pantelides, *Phys. Rev.* **B21**, 3545 (1981); G. A. Baraff and M. Schlüter, *Phys. Rev.* **19**, 4965 (1980).

19. M. Scheffler, S. T. Pantelides, N. O. Lipari, and J. Bernholc, *Phys. Rev. Lett.* **47**,, 413 (1981).

20. G. A. Baraff, E. O. Kane and M. Schlüter, *Phys. Rev.* **B25**, 548 (1982).

21. J. Bernholc, N. O. Lipari, S. T. Pantelides, and M. Scheffler, *Phys. Rev.* **B26**, 5706 (1982).

22. For a review, see J. W. Corbett and J. C. Bourgoin, in *Point Defects in Solids*, edited by J. H. Crawford, Jr., and L. M. Slifkin, (Plenum, New York, 1975), p. 1.

23. U. Kaufmann, J. Schneider, and A. Räber, *Appl. Phys. Lett.* **29**, 312 (1976).

24. For a review of the Si vacancy, see G. D. Watkins, J. R. Troxell, and A. P. Chatterjee, in *Defects and Radiation Effects in Semiconductors*, edited by J. H. Albany, (Inst. of Phys., London, 1979), p. 16.

25. G. A. Baraff, E. O. Kane, and M. Schlüter, *Phys. Rev. Lett.* **43**, 956 (1979).

26. N. O. Lipari, J. Bernholc, and S. T. Pantelides, **43**, 1354 (1979).

27. P. W. Anderson, *Phys. Rev. Lett.* **34**, 953 (1975).

28. G. D. Watkins and J. R. Troxell, *Phys. Rev. Lett.* **44**, 593 (1980); J. L. Newton, A. P. Chatterjee, R. D. Harris, and G. D. Watkins, Proceedings of the 12th Intern. Conf. on Defects in Semiconductors, Amsterdam 1982, in press.

29. S. T. Pantelides, I. Ivanov, M. Scheffler, and J. P. Vigneron, Proceedings of the 12th Intern. Conf. on Defects in Semiconductors, Amsterdam 1982, in press; I. Ivanov and S. T. Pantelides, to be published.

30. J. P. Vigneron, M. Scheffler, and S. T. Pantelides, Proceedings of the 16th International Conference on the Physics of Semiconductors, Montpellier, France, 1982, in press.

31. G. D. Watkins, R. P. Messmer, C. Weigel, D. Peak, and J. W. Corbett, *Phys. Rev. Lett.* **27**, 1573 (1971); C. Weigel *et al., Phys. Rev.* **B8**, 2906 (1973).

# SILICIDE CONTACT AND GATE IN MICROELECTRONIC DEVICES

K.N. Tu

IBM T.J. Watson Research Center

Yorktown Heights, N.Y. 10598

## ABSTRACT

A brief review about silicide metallization technology, i.e. the use of transition metal silicides as contacts and gates in large scale integrated silicon devices is presented and a short discussion about the low temperature redistribution of dopant induced by a moving silicide-silicon interface is included.

## I. INTRODUCTION

As device dimensions are reduced, device performance will become very sensitive to dimensional precision control. The smaller the device structure, the tighter the tolerance for statistical parameter variations. For example, we may not tolerate a variation of line width by more than $0.1\mu m$ in an interconnection line of $1\mu m$ wide. To process small structure with precision is challenging; not only the lateral dimensions of a device such as line width and gate width are of concern, the vertical dimensions such as gate oxide thickness and contact depth have to be scaled down as well.[1] In this paper we shall briefly review the so-called silicide metallization technology, i.e. the use of silicides as contact and gate in microelectronic devices.

In the case of a contact, a shallow junction needs a shallow contact, otherwise a deep one may punch through the junction. This is illustrated by a schematic diagram in Fig. 1. It shows that as the lateral dimensions of a contact is approaching 1$\mu$m, the vertical dimensions of the shallow P-N junction and the depth of the silicide contact in a bipolar configuration will be about 250nm and 20nm, respectively.[2] The depth of a silicide contact is measured by the amount of Si consumed from the substrate forming the contact. To control the depth to within 20nm means we have to tailor the reaction at the silicide-silicon interface so that a uniform and limited interfacial reaction can be achieved. Such a limited reaction would also reduce the sidewise reaction as shown in Fig. 1. Sidewise reaction during contact formation affects the lateral dimensions, so it is of concern in a short channel MOSFET device. Moreover, a limited interfacial reaction would cause a much smaller redistribution of dopant near the interface, a subject which we shall discuss at the end (section IV) of this paper. In section II, a general scheme of making shallow silicide contacts using refractory metal alloys and Si alloys will be reviewed, including their Schottky barrier behavior.

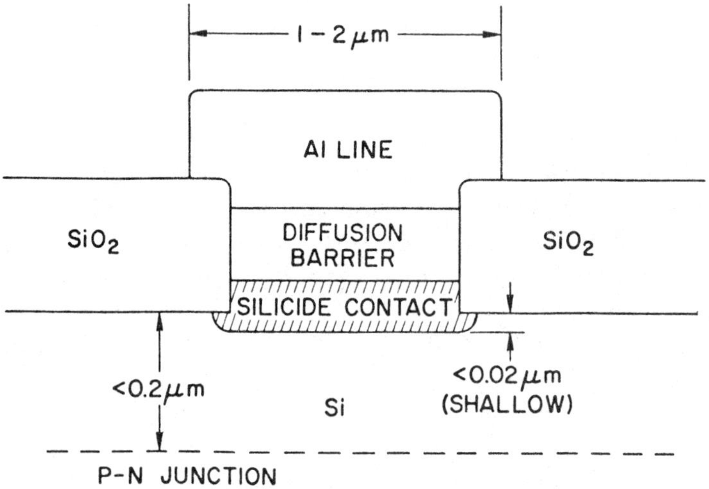


Fig. 1. Schematic diagram of the cross-sectional view of a shallow silicide contact on a shallow junction.

In the case of an FET gate structure, the low conductivity of a heavily doped poly-silicon gate has become a limiting factor in reducing the gate dimension, hence a composite of silicide and poly-silicon (the so-called polycide)[3-5] has replaced poly-silicon as gate on top of the gate oxide as shown in Fig. 2. The composite takes advantage of the high conductivity of the silicide and the thermal stability of the interface between poly-silicon and silicon dioxide. In section III, a brief review of the composite material and its properties are given. Since silicide is also used as ohmic contact to the heavily doped source and drain regions in the FET structure, it is desirable to find a silicide which can serve as gates and contacts at the same time. This means we can process them together. As a result it not only simplifies the manufacturing steps, but also reduces the lateral misalignment due to a repeated lithographic procedures for gates and contacts. The latter is of serious concern in shrinking the lateral dimensions of a device. How to achieve a self-aligned silicide gate and silicide contact of submicron dimension in an FET structure is a subject of wide interest at the present time,[6,7] and it will not be reviewed in this paper.

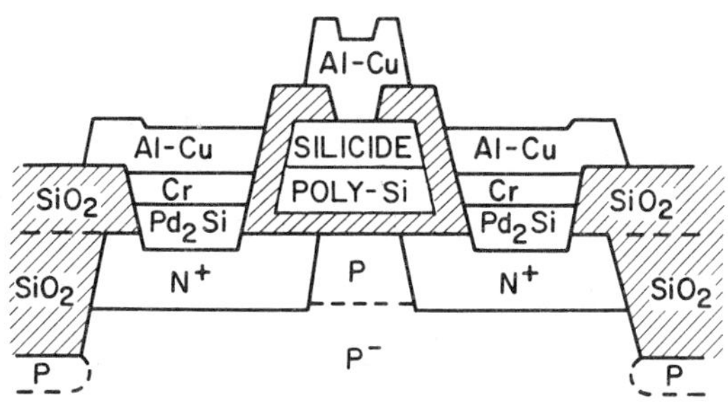


Fig. 2. Schematic diagram of the cross-sectional view of an MOS-FET device with silicides as contact and gate.

## II. FORMATION OF SHALLOW SILICIDE CONTACTS

The contact reaction between a transition metal and silicon to form silicide can be divided into three steps. First, the metal must diffuse through or dissolve the native oxide on the Si surface, then, interdiffusion and dissolution between the metal and Si occurs, and finally silicide compound formation takes place. It is conceivable that we may choose a metal which can dissolve the native oxide but forms no silicides, hence it makes an intimate contact to Si yet it consumes very little Si. We note that Al is such a metal. However, it has caused the so called "Al-penetration" problem[8] due to the dissolution reaction which is hard to avoid. Typically, a contact to Si without silicide formation always shows a wide spread in its electrical properties, unacceptable for a high performance device.

On the other hand, if a ultra-high vacuum system and a Si substrate free of any surface oxide are available and if we can deposit an epitaxi-

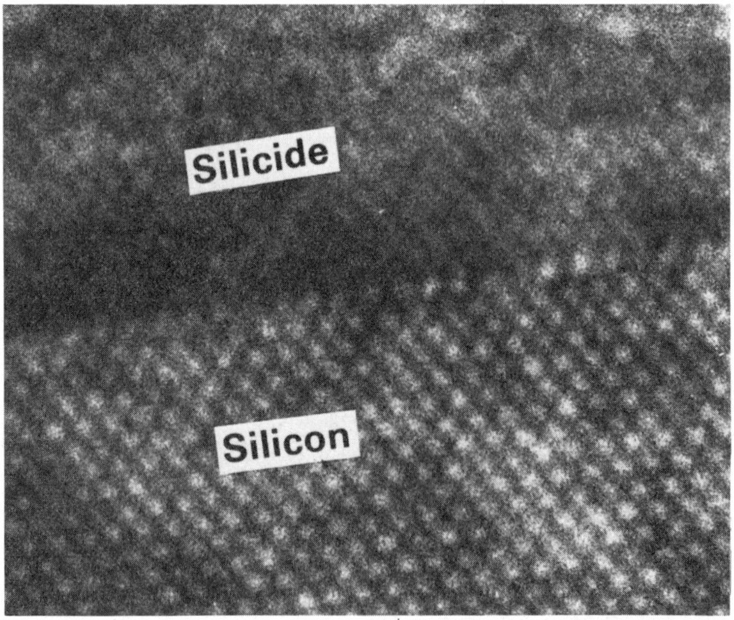


Fig. 3. High resolution lattice image of a cross-section of the epitaxial interface between $NiSi_2$ and (100) Si.

al layer of silicide by co-deposition, sputtering or molecular beam, we can obtain the ideal contact with an atomically flat interface. A high resolution transmission electron micrograph of a cross-sectional view of an epitaxial $NiSi_2$ on (100)Si is shown in Fig. 3.[9] The epitaxy occurs because $NiSi_2$ has a cubic $CaF_2$ structure with a lattice parameter a=3.54Å.[10] Obviously, Fig. 3 shows what may be the best structure we can do about a silicide contact, nevertheless, it is not practical at present to process such a contact on a device, instead a reactive contact which is uniform and limited is most desirable.

In addition to controlling the interfacial silicide formation, we also have to control electrical contact properties, such as Schottky barrier

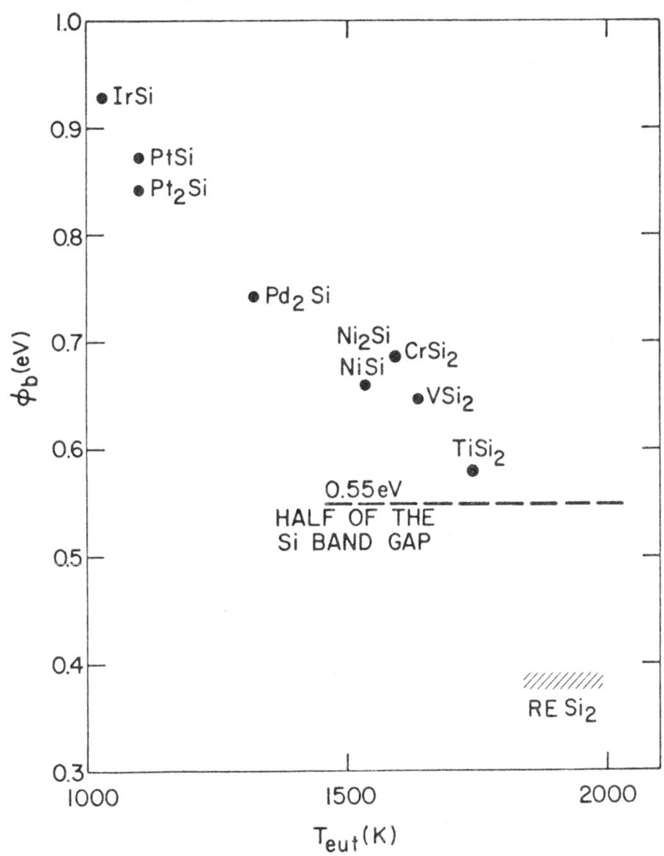


Fig. 4. Plot of Schottky barrier height against eutectic temperature of transition and rare-earth metal silicides.

height and contact resistance. We shall concentrate on the barrier height, and there are two points worth mentioning. First, the control of contact properties and the control of contact reaction are inseparable. Specifically, "formation of Schottky barrier" is directly related to "formation of silicides" in silicide-Si contacts. Electronic equilibrium across a contact which defines a Schottky barrier has to be considered in conjunction with chemical equilibrium across the contact. Second, the Schottky barrier height of silicides on Si exists over a very wide range from 0.93 to 0.37eV,[11] and its value depends on the chosen silicide rather than on a specific pinning mechanism. In Fig. 4, a plot of Schottky barrier height against eutectic temperature of transition and rare-earth metal silicides is given.[12]

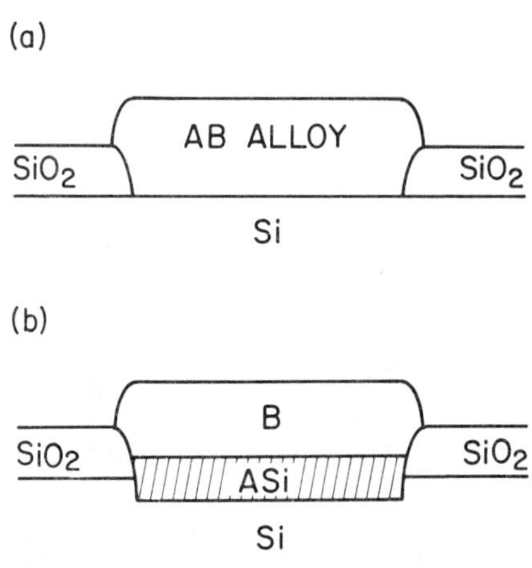


Fig. 5. Schematic diagram of an alloy AB being used to control silicide formation, where A and B form silicide and no silicide respectively. The amount of Si consumed by the reaction can be predetermined by the thickness and composition of the alloy film, and silicide composition. (a). before and (b) after contact reaction.

In the following, we shall review the use of thin alloy films to produce shallow silicide contacts of depth of about 0.01$\mu$m and correlate them to their Schottky behavior.[2] Contacts of both high barrier (0.85 to 0.75eV) and low barrier (0.55 to 0.40eV) will be discussed. Shallow ohmic contacts can be made on heavily doped Si using the same approach, so they will not be presented.

The primary effect of using thin alloy films on Si for shallow silicide contact formation is to dilute the interfacial reaction so that a limited reaction can be achieved, see Fig. 5. In selecting an element B to alloy with an element A which forms a silicide such as ASi, we require that B has to fulfill two criteria. First, B has to form solid solution with A so that a homogeneous alloy of AB can be prepared on Si in order to achieve a uniform interfacial reaction over the entire contact area. Second, B does not react with Si to form silicides, otherwise it defeats the purpose of alloying. Now, if we take A to be Pd or Pt whose silicides have a high Schottky barrier height on n-type Si, we can choose B from the following three groups, namely, (a) Au, Ag and Al, (b) Ge and Si, and (c) refractory metal elements such as W and V. We note that refractory metals form silicides such as $WSi_2$ and $VSi_2$, so it

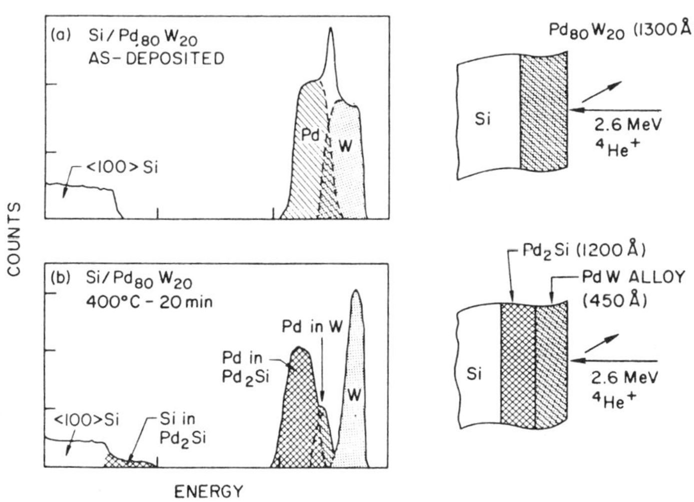


Fig. 6. Rutherford backscattering spectra of a 1000Å $Pd_{80}W_{20}$ alloy before and after the annealing at 400°C for 20 minutes on Si.

appears at first that we may not use them. However, it is known that the temperatures of silicide formation of near-noble metals and refractory metals are quite different, the former such as $Pd_2Si$ forms around 200°C and the latter such as $WSi_2$ forms around 700°C.[2] Their differences are summarized in Table I. It is conceivable that at a temperature around 400°C, which is above the formation temperature of near-noble metal silicides but below that of refractory metal silicides, an alloy of a near-noble metal and a refractory metal might react selectively with Si and result in the formation of a near-noble metal silicide alone. In other words, by making use of the wide separation of their silicide formation temperatures, we can alloy Pd and W for a limited $Pd_2Si$ contact formation. The question whether Pd and W forms solid solution or not can be easily answered by an inspection of the Darken-Gurry plot.[13]

In the following, we shall briefly review shallow silicide contact formation by using Pd-W[14,15] alloys and Pd-Si alloys.[16,17] We will not consider alloys containing Au, Ag or Al. Fig. 6 shows Rutherford backscattering spectra (RBS) of a $Pd_{80}W_{20}$ alloy deposited on Si in the as-deposited states and after an annealing at 400°C for 20 minutes.[1] After the annealing, the signal of Pd and the signal of W parted company. While the Pd signal has extended to lower energy channels joining the Si, the W signal grew at the higher end, indicating the depletion of

Table I.: Comparison Between Near-noble Metal Silicide and Refractory Metal Silicide.

| Kinetics | Near-noble metal | Refractory metal |
|---|---|---|
| 1st phase formation | $M_2Si$ | $MSi_2$ |
| Formation temperature | ~200°C | ~600°C |
| Growth rate | $x^2 \sim T$ | $x \sim T$ |
| Activation energy of growth | 1.1-1.5eV | >2.5eV |
| Dominant diffusion Species | Metal | Si |
| Mechanism of metal diffusion in Si | Interstitial | Substitution |

Pd. These results indicate the formation of a two-layer structure by the annealing. We note that the inner layer of $Pd_2Si$ can be used as a contact to the Si and the outer layer of W-rich alloy can serve as a diffusion layer to protect the contact against Al-penetration.

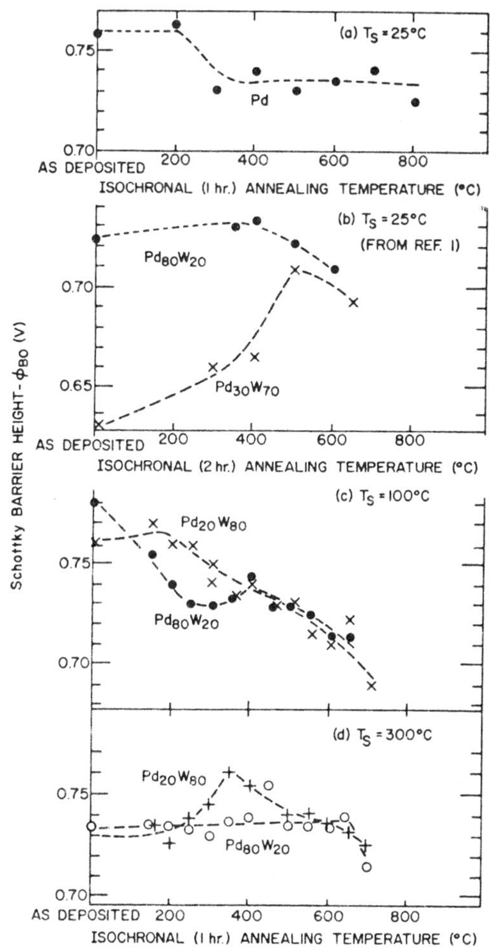


Fig. 7. Schottky barrier height of Pd and Pd-W alloys on Si as a function of isochronal annealing temperatures: (a) pure Pd deposited at 25°C, (b) alloys deposited at 25°C, (c) alloys deposited at 100°C, and (d) alloys deposited at 300°C.

In Fig. 7, Schottky barrier heights of Pd and Pd-W alloys on n-type Si as a function of isóchronal annealing temperature and deposition temperature are shown.[14] We found that by maintaining a substrate temperature of 100°C during the deposition of W-rich $Pd_{20}W_{80}$ alloys, contacts of $Pd_2Si$ on Si with a Schottky barrier height of 0.74eV can be obtained without annealing. In Fig. 8, Schottky barrier heights of Pd-Si alloys on n-type Si as a function of isochronal annealing are shown.[16] A comparison between the Si alloy and the refractory metal alloy for shallow silicide formation has already been published.[15] Since it is frequent to encounter the so-called "parallel contacts" or "mixed diodes" in a nonuniform reaction between an alloy and Si, parallel contacts have been analyzed by current-voltage, capacitance-voltage[18,19] and photo-response[15] measurements.

## III. REFRACTORY METAL SILICIDE ON FET GATE

Unlike the application in Schottky and ohmic contacts wherein the near-noble metal silicides such as PtSi and $Pd_2Si$ are commonly used, the application of silicide in FET gate has been dominated by the use of refractory metal silicides[2-4] e.g. $WSi_2$ and $TaSi_2$. To serve together with poly-silicon as a gate, the silicide must be a good conductor, stable with the poly-silicon so that the gate oxide is not affected, and capable to survive a high temperature process without oxidation. It is due to the requirements of the high temperature stability with poly-silicon and against oxidation that silicides of refractory metals rather than refractory metals themselves are selected, otherwise the latter would be favorable for their high conductivity. For these requirements, the conduction and oxidation behavior of refractory metal silicides are the two most important properties for their application as a gate. The oxidation of silicide is also interested for the formation of an isolating layer on top of the gate since an additional metallization layer for interconnection is required above the gate. Table II lists the conductivity and other properties of some of the refractory metal disilicides.

To form a layer of refractory metal silicide on poly-silicon, we can deposit a refractory metal film on the Si and followed by a high temperature annealing (>600°C) to form the disilicide, or we can codeposit the disilicide directly on the Si. It has been shown that the codeposition produces a more uniform layer on the poly-silicon,[4] besides the co-deposition can be adjusted to prepare a Si-rich film for oxidation so that the excess Si in the disilicide rather than the poly-silicon will be consumed by the oxide growth. However, the codeposition tends to produce a low conductivity film in the as-deposited state so a high temperature annealing (~900°C) is required to improve it.

When a composite of refractory metal disilicide and poly-silicon is oxidized in dry or wet oxidation atmosphere, a layer of $SiO_2$ grows on top of the disilicide. The growth consumes the poly-silicon rather than the disilicide and the latter serves as a mass transport layer for the Si so

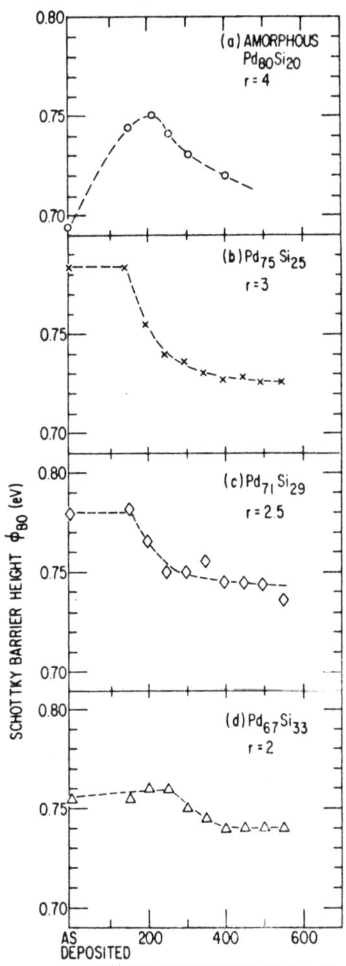


Fig. 8. Schottky barrier height of various Pd-Si alloys on n-Si as a function of isochronal (30 min.) annealing temperatures.

it is displaced inward after the oxidation. In other words, the disilicide survives the oxidation process and remains a good conductor. In this reaction, the thermal stability of the disilicide and the kinetics of mass transport across the disilicide are the two fundamental questions. The stability will depend on the formation energy of the disilicide, refractory metal oxide and $SiO_2$ and whether there is excess Si or a layer of poly-silicon under the disilicide. It is clear that not all the disilicides will behave the same, nevertheless, when excess Si is available, many refractory metal disilicides such as $WSi_2$, $MoSi_2$ and $TaSi_2$ have been found to be stable during the oxidation process. About the kinetics of mass transport, there are two possible paths. One involves the dissociation of refractory metal disilicide wherein the dissociated Si reacts with

Table II: Properties of Some Refractory Metal Disilicides*

| Property | $TiSi_2$ | $TaSi_2$ | $MoSi_2$ | $WSi_2$ |
|---|---|---|---|---|
| Resistivity ($\mu\Omega$cm) | 15-25 | 35-50 | 90-100 | 70 |
| Heat of Formation (kcal/g-atom) | -10.7 | -8.0 | -9.3 | -7.5 |
| Thermal Expansion coefficient (ppm/°C) | 10.5 | 8.8 | 8.25 | 7.9 |
| Etch rate in 10:1 BHF (Å/min) | >2000 | 100~300 | low | low |
| Formation Temp on Si (°C) | 600 | 650 | 525 | 700 |

*See M-A. Nicolet and S.S. Lau, in VLSI ELECTRONICS: MICROSTRUCTURE SCIENCE, Norman Einspruch, Series Editor, SUPPLEMENT A - MATERIALS AND PROCESS CHARACTERIZATION, Graydon Larrabee, Guest Editor, (Academic Press, New York, in press).

oxygen atoms to form $SiO_2$ and the metal diffuses inward to react with poly-silicon to form silicide again. The other path involves simply the direct transport of Si form the poly-silicon across the disilicide. Since it has been shown previously that Si is the dominant diffusing species during the growth of refractory metal disilicides,[8] the direct transport of Si across the disilicide layer for the oxidation is expected and verified. The transport will cause a flux of vacancies to condense at the

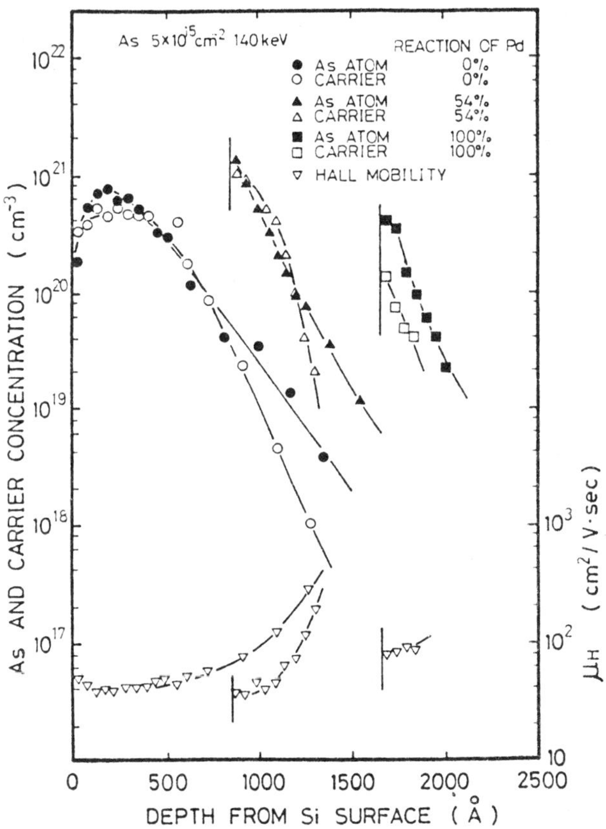


Fig. 9. Depth profiles of total As concentrations measured by neutron activation analysis and of As carrier concentration and Hall mobility measured by differential Hall effect during $Pd_2Si$ growth at 250°C. The circles, triangles and squares represent the distribution before and after annealing at 250°C for 70 and 375 minutes, respectively.

interface between the disilicide and polysilicon and in turn will weaken the bonding or adhesion across the interface. Therefore, peeling tends to occur if the disilicide film is thermally stressed, so the stress in the disilicide cannot be ignored.

## IV. REDISTRIBUTION OF DOPANCY DURING SILICIDE FORMATION

What happens to the dopant in the layer of Si consumed by silicide formation has been an intriguing question. It has been found that dopant atoms can redistribute themselves during silicide formation and a fraction of them has diffused into Si and appeared as a pile-up near the silicide-Si interfaces.[21,22] The phenomenon of As redistribution during $Pd_2Si$ formation at 250°C has been observed, and the depth of redistribution was determined to be about 100Å into the Si. The intriguing point about the phenomenon is that it is a low temperature doping processing because about 50% of the As atoms are found to be substitutional and electrically active. Since it accompanies silicide formation, a shallow silicide contact would reduce the amount of redistribution and in turn the impact on contact properties.

In Fig. 9, actual profiles of As after $Pd_2Si$ formation are shown.[23] The Si wafers were first implanted with 140KeV As atoms at a dose of $5x10^{15}$ ions/$cm^2$. They were annealed at 900°C for 30 minutes in order to remove implantation damage. Then Pd film of 2500Å was deposited onto the wafers and $Pd_2Si$ formation was carried out by annealing at 250°C for 10 to 375 minutes. The profiles of As atoms in the Si before and after the annealings were analyzed by neutron activation analysis[23] and Rutherford backscattering[24] for all the As atoms, differential Hall measurement for electrically active carriers[23], and ion beam channeling for interstitial As solutes.[24] The results of these measurements agree quite well with each other that a low temperature doping of As occurs at 250°C during $Pd_2Si$ formation. Its effects on Schottky barrier height, contact resistance, and shallow p-n junction have been studied.[25]

## REFERENCES


1. K.N. Tu, W.N. Hammer and J.O. Olowolafe, J. Appl. Phys. *51*, 1663 (1980).
2. K.N. Tu, J. Vac. Sci. Technol. *19*, 766 (1981).
3. B.L. Crowder and S. Zirinsky, IEEE Trans. Electronic Devices, ED26, 369 (1979).
4. S.P. Murarka, J. Vac. Sci. Technol. *17*, 775 (1980).
5. A.K. Sinha, J. Vac. Sci. Technol. *19*, 778 (1981).
6. C.M. Osburn, M.Y. Tsai, S. Roberts, C.L. Luchese and C.Y. Ting Proceedings of ECS meeting (1982).
7. C.Y. Ting, S.S. Iyer, C.M. Osburn, G.J. Hu and A.M. Schweighart, Proceedings of ECS Meeting (1982).
8. K.N. Tu and J.W. Mayer, in "Thin Films - Interdiffusion and Reactions" edited by J.M. Poate, K.N. Tu and J.W. Mayer, Wiley-Interscience, New York (1978).
9. H. Föll, P.S. Ho and K.N. Tu, Phil. Mag. *45*, 31 (1982).
10. K.N. Tu, E.I. Alessandrini, W.K. Chu, H. Krautle and J.W. Mayer, Japan J. of Appl. Phys., Supplement 2, Part. 1, page 669 (1974).
11. K.N. Tu, R.D. Thompson and B.Y. Tsaur, Appl. Phys. Lett. *38*, 626 (1981).
12. G. Ottaviani, K.N. Tu and J.W. Mayer, Phys. Rev. B *24*, 3354 (1981).
13. L.S. Darken and R.W. Gurry, "Physical Chemistry of Metals, McGraw-Hill, New York (1953).
14. J.O. Olowolafe, K.N. Tu and J. Angilello, J. Appl. Phys. *50*, 6316 (1979).
15. M. Eizenberg and K.N. Tu, J. Appl. Phys. *53*, 1577 (1982).
16. S. Kritzinger and K.N. Tu, J. Appl. Phys. *52*, 305 (1981).
17. M. Eizenberg, H. Föll and K.N. Tu, J. Appl. Phys., *52*, 861 (1981).
18. I. Ohdomari and K.N. Tu, J. Appl. Phys. *51*, 3735 (1980).
19. R. Thompson and K.N. Tu, J. Appl. Phys. *53* 4285 (1982).
20. T. Okumura and K.N. Tu, J. Appl. Phys. (submitted).
21. M. Wittmer and T.E. Seidel, J. Appl. Phys. *49*, 5827 (1978).
22. I. Ohdomari, K.N. Tu, K. Sugaro, M. Akiyama, I. Kimura and K. Yoneda, Appl. Phys. Lett. *38*, 1015 (1981).
23. I. Ohdomari, K. Sugaro, M. Akiyama, T. Maeda, K.N. Tu, I. Kimura and K. Yoneda, Thin Solid Films *89*, 349 (1982).
24. M. Wittmer, C.Y. Ting, I. Ohdomari and K.N. Tu, J. Appl. Phys. (in press).
25. M. Wittmer, C.Y. Ting and K.N. Tu, J. Appl. Phys. (submitted).

# PHOTOLITHOGRAPHY AND X-RAY LITHOGRAPHY

A. Heuberger

Institut für Festkörpertechnologie, Mikrostrukturtechnik

Lentzeallee 100, D-1000 Berlin 33, Fed. Rep. of Germany

Summary

Very intense technological efforts to increase the integration density of semiconductor devices have been made in the development of new and economical lithography methods for structures below 2 µm. The lithography methods now used for modern fabrication lines are optical projection systems in the wavelength range of 400 to 250 nm, based on high-resolution objective lenses or mirror optics. The resolution limit arising from diffraction is far below 1 µm, and impressive examples of optical sub-µm patterning have already been realized. With increasing resolution, however, equipment costs as well as the complexity of the process technology are growing rapidly, e.g., small step-and-repeat areas, extremely low depth of focus, multilayer techniques, suppression of interference effects, and so on. Therefore, in the practical case of application in circuit fabrication, the resolution limit will be somewhat higher; it seems possible that this will be in the range of 1 µm.

In order to realize a production process to generate sub-µm structures, X-ray lithography is the most promising approach. X-ray lithography at wavelengths between 0.5 and 5 nm is a simple one-to-one shadow-projection process, with structural resolution as good as 0.1 µm under certain conditions. The main factors limiting the resolution are Fresnel diffraction, fast secondary electrons, the relatively-low mask contrast attainable in the soft X-ray range, and - most important - the individual radiation characteristics of the X-ray source in question. For a quantitative comparison of the different X-ray sources which are applicable for lithography, numerical calculations of the generated resist patterns are necessary. The simulation model on which the calculations are based has to take into account the effects mentioned above, depending on the spectral distribution of the individual X-ray source, as well as the spectral absorption of windows, mask substrates, and mask absorbers. Important boundary conditions in this connection derive from the present state of resist technology, especially in regard to sensitivity, and from the necessity of compromising between source radiation power and tolerable mask heating during exposure. Based on these considerations, a comparison between X-ray tubes, storage rings, and plasma sources leads to the conclusion that synchrotron radiation is superior to the others.

## 1. Introduction

In connection with the increase in the packaging density of integrated circuits lithography processes have to be improved or new techniques developed to create the fundamentals for future VLSI technologies. It seems certain that production lines for VLSI circuits with sub-µm dimensions will be needed to a greater extent by the end of the eighties. The requirements for lithography are twofold: first, there is a need for serial writing techniques, e-beam or ion-beam processes, for mask making and direct wafer exposure with short turnaround times, e.g., for wafer personalization of gate array circuits; second, fast parallel mask projection methods for large-scale production of standard circuits are needed. Only the second kind of technique will be considered here. In this context, photolithography and X-ray lithography seem the most promising approaches to realize mask-related techniques with high resolution that are suitable for large-scale application. Other methods, e.g., e-beam and ion-beam shadow printing through masks, also have advantageous features, but worldwide development activities are too limired to solve all problems related to a new lithography technique in time. Therefore, the following statements only concern optical and X-ray lithography. The inclusion of both of these competing techniques in one presentation will enable a comparison and evaluation of both in regard to achievable resolution as well as process applicability.

## 2. Resolution limits in photolithography

Regarding the resolution limit, we only need to consider the so-called optical wafer stepper diagrammed in Fig. 1; all other techniques - even the Perkin-Elmer approach - will have a resolution limit in respect to fabrication conditions between 1.0 and 1.5 µm line width. Therefore, we have to compare the features of a optical wafer stepper with those of a wafer stepper using X-rays. The basic principle of an optical wafer stepper is relatively easy. A homogeneously-illuminated mask or reticle with a magnification factor between 4 and 10 is projected by a lens onto a resist-coated wafer. The image field of this lens allows, in most cases, the simultaneous exposure of a 1 $cm^2$ field. The exposure of the complete wafer is done step-by-step, using a laser-interferometric-controlled stage. The structural resolution is determined by the properties of the lens; the most important parameter is the numerical aperture.

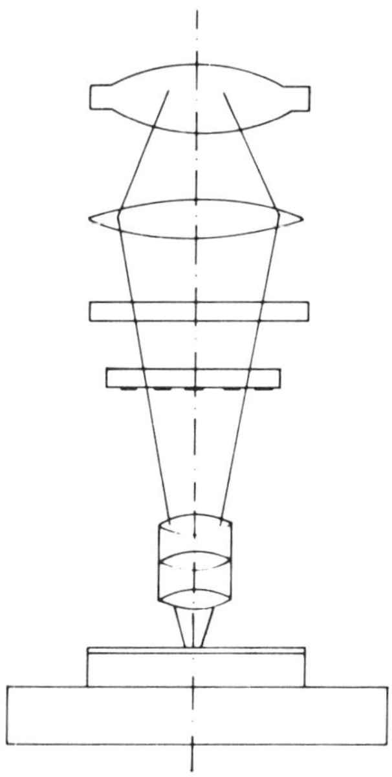


Fig. 1: Basic principle of an optical wafer stepper

A rough estimate of the resolution limit for such an arrangement can be provided using the simple "Raleigh criterion", as illustrated in Fig. 2. According to this criterion, two lines can be resolved if the maximum of one diffraction distribution corresponds to the minimum of the other. The resulting smallest line width is given by $w_l = \lambda/1.61\ A$, with $\lambda$ = wavelength and $A$ = numerical aperture. In the lower part of Fig. 2, this relationship is shown for two different wavelengths. The conclusions are that the numerical aperture is the dominant parameter (the wavelength can only be varied with a relatively small range), and that there is no physical reason not to have a resolution down to the submicron range.

However, the resolution is only one problem in lithography; there are quite a few other questions which have the same importance as the resolution, e.g., depth of focus, influence of substrate reflectivity, edge quality of the resist patterns, etc. To

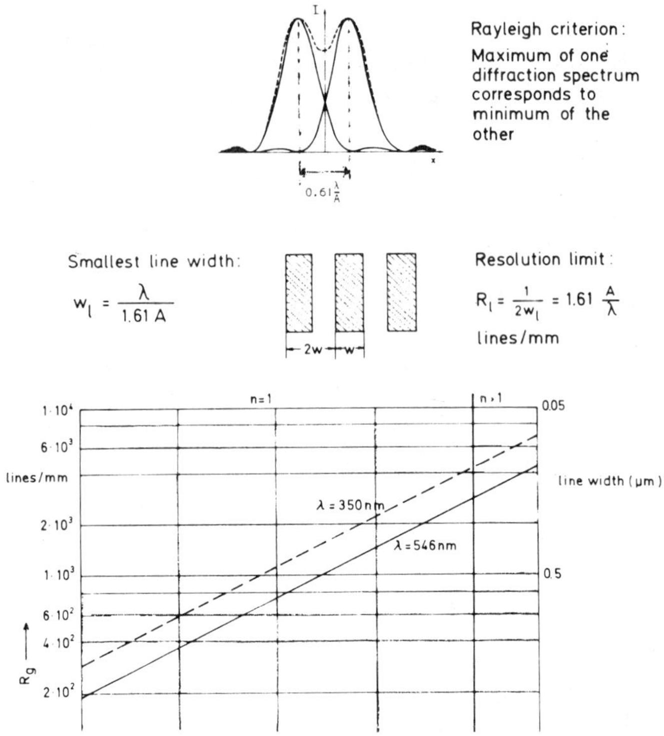


Fig. 2: Resolution limit due to diffraction

provide information on these questions, even for the submicron range where experimental data are not available in full scope, physical models and numerical computer programs are needed. The best simulation model for optical lithography which has been published to date is the "SAMPLE" program from Berkeley /1, 2/. To give an impression of the performance of this program and the underlying physical parameters, the program inputs and outputs are listed in the following table:

| Inputs | | Outputs |
|---|---|---|
| | A. Image Simulator | |
| Lens numerical aperture | | Image profile |
| Illuminator coherence | | Contrast |
| Focus error | | Slope |
| Mask line and space width | | |

Table 1: The photolithographic simulator "SAMPLE" /1/

| | |
|---|---|
| **B. Exposure Simulator** | |
| Substrate properties | 2-dimensional array |
| Thin film properties and thickness | of resist exposure |
| Resist properties and thickness | versus position and depth |
| Exposure dose | |
| **C. Development Simulator** | |
| Resist development properties | Resist slope |
| Development time | Resist line width |
| | Resist line-edge profile |

Naturally, the resolution limit in optical-projection lithography can be calculated very precisely using this program, including the influence of the resist properties, especially that of resist degradation. The output parameter suitable for characterizing the resolution of a given optical system is the contrast function $C = (I_{max} - I_{min})/(I_{max} + I_{min})$ where $I_{max}$ and $I_{min}$ denote the respective intensities at the center of a line and between two lines which are to be resolved. Fig. 3 shows an example for a lens with a numerical aperture of 0.42 /3/. For sufficiently,good pattern definition, most of the resists applied in IC fabrication need at least a contrast of 80%. Using this condition, the resolution limit (line width) can be estimated by $w_l = \lambda/2A$, which is nearly the same expression as that resulting from the Raleigh criterion. The contrast exhibits a typically sharp degradation below a line width of $w_l$.

The best lenses in currently-available optical wafer steppers have apertures of about 0.3 at a wavelength of 436 nm. This leads to a resolution limit of about 0.7 µm, using the above expression. To improve this value, the wavelength must be reduced and/or the aperture enhanced, as indicated in Fig. 4. A wavelength reduction down to about 200 nm seems feasible, but asides from the other problems indicated in Fig. 4, it is doubtful whether numerical apertures can be realized as high as the range of 400 nm. Another question is that of the best realizable apertures in the future, due to further improvements in lens-production techniques combined with better numerical methods for the calculation distortion correction. Some hints are given by research prototypes now in development: for example, outstanding laboratory results have been published recently by the Siemens Research Laboratory in Munich /3/. It was demonstrated, by generating real device structures instead of the grids normally used for high-

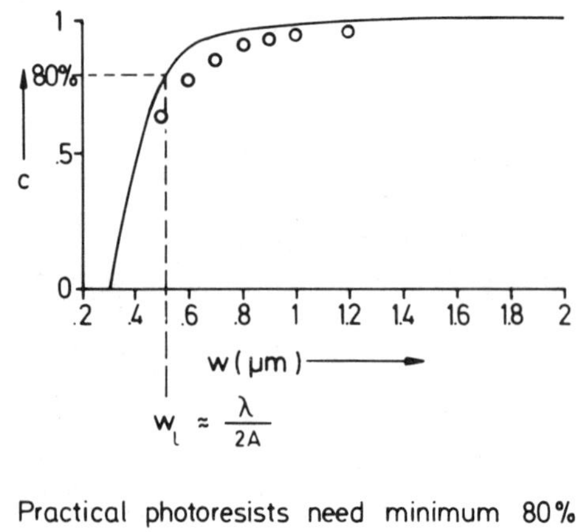


Fig. 3: Example for a contrast function of an optical system with numerical aperture A = 0.42 (calculated with SAMPLE; circles represent experimental results) /3/.

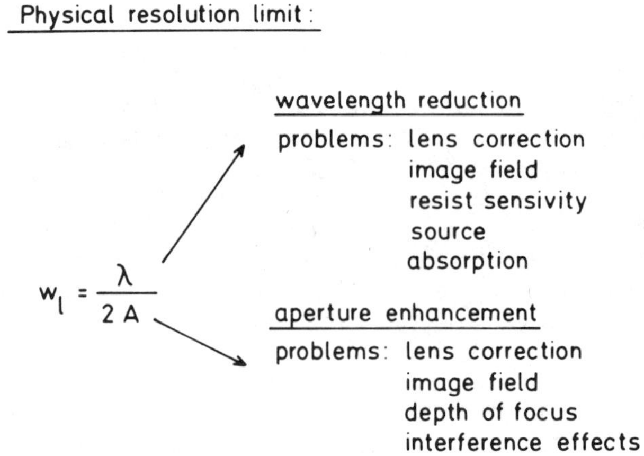


Fig. 4: Basic methods to enhance the resolution of optical lithography systems.

resolution tests, that a submicron technology with 0.5 µm features can be established with optical 10:1 projection printing. Fig. 5 shows, as an impressive example, a static memory circuit from a currently-used polysilicon-gate MOS process shrunk by a factor of three. The basis for these results is a special projection lens (no. 10-78-02) with a numerical aperture of 0.42, which was a custom development by Zeiss for Siemens. Its contrast function is illustrated in Fig. 3; the smallest resolvable line width $w_l = 0.5$ µm.

## 3. Application problems in optical projection printing

The most important disadvantage of this lens is its small image field of only 7 mm Ø, which is unsuitable for applications in economical IC production. However, most specialists in optics agree that lenses currently under development will combine large apertures and image fields of 15 mm Ø (suitable for 1 cm² step-and-repeat fields).

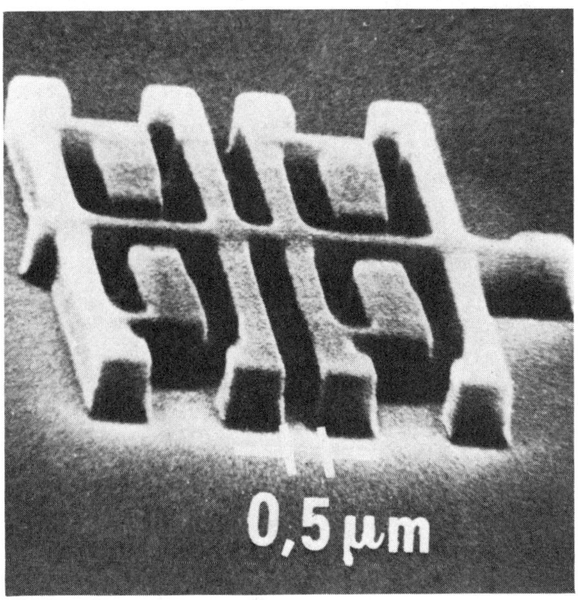


Fig. 5: Resist structures of a static memory cell with 0.5 µm dimensions (1.5 µm AZ 1450 on polysilicon) /3/.

Other serious problems arise from the extremely-small depth of focus inherent in such lenses. Here, a physical law states that the depth of focus is reduced quadratically with the numerical aperture. In the case of the so-called "Siemens lens", this parameter is only about 1.2 µm; a defocus of only 1 µm already leads to a considerable contrast degradation /3/. Therefore, the wafer and chip topography has a strong influence on the pattern quality, even in case of an autofocus adjustment at each step.

Several other problems are related to lens correction. The corrected wavelength range decreases with increasing aperture and image field. For example, the "Siemens lens" is corrected for a wavelength of 436 nm within only 1 nm. Due to this monochromatic exposure, the troublesome interference effects caused by light reflection from the various optical interfaces of the semiconductor-layer sequence (especially, resist-to-air and resist-to-substrate) will become more important. The best-known effect arising from interference are standing light waves, for which an impressive example is shown in Fig. 6. The reflected light intensity, and therefore the quality of the generated pattern, are greatly affected by local variations due to wafer topography and processing, e.g.:

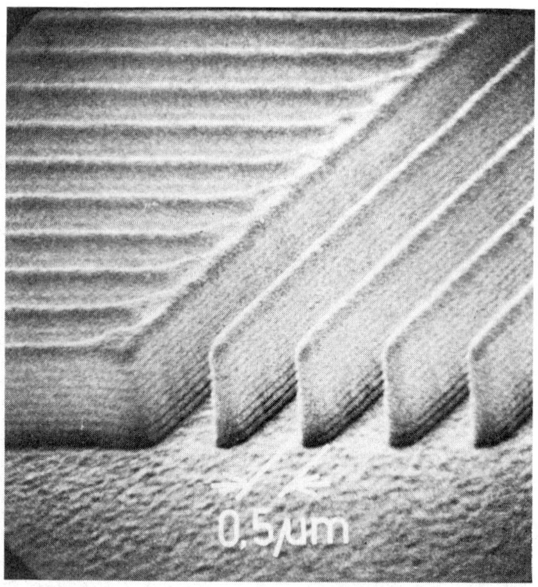


Fig. 6: Resist pattern with 0.5 µm dimensions and high aspect ratio /3/. The ripples on the resists walls are caused by standing waves.

Unavoidable differences in reflectivity between the various layer materials (e.g., the intensity ratio of reflected and incident light for aluminum can be twice as high as that of a polysilicon substrate /4/).

- Local variations in resist thickness over the wafer area (striations, vertical steps). Thickness variations in the range of only 0.1 μm can influence the intensity ratio up to a factor of 5 /5/.

- Fluctuations of the resist-absorption coefficient due to processing (aging, prebake, etc.) and bleaching during exposure.

- Disturbances of the optical symmetry, shadow effects at edges, and profile steps of the wafer topography.

- Technological influences on the reflectivity of a layer material due to fluctuation or alteration of the processing parameters (e.g., variation of the grain size in polysilicon caused by changes in vacuum, gas flow, or temperature).

This list should give an impression of how sensitive optical submicron lithography is to relatively-small variations in technological and geometrical parameters.

Summarizing the performance of optical lithography for submicron patterning, the following conclusions can be drawn: In principle, optical lithography can be applied down to the deep submicron range. The resolution limit for advanced systems will be about 0.5 μm. (Lenses with numerical apertures of 0.4 and adequate image fields will be available within a few years.) However, the technological problems for application in large-scale circuit production increase very rapidly in the range below 1 μm. A solution to these problems can only be obtained by using "multilayer-resist techniques" which introduce additional costs and new technological problems, especially in regard to increased defects.

## 4. General principle of X-ray lithography

As mentioned in section 2, the wavelenght reduction beginning from the optical range, in order to eliminate all disturbing effects such as diffraction and interference, stops very soon at about 200 nm. For high-resolution lithography, shorter wavelengths would be more suitable. This leads to X-ray lithography, since soft X-rays between about 4 and 0.5 nm represents the only shorter-wavelength range which can be used for a mask-related lithography process. At these wavelengths, no problems with diffraction or interference arise down to the order of magnitude of 0.1 μm. However, there are two important physical facts which make X-ray lithography much more difficult than the optical process:

- In the wavelength range of X-rays, the difference in transparency is not as high as in the visible wavelength region, which means that for the X-ray region at a given wavelength, there are no materials available which would be fully transparent in thicker dimensions (such as glass in the optical case), or which would fully absorb the radiation in very thin layers (such as chrome in the optical case).

- There are no imaging optics available possessing a useful efficiency, which means that a condenser for homogeneous illumination of the wafer is not realizable.

These two points already summarize the general problems of X-ray lithography. As a result of the first point, mask technology has to be modified. In order to obtain a sufficiently-transparent mask substrate in this wavelength region, a light element with low atomic number and low absorption has to be selected. Through use of a thin foil covered with a relatively thick absorber structure, one can obtain the required mask contrast.

The wavelength region usable for the purpose of lithography is determined by the absorption properties of the mask substrate as well as those of the resist; this region corresponds to approx. 0.5 - 4 nm. For longer wavelengths, the limit is determined by the high absorption in the thin mask substrate; for shorter wavelengths, the limiting factor is the decreasing absorption in the resist layer, which becomes transparent in the case of harder radiation.

The second point mentioned above means that the only way to realize X-ray lithography is one-to-one shadow projection of an X-ray-transparent membrane with absorber structures onto a resist-coated wafer, as shown in Fig. 7. Between maskand wafer is a small gap, the proximity distance (typically 50 µm), to protect the mask against mechanical damage. Projection printing with a magnified reticle, as in the case of an optical wafer stepper, seems to be impossible with X-ray lithography. Furthermore, the lack of useful optics such as lenses in this wavelength range means that the radiation has to be used in the same form (i.e., wavelength distribution and geometrical characteristics) as it is emitted from a given X-ray source. It is true that several research groups are investigating techniques for focusing soft X-rays, but realization of a condenser having high efficiency and satisfying stability which would be applicable for production purposes does not appear feasible. Therefore, the successful application of X-ray lithography depends critically on the question of what kinds of X-ray sources are available.

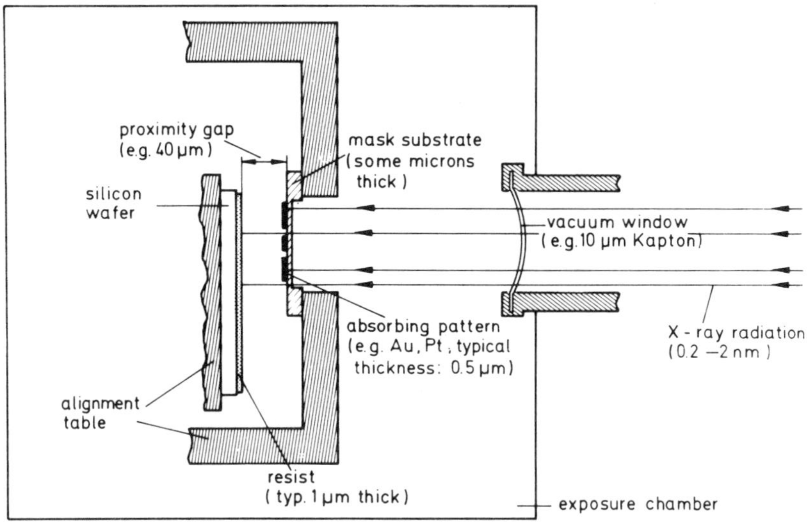


Fig. 7: Schematic exposure arrangement for X-ray lithography using synchrotron radiation

## 5. Technology of X-ray masks

The most decisive consideration for an X-ray mask is the selection of the material for the X-ray transparent mask substrate. This material must have a low atomic number, and must be suitable for preparation of a thin but stable membrane with a thickness of only several microns. Since the beginnings of X-ray lithography, nearly all light elements and their compounds have been taken into consideration. However, only a few of them remain as promising materials to fabricate satisfying X-ray masks. From the present point of view, the most important are these: silicon, silicon nitride, silicon carbide, and polyimide. The mask concept of Bell /6/, the boron-nitride/polyimide sandwich mask, is only applicable with a palladium source; for softer radiation, the absorption of this membrane is too high. Furthermore, it is probable that the requirements for X-ray masks can be met by single layers, which are easier to fabricate. Light metals too are very interesting for substrate preparation (e.g., titanium /7/), but there are significant disadvantages related to this concept. First, the fabrication uses no batch process, and second, it is very difficult in this case to generate areas within the metal foil which are completely transparent to visible light for supporting the alignment marks. This is very important, since all specialists agree that alignment in X-ray lithography has to be done using visible light, as opposed to the use of X-rays or e-beams /8/.

At the present state of the art, a statement as to what kind of substrate material will prove best cannot yet be made. To answer this question, more experience is required from applications in pilot lines, especially regarding defect densities and fabrication expenditures. On a laboratory scale, the following fundamental requirements for X-ray masks can be fulfilled with the above-mentioned materials:

| | |
|---|---|
| X-ray transparency (10 Å) | ⩾ 60% |
| Contrast | ⩾ 10-20 |
| Optical transparency | ⩾ 50% |
| Step-and-repeat field size | up to 5 x 5 cm² |
| Geometrical stability (against environmental influences, esp. radiation damage) | $10^{-6}$ |
| Membrane flatness | ⩽ 1 µm |

The last requirement, for a flatness of better than 1 µm, is made assuming the central projection of an X-ray tube. In the case of parallel synchrotron radiation, the unflatness can be as high as, e.g., 5 µm without causing unacceptable pattern distortions. The main problems of X-ray-mask technology which have to be solved in the next phase of investigation are a consequence of the fact that defects in 1:1 masks with submicron features cannot be repaired easily. Therefore, a mask technology must be developed to provide masks with zero defect density as well as acceptable yield. Furthermore, a fast detection method for submicron defects is also needed.

To illustrate the technological state of the art, current silicon masks will now be described in more detail. Silicon is certainly the most important mask material, since the unique and manifold tools of silicon processing make silicon suitable for the membrane itself, as well as for the starting substrate or frame for other substrate materials, especially silicon compounds (preferential and self-stopping etching, layer deposition by epitaxy, CVD and plasma, batch-processing, etc.) Using silicon membranes with a 4" diameter (in a 125 mm wafer) and a thickness of 2-3 µm can be made easily; an example is shown in Fig. 8. The stability of inorganic materials such as silicon against environmental influences (e.g., radiation and humidity) is very high.

Fig. 9 illustrates the essentials of the currently-used fabrication procedure, which is much simpler than in the beginnings of X-ray lithography. The depth of the boron doping - generated by diffusion or epitaxy - determines the thickness of the silicon membrane via etch stopping. The opening of the area to be etched is done using a simple mechanical aperture. Due to the properties of the ethylene-diamine etchant the absorber patterning can be

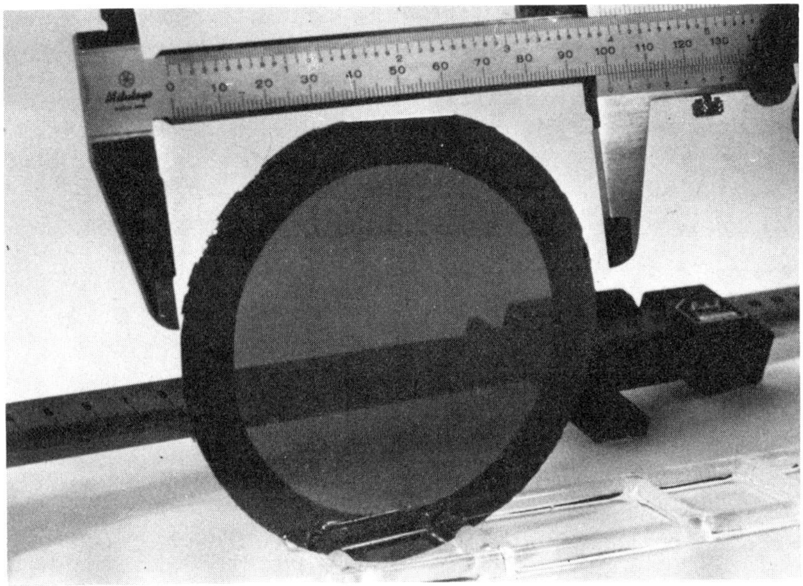

Fig. 8: Etched silicon substrate for X-ray masks (3" window with 2 µm thickness in a 100 mm wafer) /8/.

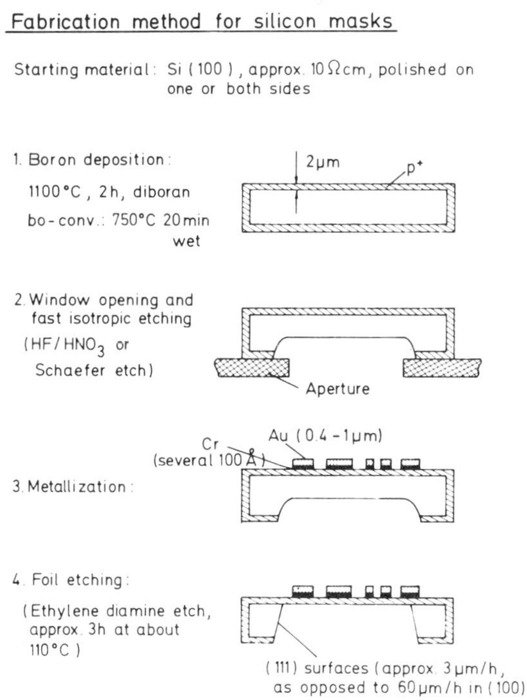


Fig. 9: Example for a process sequence for silicon substrates

done before the final thinning since the metal layer remains unetched. However, there are some problems with pattern distortions at the end of the thinning process, caused by membrane stress induced by the heavy boron doping. Silicon technology, though, provides solutions to control the membrane stress, e.g., out-diffusion, Ge doping, epitaxy, etc. (A small amount of stress is necessary to make the membrane smooth and flat.) Furthermore, it is also possible to write with e-beams directly onto the thin membrane, which means that the last two steps in Fig. 9 are reversed. Stress control is necessary in all cases because the membrane flatness depends on this very greatly.

## 6. Radiation characteristics of the various X-ray sources

For lithography purposes, X-ray tubes, synchrotrons or storage rings, and plasma sources are currently in use to greater or lesser degrees. Most frequently used are X-ray tubes, since they involve a relatively well-known technique and are comparatively inexpensive. However, in addition to several other disadvantages such as low intensity and difficult-to-change wavelength, they involve very unsatisfactory radiation characteristics. Specifically, they emit isotropic radiation, in combination with an extended area from which the radiation is emitted. The spot size of the source typically amounts to at least several mm. This leads to the well-known problems illustrated in the upper part of Fig. 10. The finite spot size denoted by 2r, acting together with the proximity distance s, causes blurring ($\delta$) and run-out (b). The term run-out refers to a systematic alteration in the lateral dimensions of the projected resist patterns. This effect is not as problematical as blurring, since it can be compensated for in principle by the mask layout, with a magnification scale which depends on the proximity distance as well as the distance between source and mask. In addition, consideration has been given to the possible use of this effect for compensation of process-induced linear wafer distortions. If there are uncontrolled variations of the proximity distance and unflatness of the mask membrane, however, the resulting pattern distortions will enlarge very rapidly to values of 0.5 µm and greater. Clearly, a professional sub-µm technology is very hard to realize under such conditions. Attempts to improve the situation by increasing the distance between source and mask are very soon thwarted by the fact that the radiation density decreases quadratically with increasing distance.

Therefore, it is necessary to consider the possibility of an alternative X-ray source offering better radiation characteristics. In the lower part of Fig. 2, three types of characteristics are shown which are more applicable in lithography: the

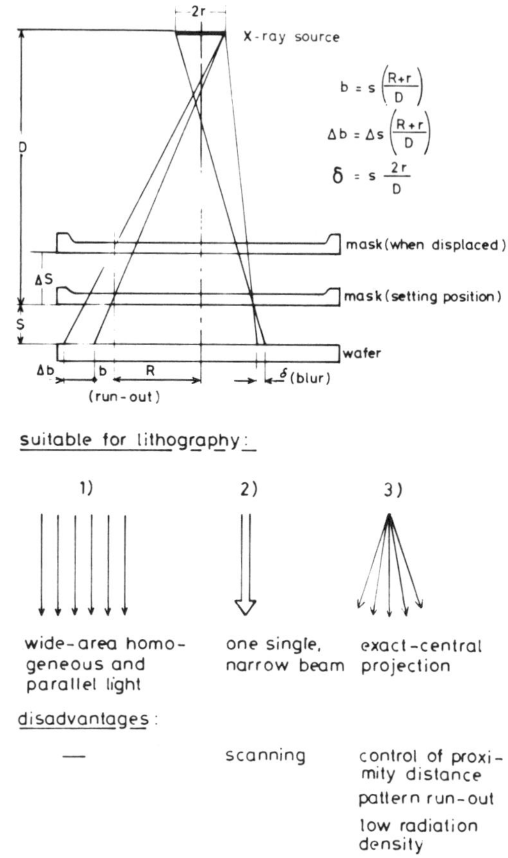


Fig. 10: General principles of radiation characteristics for X-ray sources

best one would be the parallel and homogeneous floodlight of at least the size of the applied step-and-repeat field. As already said, however, there is only an extremely small chance of its realization.

The second case, the single beam with small aperture, is of the same quality in regard to the projection accuracy as the first case; here, though, there is the problem of a time-consuming and relatively-difficult scanning process. In addition, local mask heating is somewhat higher than in the case of homogeneous illumination, depending on the scan velocity.

The third case, exact central projection with extremely small spot size, solves the problem of blurring, but all the other problems of the X-ray tube are still present, especially the exact control of proximity distance and mask flatness. Only in combination with a very high radiation density, to allow enlargement of the distance D between source and mask, would this case be suitable for sub-μm lithography. A fairly good approximation to this third case is represented by the various kinds of plasma sources, which provide a very small focus spot and a high intensity in comparison to X-ray tubes.

Synchrotron radiation offers a mixture of the first two types of characteristics: in the direction parallel to the orbit plane of the circulating electrons in a storage ring, the radiation is similar to floodlight illumination, with a broadness which is large compared to a silicon wafer. In the direction perpendicular to the orbit of the electrons, the synchrotron radiation is collimated very narrowly. The spatial arrangement is illustrated by the highly-simplified diagram of a storage ring shown in Fig. 11.

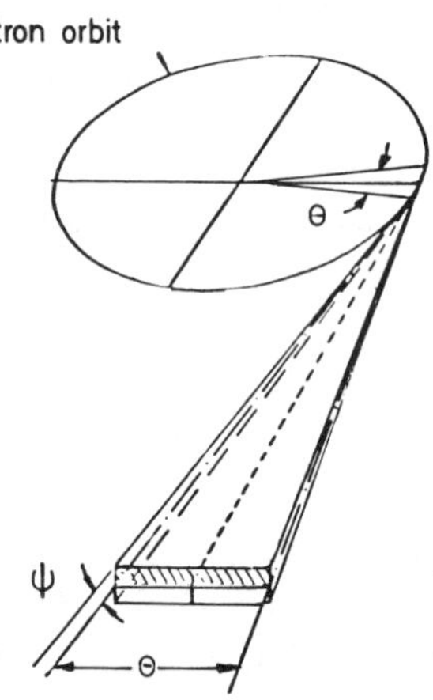


approximation of the angular spread $\langle \Psi \rangle$

for wavelengths near $\lambda_c$ :

$$\langle \Psi \rangle \approx \frac{m_0 c^2}{E}$$

E = electron energy

$m_0 c^2$ = rest energy of an electron

Fig. 11: Emission characteristics of synchrotron radiation

The angle $\psi$ represents the direction vertical to the orbit plane. The radiation power decreases rapidly with decreasing $\psi$; one can see that the vertical collimation of the synchrotron radiation is better than 1 mrad. At a typical distance from the source point of 10 m, the soft X-ray component of the synchrotron radiation illuminates a horizontal stripe having a height of less than 1 cm. For purposes of lithography, the synchrotron radiation can be assumed as absolutely parallel. In contrast to the monochromatic emission from an X-ray tube, synchrotron radiation provides broad-band emission, similar to black-body radiation. The maximum of the radiated power can be chosen freely by altering the energy of the circulating electrons and the magnetic field in the bending magnets at a given bending radius.

Fig. 12 summarizes the properties of the X-ray sources under discussion. As opposed to the spectrum of a storage ring, the other sources are more-or-less monochromatic, depending on the target material as well as the gas within the discharge volume. The wide-band spectrum has some advantages in respect to the Fresnel diffraction. Furthermore, it is very important that the spectrum of a storage ring can be changed easily, since the maximum of the radiation can be adjusted to maximum of sensitivity (transparency of mask and absorption of resist), entirely independent of the characteristic line of a target material.

| Parameter | Storage ring | X-ray tube | Plasma source |
|---|---|---|---|
| spectrum | wideband typically 4 - 40 Å | monochromatic 8.3 Å (Al) 6.7 Å (Si) 4.4 Å (Pd) | wideband with monochromatic lines 7 Å, 12 Å |
| focus spot | 0.5 mm | 5 mm | ≤ 1 mm |
| aperture | 1 mrad (with oscill. ≤ 5 mrad) | isotropic | isotropic |
| distance source/mask | 10 m | 30 cm | > 50 cm |
| blurring (prox. d. 50 µm) | < 0.01 µm | ~ 0.5 µm | ~ 0.1 - 0.2 µm |
| homogeneity | horiz: homogen. vert: Gaussian | homogeneous | homogeneous |
| pattern distortion | Fresnel | Fresnel + blurring | Fresnel (blurring) |
| intensity | $10^{-1}$ W/cm$^2$ | $10^{-3}$ W/cm$^2$ | $10^{-2}$ W/cm$^2$ |
| time dependence | pulsed with high repetition rate (MHz) vert. scan : 1 Hz to 100 kHz | constant | pulses with $10^5$ - $10^6$ W/cm$^2$ (~20 ns, 0.1 - 1 Hz) |

Fig. 12: Comparison of the various X-ray sources

The focus spot and the aperture have already been discussed; here too, all advantages are on the side of the storage ring. In this case, the highest throughput is also achievable /9/. Only plasma sources, with their relatively-small focus in combination with especially-high intensity /10/, and thus an increased distance between source and mask, are also of interest for sub-µm pattern generation. With both interesting principles of plasma sources, laser-induced plasma and plasma focus, spot sizes down to 100 µm have already been realized /11/; adding the lateral instability by averaging over many pulses, the effective spot size should be lower than 1 mm. Therefore, the blurring effect is much lower than in the case of X-ray tubes. The overall pattern distortion will be discussed in sect. 7, referring to calculations of the resulting resist patterns. The intensities shown in the table of Fig. 12 indicate the power density utilizable in lithography; these are related to the source/mask distance shown above, and represent the maximum values realizable with currently-available equipment. The specifications of the plasma source have been chosen according to the plasma-focus system published this year by Physics International /12/.

Very important in connection with mask heating is the time dependence of the radiated power, as indicated in the last row of Fig. 12. The synchrotron radiation is pulsed because the circulating electrons assemble into bunches; the pulse frequency is determined by the number and circulation time of the bunches, and ranges up to several hundred MHz. In respect to lithography, however, synchrotron radiation can be assumed as uniform in time, due to this high repetition rate. This assumption is not affected by the vertical scanning motion which is necessary to provide a homogeneous illumination in the vertical direction too, if the scan velocity is properly chosen /13/. However, the time dependence of plasma sources is very inhomogeneous, even in respect to lithography applications. All types of plasma sources provide short, very strong pulses having power densities up to $10^6$ W/cm, with typical pulse lengths of 20 nsec. Compared to this, the repetition rate is very slow, amounting to about 0.1 - 1 Hz, depending on the expense of the equipment. This fact can cause severe problems in connection with mask heating /13/.

## 7. Pattern resolution with different X-ray sources

For calculations of the structural resolution achievable with the various types of X-ray sources, a computer model has been developed which is described in detail in Ref. /14/. This program is able to calculate the generated resist patterns for the three X-ray sources already considered and for the most common X-ray resists, such as PMMA, FBM, XXL15, etc., starting from the basic parameters such as type and thickness of absorber, mask membrane, window and resist, current and electron energy in the case of a storage ring, target material, primary power, source area in the case of an X-ray tube, etc.

The following effects can be calculated by means of a mathematical solution:

- spectral absorption of all materials involved, depending on the spectral characteristics of the source,
- Fresnel diffraction in respect to the complete spectrum,
- Fresnel diffraction including non-fully-opaque absorbers,
- extension of the focus spot of the source,
- range of the backscattered photoelectrons,
- beveled absorber edges.

The relationship employed between dissolution rate and dose, for the most important X-ray resists, is based on empirical results.

To compare the high-resolution capabilities of the various X-ray sources, Figs. 13 - 15 show generated resist patterns for identical exposure conditions: 1-µm-thick and 0.5-µm-wide gold absorbers on a 2-µm-thick silicon membrane, a proximity distance of 50 µm, and PMMA resist exposed with a dose of 1000 J/cm$^3$. The pattern generated by an X-ray tube with an aluminum target is reproduced in Fig. 13. This shows the geometrical shape of the different resist patterns, i.e., resist depth vs distance from the middle of the absorber, due to the progress of the development procedure at a fixed exposure dose. The parameter at the right side of each resist profile refers to the development time. The resist has a initial thickness of 1 µm; zero in the scale at the left side indicates the initial surface of the resist layer. It is easy to see that there is no way to realize submicron patterns.

Using synchrotron radiation, however, conditions can be found for an accurate transfer of a 0.5-µm-wide absorber structure (Fig. 14). Fig. 15 represents the plasma source: the source parameters have been chosen according to the data of Physics International; the source/mask distance is 60 cm, in order to obtain the same exposure time as that of the X-ray tube (Fig. 13). Here, a pattern definition is obtained which is not as good as that of synchrotron radiation; however, it would be sufficient for submicron lithography.

These statements can be affirmed by some examples of achieved resist patterns, which show that the use of synchrotron radiation is the best method for carrying out X-ray lithography. These

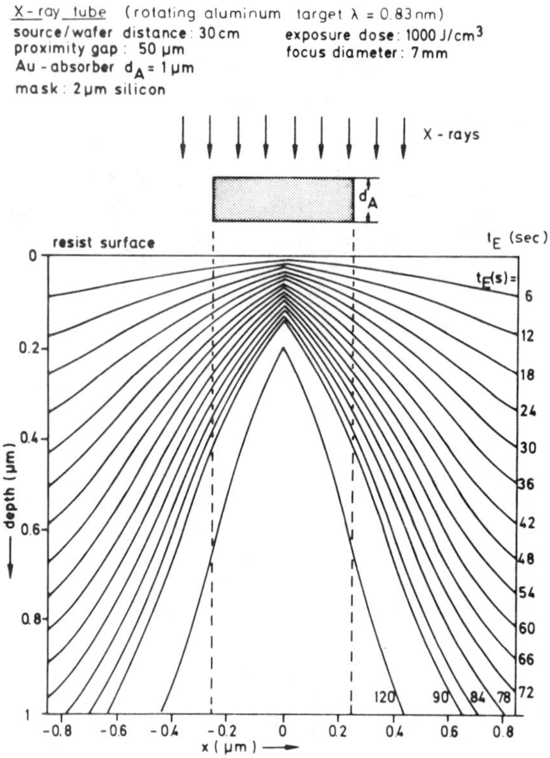


Fig. 13: Calculated resist profile in the case of an X-ray tube

examples are not connected with an application in microelectronics, where only thin resist layers, mostly below 2 μm, are needed. The problem to solve here was to generate extremely thick resist patterns of several 100 μm, with lateral dimensions down to 1 μm, for applications in micromechanics /15/. Fig. 16 shows needles of PMMA with a height of about 130 μm and diameters down to several microns. Fig. 17 illustrates the complete micromechanical structure, with a height of about 350 μm, which has been achieved up to now. Fig. 18 enlarges the area of Fig. 17 where the smallest dimensions (down to 1 μm) are located. The steepness of the walls is surprisingly good; the deviation from the vertical can be kept to below 1 μm over the total height.

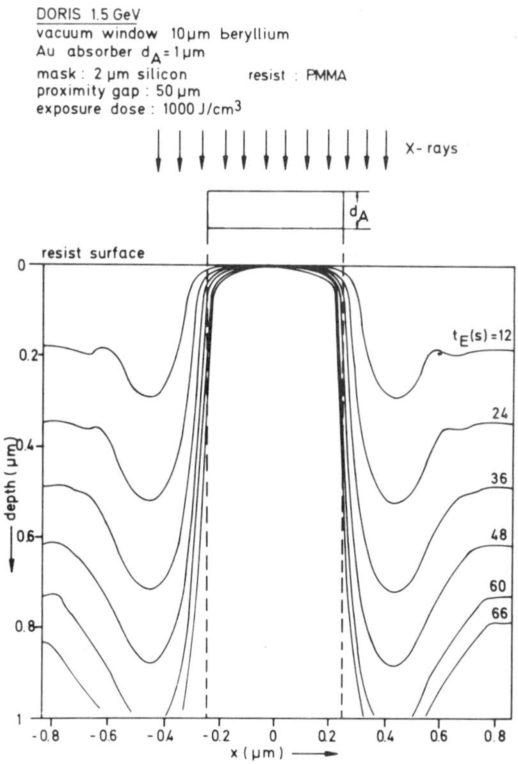


Fig. 14: Calculated resist profile in the case of a storage ring

## 8. The compact storage ring

Synchrotron-radiation sources providing facilities for lithography experiments in Germany include the electron-storage ring DORIS in Hamburg and the electron synchrotron BONN II in Bonn. The 800 MeV storage ring BESSY provides nearly-ideal conditions for lithography experiments. Outside of Germany, lithography activities are being conducted at several locations. To be mentioned first is the small storage ring at Brookhaven, where IBM is doing intensive research; further, there are Orsay (ACO), Stanford (Spear), Dawesbury, and the photon factory in Japan. However, all these facilities are only well-suited for R+D activities; an application to circuit production does not seem practical.

Therefore, if X-ray lithography using synchrotron radiation is to have a chance for broad application, it is necessary to deve-

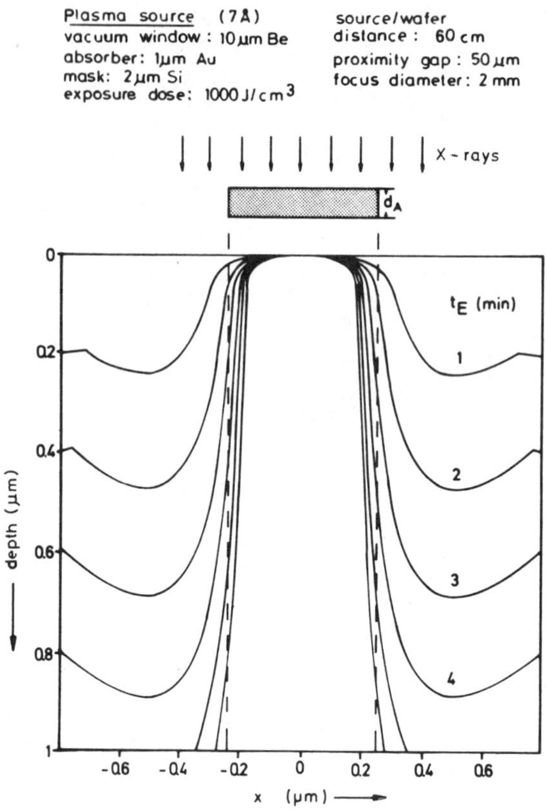


Fig. 15: Calculated resist profile in the case of a plasma source

lop special storage rings which are optimized in respect to costs and performance for lithography applications only. Currently-existing storage rings are specialized for purposes of fundamental research, with high demands on electron-beam focusing, time dependence, spectrum and beam stability, and so on. As opposed to this, a storage ring or synchrotron for lithography is a relatively simple machine, in which the beam size of the circulating electrons is artificially enlarged by stimulation of oscillations.

The smallest version of such a compact storage ring is the so-called "Klein-Erna" concept, which has nearly the dimensions of a table-top machine with a outer diameter of about 2 m /16/. The optimal size, though, is still a subject for discussion. Here, a compromise has to be found between costs, applicable number of beam lines, and needed capacity. Many fundamental considerations

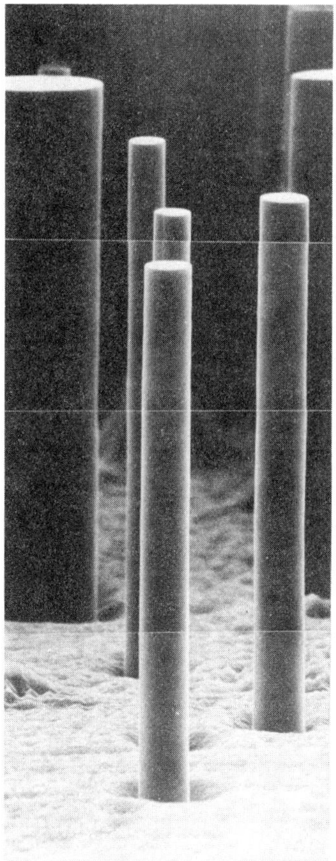

Fig. 16: PMMA needles with a height of approx. 130 μm

in regard to circuit fabrication are involved in this matter, e.g., the question of whether it is acceptable for a fabrication line to depend on only one machine with its limited up-time, the problem of the best configuration for a reasonably-clean room arrangement, etc.

In Germany, the decision has been made - based on the early start already taken in this direction - to develop a storage ring offering minimum size and costs. The basic principles to fulfill these requirements are the application of superconducting techniques as well as so-called weak focusing. In cooperation between the Fraunhofer-Gesellschaft and BESSY, a project has been started in Berlin to fabricate a prototype which will be ready for the first lithography experiments in 1985. The proposed specifications are listed in Fig. 19.

Fig. 17: Micromechanical structure with a height of approx. 350 µm

9. Conclusion

Fig. 20 shows a comparison between a future optical wafer stepper with highly sophisticated lenses and an X-ray stepper which will be available in the second half of the eighties. Aside from mask inspection and repair, all advantages are with X-ray lithography. However, most of these advantages, such as the exposure of very thick resist layers without the necessity of multilayer techniques, noncritical proximity distances (e.g. 100 µm), no problems with depth of focus, high throughput even in the case of insensitive but stable resists, etc., can only be obtained using parallel X-ray light of high intensity, with the possibility of adjusting the source spectrum to the mask-and-resist combination. Synchrotron radiation offers the most advantageous preconditions by far for this purpose. This means that for an economical application of X-ray lithography, a compact storage ring must developed which is related to the same range of necessary investment costs per exposure station as in the case of the X-ray tube. With appropriately-designed storage rings, power densities of more than 100 mW/cm² are achievable, which is about 2 orders of magnitude higher than in the case of an X-ray tube. Assuming the availability of such a compact storage ring, it is very possible to suppose that the process of X-ray lithography will represent the simplest and most economical approach for the fabrication of

Fig. 18: Detail of the pattern shown in Fig. 17 (some features are in the submicron range)

device structures in the sub-µm region, in spite of the fact that optical techniques can also be extended to the submicron range. It is very expensive and problematical, though, to work near the physical limits; X-ray lithography is far away from its limits even in the case of 0.1 µm features. Therefore, the activities in Germany are aiming towards a sophisticated X-ray source, and not towards a high-sensitivity resist which would necessarily be unstable against etching, etc. This approach is quite different from that of Bell, e.g., as indicated in Fig. 21.

Plasma sources extrapolated to the future could represent the best compromise between performance and price; in this connection, though, there are still considerable development problems of a basic nature to be solved. Due to the huge pulse power, mask

| Parameter | Proposed value |
|---|---|
| $\lambda_c$ | 20 Å |
| current | 300 mA |
| aperture (including betatron oscillation) | 5 mrad |
| vertical movement | linear |
| lifetime | 20 h |
| outer dimensions (including shielding) | 2 m |
| number of beam lines | 5 - 10 |
| weight | < 10 t |
| power consumption | 50 -100 kW |
| price without beam lines etc. | < 5 Mill. DM |

Fig. 19: Specifications for a prototype compact storage ring to be developed in Berlin (Bessy and Fraunhofer-Gesellschaft)

heating problems can arise. All this does not mean that X-ray tubes would not be a good interim solution for an early start with small-scale or laboratory activities; in the future, though, meaning at the end of the eighties, the compact storage ring will win the race.

| | photo | X-ray |
|---|---|---|
| resolution | 0.5 µm | 0.1 µm |
| resist thickness | < 1 µm | <10 µm |
| depth of focus | <1 µm | 100 µm |
| influence of substrate | extremely high | low |
| sensitivity to process variation | extremely high | low |
| sensitivity to dust particles | extremely high | low |
| technological stability of resists | good | good |
| multilayer technique | necessary | not necessary |
| mask inspection | rel. easy (10x) | difficult (1x) |
| mask repair | possible | not possible |
| step-and-repeat | 1 cm $^2$ | ~ 25 cm $^2$ |
| throughput | ~25 wafer/h | ~ 50 wafer/h |

Fig. 20: Comparison of photolithography (wafer stepper) and X-ray lithography (synchrotron radiation)

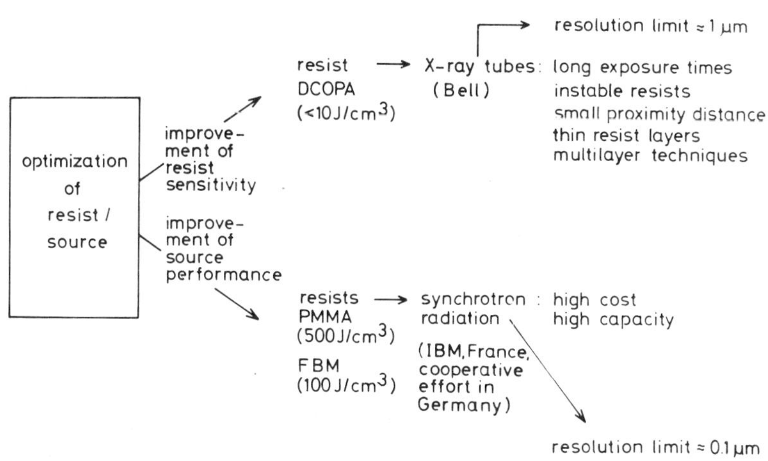


Fig. 21: Different approaches for X-ray lithography application

REFERENCES

/ 1/ W.G. Oldham, A.R. Neureuther, Solid State Technology, May 1981, p. 106

/ 2/ A.R. Neureuther, P.K. Jain, W.G. Oldham, SPIE Vol. 275, Semiconductor Microlithography VI, 1981, p. 110

/ 3/ W. Arden, Siemens Forsch.- u. Entwicklungs.-Ber. Bd. 11 (1982), Nr. 4, Springer-Verlag 1982

/ 4/ W.G. Oldham, Limits of Optical Lithography, Summer Course at Uny. of Leuven, June 1980

/ 5/ D. Widmann, H. Binder, IEEE Trans. ED 22 (1975), p. 467

/ 6/ D. Maydan, G.A. Coquin, J.J. Levinstein, A.K. Sinha, D.N.K. Wang, 15th Symp. Electron, Ion, Photon Beam Technology, Boston, 1979

/ 7/ W.D. Buckley, J.F. Nester, H. Windischmann, Proceedings Microcircuit Engineering 80, Int. Conf. on Microlithography, Amsterdam, 1980, p. 327

/ 8/ A. Heuberger, H. Betz, S. Pongratz, "Present status and problems of X-ray lithography", Festkörperprobleme XX, Ed. J. Treusch, Vieweg, Wiesbaden, 1980, p. 259

/ 9/ W.D. Grobman, "Status of X-ray lithography", CH 1616-2/80/ 0000-0415, 1980, IEEE

/10/ H.I. Smith, Proc. Ninth Int. Conf. Electron and Ion Beam Science and Technology, St. Louis, Ed.: R. Bakish, The Electrochemical Society, May 1980

/11/ B. Wende, Physikalisch Technische Bundesanstalt, Berlin, private communication, and: G. Herziger, H. Krompholz, W. Schneider, and K. Schönbach, "A steady-state fluid model of the coaxial plasma gun", Physics Letters 71A (1979), p. 54

/12/ Physics International, PIXI-10 pulsed plasma soft X-ray laboratory source; preliminary specifications, San Leandro, March 1982

/13/ K. Heinrich, H. Betz, and A. Heuberger, "Heating and temperature-induced distortions of silicon X-ray masks", to be published in IEEE ED

/14/ K. Heinrich, H. Betz, A. Heuberger, and S. Pongratz, "Computer simulations of resist profiles in X-ray lithography", J. Vac. Sci. Technol. 19 (1981), p. 1254,

/15/ E.W. Becker, H. Betz, W. Ehrfeld, W. Glashauser, A. Heuberger, H.J. Michel, D. Münchmeyer, S. Pongratz, and R. v. Siemens, "Production of separation nozzle systems for uranium enrichment by a combination of X-ray lithography and galvanoplastics", to be published in Naturwissenschaften

/16/ U. Trinks, F. Nolden, A. Jahnke, "The table-top synchrotron radiation source "Klein-Erna", Nucl. Instr. and Methods 200 (1982), p. 475

/406/podst/21.3.1983/

## ELECTRON BEAM LITHOGRAPHY

W. Kulcke

German Manufacturing Technology Center

IBM Deutschland GmbH, Sindelfingen

### INTRODUCTION

Over the past decade, the trend in integrated circuits for computers always has been a doubling of the circuit density every year. To-day's products have 64.000 memory cells or 700 logic circuits on a 5 x 5 $mm^2$ sized chip. For to-morrow, we can expect 288.000 and 516.000 memory cells or 5000 logic circuits on a chip.

In the past, this density increase was mainly achieved by design cleverness all at geometries $\gtrsim$ 2 μm. To-day, a dynamic RAM consists merely of one transistor and one capacitor. Thus, there is no way to further reduce the circuit components. Future circuit densification can only be achieved through smaller dimensions of the components. The trend leads from to-day's 2 μm wide lines (with 0.8 μm layer to layer overlays) toward 1 μm wide lines (with 0.4 μm layer to layer overlays) and smaller.

### LIMITATIONS OF PHOTOLITHOGRAPHY

To-day's technology to generate these fine geometrical structures is photolithography. Several photolithographic steps are required as they are shown in Fig. 1.

According to that figure, the structural design data for the integrated circuits of one chip are provided on a magnetic tape. Opto-mechanical pattern generators compose from these data a first physical mask. For reasons of achievable precision, this so-called Single Segment is generated with structures of 10 x their final size. Subsequently, the Single Segments are 10 x demagnified and

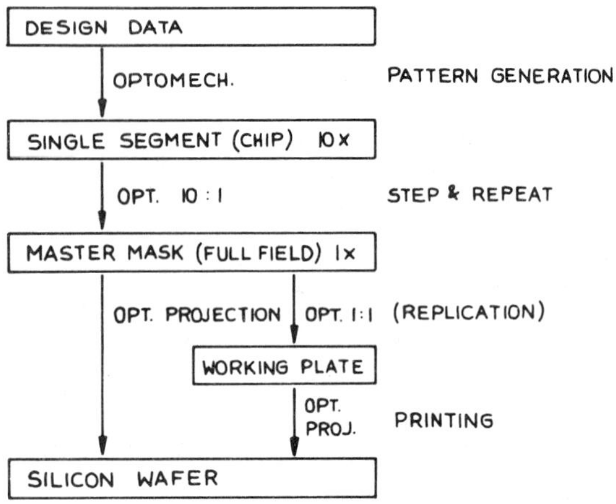


Fig. 1. To-day's photolithographic process steps

stepped and repeated to yield the Master Mask with the structures for a whole wafer in 1 x size. The pattern of this master mask - or may be a copy of it, a so-called Working Plate - is optically printed to the wafer by 1 : 1 full field projection.

This photolithographic process is very economic. Once, the mask is generated, it can be repetitively printed to the wafers at high throughput and low cost. Fifty wafer exposures per hour are achievable with to-day's equipment.

The economy of this process is very good for the fabrication of memory chips and for that of the Master Slices for logic circuits. Here a large volume of identical patterns is to be printed and one mask can be used for many prints.

One bottleneck in the photolith. process comes up, however, if the Master Slices of the logic circuits are to be personalized for the different individual logic functions needed in a computer (Fig. 2). Then each print requires its own individual mask. The photolithography becomes uneconomic. It were much more economic, if the need for a mask could be eliminated and the personalization pattern be printed to the wafer - element by element - directly according to the design data from the tape. As will be shown later: an interesting application for electron beam systems.

PERSONALIZATION OF MASTERSLICES

LIMITATIONS :

PHOTOLITH. REQUIRES MASKS

PROCESS BECOMES TOO SLOW, TOO EXPENSIVE

FUTURE REQUIREMENTS :

PATTERN VERSATILITY

MIN. GEOMETRIES: 1-2 μm

OVERLAY : 0.3 - 0.6 μm

THROUGHPUT: ~ 10 WAFER EXP./HOUR

Fig. 2. Key bottlenecks in to-day's photolithography (I)

Another bottleneck of to-day's photolithography with regard to future requirements relates to the mask making process (Fig. 3). Already to-day a chip mask for a 64.000 cell chip will have more than 200.000 structural details. To compose these by an opto-mechanical pattern generator takes about 50 hours of failure free operation. Generation of patterns for 288.000 or even 512.000 cells per chip certainly will exceed the Mean Time to Fail of such a machine and thus be impractical. So, a faster pattern generator for finer structures is another requirement for the future lithography.

Also, the photolithographic printing process itself becomes a bottle neck (Fig. 4). It is limited with regard to the minimum geometries of the structures to be printed to the wafer and with regard to the achievable overlay. Several causes care responsible.

o With every step in the photolithographic process occurs a degradation of the pattern resolution as well as one of the overlay. Therefore, for a future micron and submicron lithography as many process steps as possible should be eliminated from the sequence of Fig. 1.

o Thermal expansion and bending of mask and wafer set limits to the overlay achievable on the wafer surface - even if perfect masks were used. Thus, one should avoid printing the wafer in one blow and rather print the pattern chip by chip to the wafer, or may be die by die - if one die had the area of a few chips.

o Last but not least, the wavelength of the visible light is too long. Diffraction at the mask structures limits the minimum size of geometries and the depth of focus within which they can be printed.

The depth of focus D is governed by the Flesnel number F and depends on the minimum geometry W to be printed and on the wavelength $\lambda$ of the exposure radiation.

Tolerable line edge qualities are obtained at F = 1. The table of Fig. 5 shows for blue light of $\lambda$ = 0.4 $\mu$m wavelength, that 2 $\mu$m wide lines can be printed over a depth of focus of $\pm$ 10 $\mu$m - the required flatness of the wafer area exposed in parallel. Lines of 1 $\mu$m width would tolerate only $\pm$ 2.5 $\mu$m focal depth, a flatness tolerance for the wafer surface which cannot be achieved over a diameter of - let us say - 100 mm.

The table in the middle of Fig. 5 shows a much better situation for light of 1.0 nm wavelength - as it is applicable in x-ray radiation. Here, the flatness tolerances of the exposure area are considerably looser and acceptable.

So, aside from few process steps and from die stepping exposure, the wavelength of a future lithographic printer should be short compared to that of visible light.

Flexibility in pattern shapes is not specifically needed for volume printing. However, throughput and good economy are main requirements. Very fortunately, it appears that throughput can be gained if one abandons flexibility.

**MASK MAKING**

LIMITATIONS :

- PRESENT PATTERN GENERATOR TOO SLOW
- PRESENT MASK MAKING PROCESS TOO MANY STEPS

FUTURE REQUIREMENTS :

- PATTERN VERSATILITY
- MIN. GEOMETRIES : $0.5 - 2\ \mu m$
- PATTERN PRECISION : $\leq 0.1\ \mu m$
- THROUGHPUT : $\geq$ 1 MASK / 5 HOURS

Fig. 3. Key bottlenecks in to-day's photolithography (II)

**PHOTOLITHOGRAPHIC PRINTING**

LIMITATIONS :

- TOO MANY PROCESS STEPS
- FULL FIELD PROJECTION
- DIFFRACTION EFFECTS

FUTURE REQUIREMENTS :

- THROUGHPUT : $\gtrsim$ 50 WAFER EXP. / HOUR
- MIN. GEOMETRIES : $0.5 - 1.0\ \mu m$
- OVERLAY : $0.2 - 0.4\ \mu m$
- DIE BY DIE EXPOSURE

Fig. 4. Key bottlenecks in to-day's photolithography (III)

$$\frac{W^2}{\lambda \cdot D} \approx 1$$

W = LINE WIDTH , D = DEPTH OF FOCUS , λ = WAVELENGTH

λ = 0.4 μm

| W | D |
|---|---|
| 3 | 23 |
| 2 | 10 |
| 1 | 2.5 |
| 0.5 | 0.6 |

λ = 10 Å

| W | D |
|---|---|
| 1 | 1000 |
| 0.5 | 250 |
| 0.25 | 62 |
| 0.1 | 10 |

λ = 0.15 Å

| W | D |
|---|---|
| 1 | 67000 |
| 0.5 | 17000 |
| 0.25 | 4000 |
| 0.1 | 660 |

W, D [μm]

Fig. 5. Depth of focus

## ELECTRON BEAM LITHOGRAPHY SYSTEMS

Several contending technologies try to open up the bottlenecks of future lithography: the use of deep ultraviolet rather than blue light, or even the use of x-rays, both in combination with the exposure on ultrafast stepping tables are typical examples for future volume printing approaches.

Electron beam technology with its very short wavelength of radiation and its capability of easily deflecting the beam offers both: high throughput systems for volume production with the electron beam projection systems - and flexibility in pattern generation with the electron beam scanning systems. The latter achieve this flexibility by directly writing the pattern according to the design data. The throughput of the former is facilitated by sacrificing this flexibility and, instead, projecting a fixed mask pattern to the wafer in one blow.

### Electron Beam Scanning Systems

In the simplest version of electron beam scanning systems (Fig. 6), the target is raster scanned line by line by a focussed electron beam. The pattern is generated much like in a TV set.

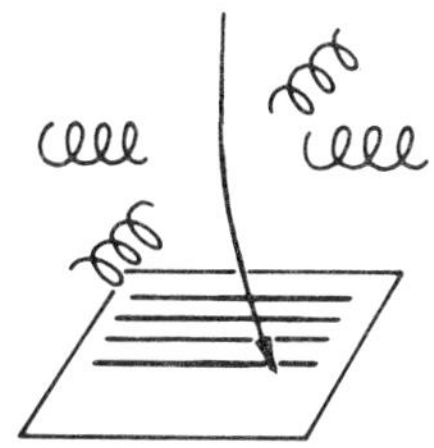

D/A CONVERSION (100 MHz) : 1 : 65000

| |
|---|
| POSITIONING ACCURACY : |
| 0.1 µm ADDRESS INCREMENTS IN 6 mm FIELD |

$10^6 \times 10^6 = 10^{12}$ ADDRESSES IN 100 mm × 100 mm FIELD

| |
|---|
| SCAN TIME FOR WAFER |
| $10^4$ sec. ≈ 3 HOURS FOR 100 mm × 100 mm FIELD |

Fig. 6. EB raster scan systems

The maximum size of a print field which can be covered by electron beam deflection may be considered disregarding the resolution of the beam column. It depends on the required lateral address increments and on the number of increments of the beam deflecting current which can be reproducibly achieved at MHz switching rates. For a 1 µm pattern capability, the lateral address increments should be about 0.1 µm. The state of the art in generating analog signals permits signal increments of 1 in 65.000. This yields a 6 mm large field, i.e. a chip field. Thus, electron beam scanning systems require the same chip size fields as called for in the future lithography requirements.

Chip field after chip field is completed by placing the wafer on an x-y table and moving it into the corresponding positions. The electron beam is registered to the chip sites by the aid of precisely located individual alignment marks.

Another way to accurately position the electron beam to the desired addresses is based on measuring the wafer position by laser interferometers attached to the x-y table. The table is moved to approximately the desired address and the laser interferometer provides a deviation signal for deflecting the beam to the precise address. Counting the interference fringes is a purely digital process. With regard to accuracy this is superior to the generation of analog beam deflecting signals. On the other side, accelerations and decelerations of the table limit the achievable speed.

To-day's widest employed electron beam scanning system for the production of photolithographic masks, the MEBES[1], employs a combination of electronic beam deflection and laser interferometer measurement (Fig. 7). Only 512 beam addresses along a very short line are accessed by beam deflection. Line under line to print a strip, and strip aside of strip to complete the mask are placed under control of a laser interferometer.

Another important characteristic of scanning systems is the achievable throughput.

If there are $10^6$ addresses per dimension of a 100 mm diameter wafer, then a total of $10^{12}$ addresses have to be scanned during one wafer exposure. At a - rather high - addressing speed of 100 MHz $10^4$ sec. or almost 3 hours are required to complete a wafer. Considerable extra time is required for the overhead, like chip stepping, alignment etc.

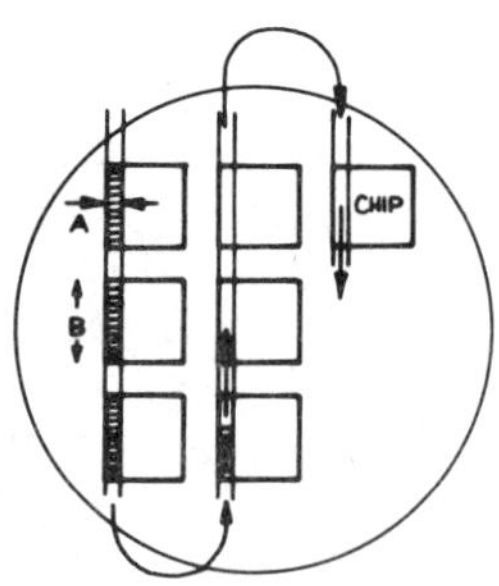


A : 512 ADDRESSES , 0.25 μm INCREMENTS

→ STRIP WIDTH 0.128 mm

B : CHIP LENGTH 30000 ADDR. ~ 8 mm

Fig. 7. MEBES principle

With a throughput of about 1 wafer per 5 hours, an electron beam scanning system is very slow. When compared, however, to the afore-mentioned optomechanical pattern generator for making Single Segments for masks, it offers a considerable speed improvement. An additional advantage is the capability to write finer geometries. Also, the deteriorating Step & Repeat process is eliminated. These advantages cause that such systems find more and more entrance to the mask houses.

For wafer exposure, however, the throughput of raster scanning systems is much too low as to be economic. Two development directions took place to achieve higher throughput (Fig. 8).

One was staying with raster scanning but using a square spot instead of a merely focused beam with a round spot. The square spot has the dimensions of the smallest geometry to be printed - however, it corresponds to a large number of simultaneously printed round beams. The example of Fig. 8 shows an improvement of the printing speed by a factor of four. In practical cases, however, the square spot corresponds to about 25 round spots.

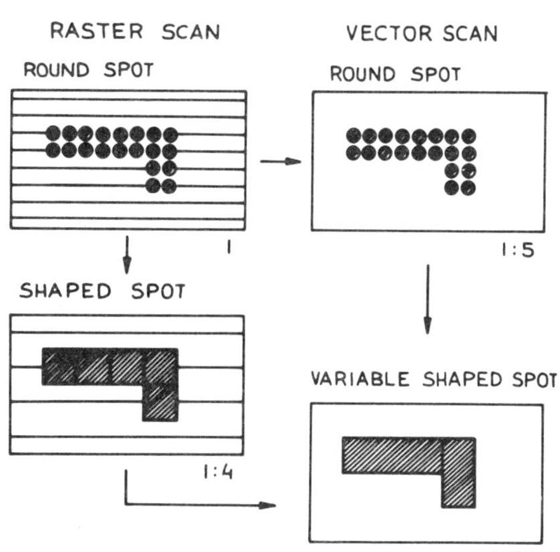


Fig. 8. Reduction of scanning time

Another direction leads to the vector scan systems. Rather than to scan through all addresses, whether or not to be exposed, these vector scan systems skip all those addresses which are not to be printed. The example of Fig. 8 shows a speed improvement by a factor of 5 corresponding to the ratio of blank and printed spots.

The present ultimate of electron beam scanning systems marries vector scan characteristics and shaped spots. In the variable shaped spot systems[2], the spot can be shaped as variably sized squares or rectangles extended in x- or y-direction. To that end (Fig. 9), one square aperture is imaged onto another square aperture. The image of the first aperture can be sidewise shifted. The effective beam shape then is determined by the overlapping parts of the two apertures.

Although several companies have variable shaped spot systems under development, the EL3 of IBM is probably the most advanced system. With its very high throughput of 10 to 20 wafers per hour for structures with down to 1.2 µm geometries, this system fits perfectly the requirements of Master Slice personalization.

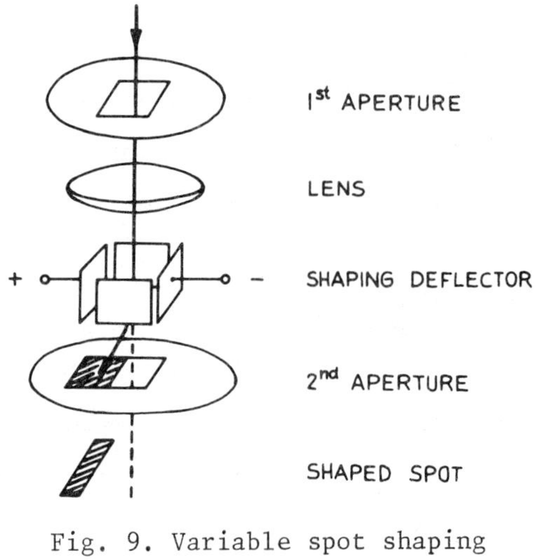


Fig. 9. Variable spot shaping

Electron Beam Projection Systems

A throughput of 10 to 20 wafers/hour (as achieved in the fastest electron beam scanning systems) can - economically - not compete with the 50 wafer/hour throughputs of to-day's photolithographic printers. This becomes even worse when, at smaller geometries and higher component densities, a scanning system has to write more individual structures.

The throughput can, however, be considerably improved, when - at the expense of flexibility - the wafer pattern or part of it is provided in a mask and this mask pattern then printed to the wafer in parallel.

In the early days of electron beam technology it has been attempted to project the image of a mask to the wafer by electron-optical lenses. These developments have been discontinued due to problems with the patterned masks, to difficulties with stepping and aligning the dies on the wafer and, with the pattern distortions originating from several kinds of disturbing electrical fields influencing the projecting electron beam.

In contrast to the case of electron beam scanning systems the development direction of projection systems lead from more complicated to more simple versions. To-day, two systems prevail, the photocathode projection system and electron beam proximity printing.

In the photocathode projection system (Fig. 10), a whole wafer pattern is projected from a cathode to the wafer in a rather unique way. The electrons originate from an areal photocathode. They are accelerated by a uniform electric field from the cathode towards the wafer. An, also, areal uniform magnetic field takes care, that all lateral motions of the electrons are transformed into spirals. The resulting electron trajectories find a focus on the wafer surface. So, every point of the photocathode is imaged to a corresponding point on the wafer, no matter what the direction of the emitted photoelectron was.

To generate the pattern for printing the entire wafer, a chromium layer is provided as a mask on a quartz plate. A CsI layer on top is illuminated through the chromium mask by ultraviolet light. It emits the photoelectrons structured according to the pattern.

The most advanced EB photocathode projection system[3] is being worked on at Philips Research Labs. in England. It was shown, that 0.5 $\mu$m lines and spaces can be printed at a throughput of 25 wafers/hour. The group claims, that even a mismatch of the mask pattern with structures already on the wafer can be compensated by slightly diverging or converging the imaging magnetic field. This facilitates a slight change of the projection scale.

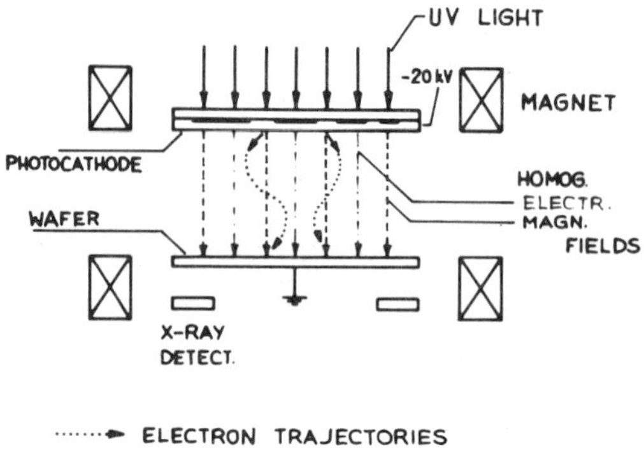


Fig. 10. EB photocathode projection

Electron beam proximity printing[4] is the newest electron beam projection system for high throughput micron and submicron lithography. It is characterized by special simplicity and freedom from distortions.

This characteristic is achieved in an arrangement, where electric and magnetic fields do not need to and hardly can act on the electron beam after the beam is spatially modulated by the pattern. According to Fig. 11, a collimated electron beam illuminates a mask and casts its shadow image on to the wafer. The mask has the size of a chip. After each chip exposure, the wafer is stepped to the next chip site by an extremely fast x-y table.

At 10 kV acceleration voltage, the electron beam has a wavelength of only 0.015 nanometers. At this short wavelength, the diffraction of the beam at the mask apertures becomes negligible down to structures of less than 0.2 μm width. According to the Fresnel equation, the useable depth of focus exceeds 1000 μm at these geometries. Fig. 12 shows the shadow print of three 0.5 μm wide lines on a wafer. The mask to wafer distance was one half millimeter. Yet, there is no blurring of the pattern, and even much finer edge structures are reproduced in all details.

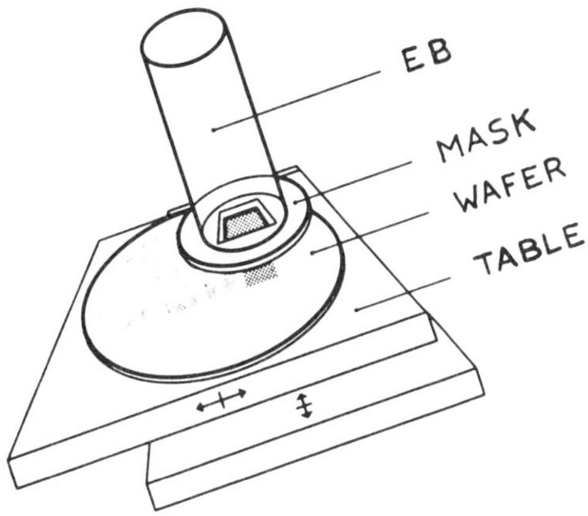


Fig. 11. EB proximity printing - schematic

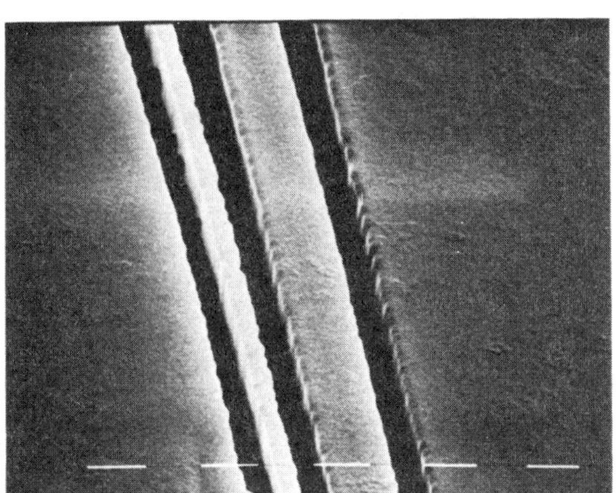

LINE WIDTH 0.5 μm

Fig. 12. EB proximity printing - printing capability

The large distance between mask and wafer opens up the possibility for another, very important feature. A slight inclination of the collimated illuminating electron beam shifts the shadow image in a corresponding direction. This feature is of special importance for exact positioning of the patterns to their nominal sites if the x-y table did not stop precisely at the correct position. In that case, the laser interferometer measures the off-position of the wafer, and a closed loop control system inclines the beam such that the shadow image is projected to its correct position.

The achievable result of positional correction is demonstrated in Fig. 13. Here, a pattern with lines of 1 μm width was printed two times to a wafer with 50% dose each. Between the two exposures, the x-y table was moved around and then returned - within its tolerances - to the original position. Fig. 13a shows that in the two table positions the patterns do not match by about three tenths of a micron. In Fig. 13b, the same was done, however, with the closed loop positional correction switched on. In this figure, no displacement remains visible, the registration of the two patterns is better than 0.1 μm.

a

b

Fig. 13. EB proximity printing - precise pattern positioning. (a) without (b) with beam inclination for positional correction

This method of positional correction is an important feature with regard to the achievable throughput of electron beam proximity printing. It permits the employment of fast stepping tables with positional tolerances up to ± 2 μm.

Table step times of 160 msec are achieved in a laboratory prototype of an EB Proximity Printer. Together with an exposure time of 50 msec per chip, this facilitates a throughput of 50 wafers per hour for pattern structures down to about 0.5 micron width.

The key to electron beam proximity printing was the availability of electron beam transmission masks[5]. These masks must have physical holes for the electrons to penetrate. A technology has been developed to make such masks which have very high precision and extreme mechanical and thermal stability. In this technology, a normal silicon wafer is etched down in a window to a thin silicon foil. The pattern to be printed is dry etched into this foil.

The favorable properties regarding accuracy and stability of these masks result from the crystalline nature of the silicon foils: they are extremely rigid and stable. Being under tensile stress from their fabrication process they do not sag in their thick silicon frame even when warmed up to 100°C.

Fig. 14 shows an electron beam transmission mask in a schematic and as a real mask at about 5 x magnification.

Masks with physical holes normally have the "mask stencil problem": islands surrounded by a line fall out. In electron beam proximity printing this problem is eliminated by the use of complementary masks. As Fig. 15 shows, the patterns are split into complementary sections and distributed to two mask halves. This split gives extra stability to the masks and provides for good dissipation of the absorbed power of the exposing electron beam.

The use of complementary masks easily demonstrates the achieved mask to mask overlay in electron beam proximity printing. Experiments yielded that patterns superimposed from their complements do not show scalloping nor blooming at the segment butts if the overlay is better than ± 0.2 μm. In Fig. 16, figures composed of the complements A and B are shown. The line widths are 2 μm. The overlaid complements A + B show figures with straight lines.

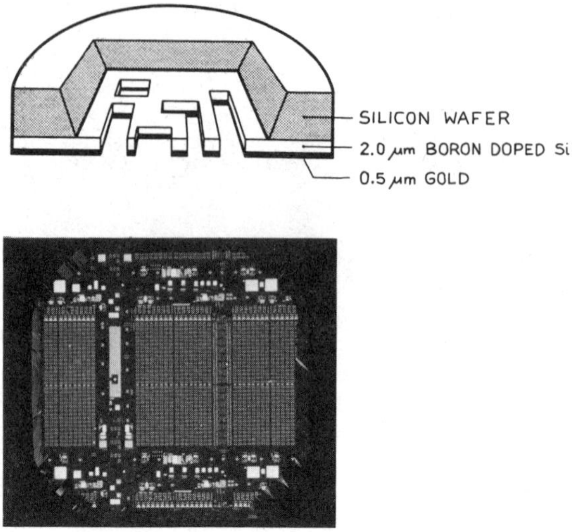


Fig. 14. EB proximity printing - mask, schematic and real

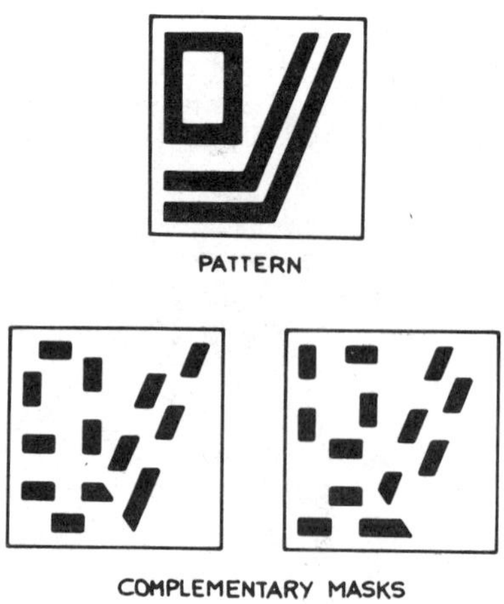


Fig. 15. EB proximity printing - complementary masks

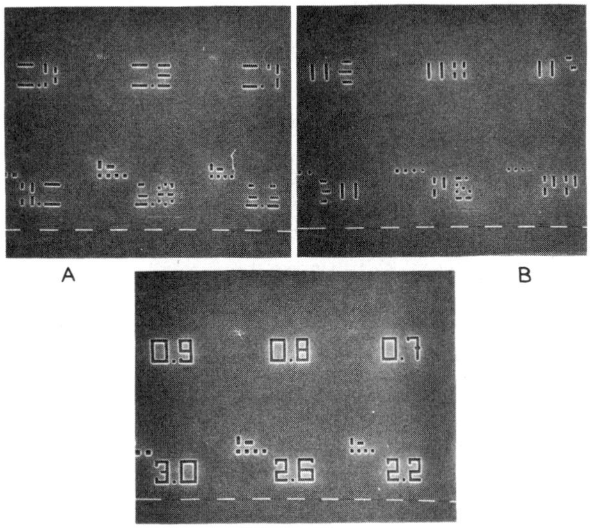


A + B

Fig. 16. EB proximity printing - overlay A + B of complementary figure segments A and B

CONCLUSION

In summing up the capabilities of electron beam scanning and projection systems, one may disregard all other competing technologies and discuss a future sole electron beam lithography (Fig. 17). This lithography would open up the bottlenecks of to-day's photolithography and would allow to economically fabricate wafer structures with geometries down to about 0.5 $\mu$m.

a. Relatively slow electron beam scanning systems with high pattern accuracy generate - directly from the design data - mask patterns in either full wafer size, die or chip size.

b. High throughput electron beam projection systems print these mask patterns to the wafer. Full field projection would be employed in a photocathode projection system, chip or die sized fields in electron beam proximity printing.

c. Electron beam variable shaped spot systems offer pattern flexibility at medium throughput rates. They are specifically suited for low volume "personalization" of Master Slices, but may be also used for mask making or volume production.

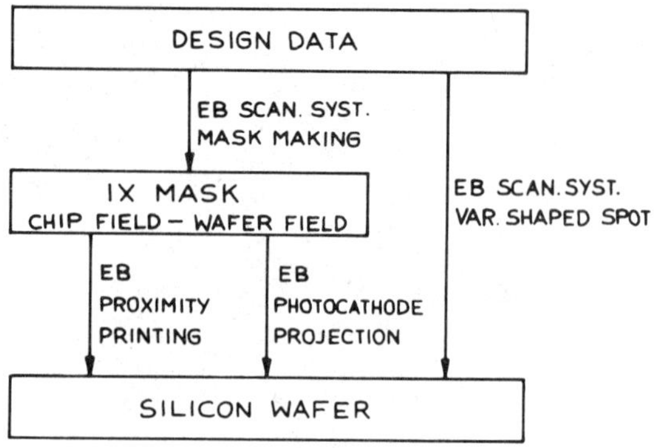


Fig. 17. A sole electron beam lithography

While (a) and (c) presently are being implemented in several companies, (b) is not of immediate need yet: the capabilities of photolithographic printers still fulfill to-day's requirements. The future will show an interesting race between photo-electron- and x-ray lithography.

Compared to Electron Beam Lithography, the other technologies have the following privileges and handicaps:

Photolithography

Privileges: building up on long-term technological experiences, no vacuum.

Handicaps: diffraction and interference effects of exposure radiation.

X-Ray Lithography

Privileges: no scattering of beam in resist and silicon wafer (Proximity Effect).

Handicaps: complicated radiation source, overlay more difficult to achieve.

REFERENCES

1. D. R. Herriott, R. L. Collier, D. S. Alles, J. W. Stafford: IEEE Trans. Electron Devices, Vol. ED-22, No. 7, p. 385, July 1975.
2. H. C. Pfeiffer, 14th EIP Beam Symp., Palo Alto, Calif., May 1977.
   E. Goto, T. Soma, M. Idesawa, ibid.
   M. G. R. Thomson, R. J. Collier, D. R. Herriott, ibid.
   J. Trotel, ibid.
3. R. Ward, A. R. Franklin, P. Gould, M. J. Plummer: J.Vac.Sci.Technol., 19 (4), Nov/Dec 1982
4. H. Bohlen, J. Greschner, J. Keyser, W. Kulcke, P. Nehmiz, IBM J.Res.&Dev., Sept. 1982
5. J. Greschner, H. Bohlen, H. Engelke, P. Nehmiz: Proc. 9th Intern. Conf. on EI Beam Science and Techn. p. 152-160, 1980.

# MICROLITHOGRAPHY FOR VLSI AND BEYOND

A. N. Broers

IBM General Technology Division
East Fishkill, New York

## INTRODUCTION

The well known drive for smallness in integrated circuits stems from the potential for smaller devices to be faster, lower power and cheaper. Lithography is the most universal integrated circuit process and is of great influence in gating progress towards smallness, but other areas are of equal importance. They include device design, material deposition (epitaxial films, oxides, silicides, *etc.*), and development of processes for transferring resist structures into real devices (ion implantation, dry etching, metallization, *etc.*). Because these other areas are complex and numerous progress has been slower than it would have been had lithography alone been the gate. As a result, the smallest dimensions in production today ($\sim 1\mu$) are still within the capabilities of optical lithography. In addition, it appears optical lithography can be extended to dimensions of $\sim 0.5\mu$, so it is likely to dominate for many years, and the resolution advantage of electron beam and X-ray methods will not be needed except in research and development. An exception to this is when fast turnaround is an asset such as in mask making and gate-array personalization. Here scanning electron beams are very valuable. They are also best for device research because, in addition to quick turn-around, they offer highest resolution (10 nm-20 nm) and best registration.

## RESOLUTION AND ACCURACY

Minimum linewidth in a manufacturing process is set by the ability to control linewidth and not simply by the ability to reproduce a given linewidth. Linewidth control must be maintained on the nonflat surfaces, encountered in typical device processes. To establish minimum linewidth, test patterns are exposed on the many different wafer surfaces encountered in the process, and a large number of measurements made for each of the widths under consideration. The distribution of widths for each nominal linewidth is fitted to a Gaussian distribution and the standard deviation ($\sigma(1)$) determined. The device designer then chooses the minimum linewidth that gives him acceptable ratio of linewidth to standard deviation. This ratio is usually 10-15:1. Devices are typically made tolerant to $\pm 3\sigma(1)$ variations in linewidth. Uniform deviations from the nominal linewidth can be compensated for by altering the linewidth in the mask, or, with direct-write electron beam exposure, by altering the written linewidth.

Errors in the position of one pattern with respect to another are known as overlay errors, and must also be treated statistically. Any systematic offset error ($\delta$) due to faulty adjustment of the alignment system is determined and subtracted to leave the random errors. The error distribution is again fitted to a Gaussian distribution and the standard deviation ($\sigma(a)$) calculated. In order to get a satisfactory yield of good devices, alignment errors up to $\pm(\delta+3\sigma(a))$ must be tolerated. Many different kinds of error contribute to overlay errors; some contributed by the exposure tool, such as alignment errors and image distortions, and some by similar errors in the mask.

The end result of this type of statistical analysis is that devices must be designed to be tolerant to errors two to three times greater than might be measured in a few laboratory experiments.

## CONTACT/PROXIMITY PRINTING

Contact or proximity printing is no longer widely used because of defects caused by mask/wafer contact. However, it is still used in noncritical situations when such defects can be tolerated, or when the resolution required is low enough to allow an adequate gap to be left between mask and wafer.

Resolution in a contact image is set by diffraction between the mask and the bottom of the resist. Thick resists, or gaps between mask and resist, degrade resolution. In practice, the minimum usable linewidth W(m) can be approximated from

$$W = 15\sqrt{(\lambda S/200)} \qquad (1)$$

where $\lambda$(m) is the wavelength of the radiation used to expose the resist (it is assumed that the wavelength is the same in the resist as in the gap), and S(m) is the distance between the mask and the bottom of the resist. This is the condition where the intensity at the center of an isolated line matches the intensity in a large area. This criterion is derived from the degradation in resolution due to Fresnel diffraction.[1] More rigorous computations of exposure profiles for proximity printing have been made by Lin.[2]

Linewidth versus gap for deep UV radiation and soft X-rays is plotted in Figure 1. Ultimately, linewidth is limited only by the thickness of the imaging layer. For example, it should be possible to produce 250 nm dimensions in 100 nm thick resist provided intimate contact is maintained between mask and resist. Multilayer resist schemes allow this resolution to be transferred into thicker resist layers.

## OPTICAL PROJECTION

Two forms of optical projection cameras have been used for integrated circuit fabrication; 1:1 full-wafer scanning and step-and-repeat (S/R) reduction. The main advantage of both types over their predecessor, contact printing, is that mask and sample are out of contact and the potential for damage eliminated. To date, the scanning cameras have used reflecting lenses and the S/R cameras refracting lenses, but there is no particular reason for this, and ultimately it may be best to combine mirror lenses with S/R. This has been done already in the Ultratech S/R camera[3] which uses a modified Dyson mirror lens (0.35 N.A.). The lens combines a primary mirror with an achromatic refractive element that corrects for astigmatism. Ultimately, mirror lenses should produce higher resolution than refractive lenses because they can be used at shorter wavelengths.

Sixty percent contrast in an optical image is generally considered adequate for satisfactory resist exposure, although 80% is needed for image size control of a tenth of a minimum linewidth. For an incoherently illuminated and diffraction limited optical system, contrast of 60% is obtained for a linewidth of $[\lambda \div (1.28 \times \mathrm{N.A.})]$. Contrast is expressed as the modulation transfer function (M.T.F.) where

$$\mathrm{M.T.F.} = 2 \div [\pi \times (\phi - \cos\phi \, \sin\phi)]$$

$$\phi = \arccos \{\lambda \div [4L \times (\mathrm{N.A.})]\}$$

and L(nm) is the linewidth. Higher contrast can be obtained for relatively large linewidths by using partially coherent illumination; that is by arranging that the image of the source only partially fills the pupil of the projection lens. Partial coherence produces higher contrast for linewidths above that at which 35% contrast is obtained in the incoherent case. Thirty percent to 70% filling of the pupil has proven optimum, and for example, can increase contrast from 60% for the incoherent case to more than 80% for an acceptable degree of

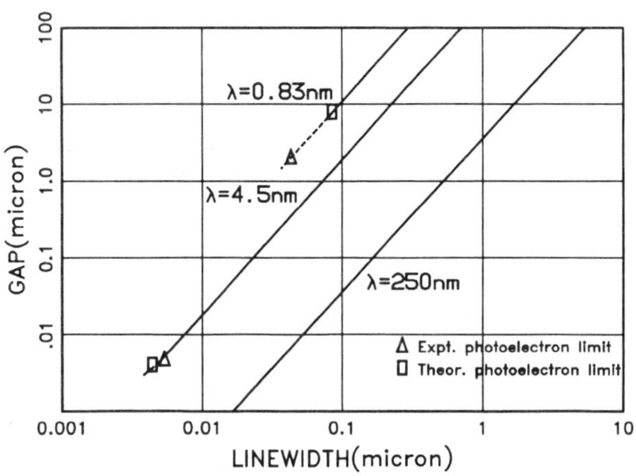


**Figure 1.** Linewidth versus gap for deep UV and X-ray proximity printing. Theoretical points correspond to the Gruen range for the maximum energy photoelectrons. Experimental points were measured by Feder and Spiller.[23]

partial coherence. Contrast for the Micralign (0.16 N.A.) optical system and a 0.35 N.A. refractive lens are shown in Figure 2. Partial coherence increases the depth of field but increases the exposure time. Too high a degree of partial coherence gives rise to undesirable interference effects between lines.

Depth of field depends on the substrate reflectivity, the degree of partial coherence and the minimum feature size.[4] An approximation can be made from the depth of field for the incoherent case which is given by $\pm\lambda \div 2(N.A.)$.[2] Two layer resist processes in which the image is formed in a thin, flat resist layer on top of a much thicker planarizing layer, alleviate the need for a large depth of field and make it easier to form high resolution, high aspect ratio, resist patterns.[5,6] These resist systems also make it possible to obtain satisfactory results at contrast levels as low as 40%.

## Scanning Mirror Systems

Several mirror lens scanning projection cameras are commercially available. The most common are the Micralign series built by Perkin Elmer,[7,8,9] although the MPA500 built by Canon is widely used in Japan. All of these cameras have mirror lens systems with a numerical aperture of 0.16, and operate at unity magnification. In the Micraligns, a crescent on the mask is critically illuminated for a curved capillary arc lamp and the illuminated portion of the mask imaged onto the wafer. Full wafer exposure is accomplished by scanning mask and wafer past the optical system.

The basic optical system of the Micralign 100, 200 and 300 series is shown in Figure 3. These cameras can be operated in three wavelength regimes, conventional ultra-violet (UV $\lambda$=300 nm) and deep UV ($\lambda$=250 nm). To optimize performance in the three regimes, however, it is necessary to change the multi-layer coatings on some of the optical elements. In particular, coatings must be changed on the mirror that follows the mask, and on the smaller of the spherical mirrors. The dichroic coatings on these components transmit green light for alignment, but reflect UV for exposure. The spacing between the layers of these coatings must of course be changed when the exposing wavelength is changed.

Optical component changes are not necessary with the model 500, shown in Figure 4, where the roof prism of the earlier systems is replaced with a second set of spherical mirrors. The second pair of mirrors reverses the parity of the image in the same way the roof prism reversed parity in the earlier systems allowing the 500 to use the same masks as the other cameras, which in turn used the same masks as contact printers.

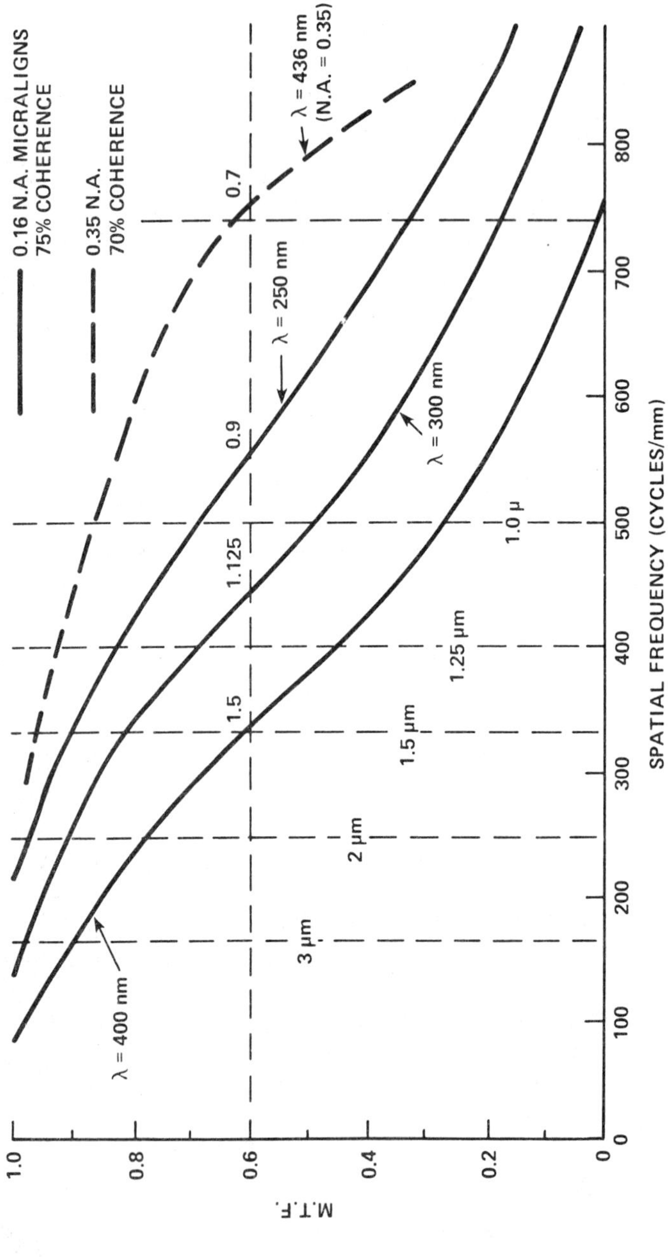


**Figure 2.** M.T.F. for Perkin Elmer Micralign systems, and for a refractive lens with a numerical aperture of 0.35.

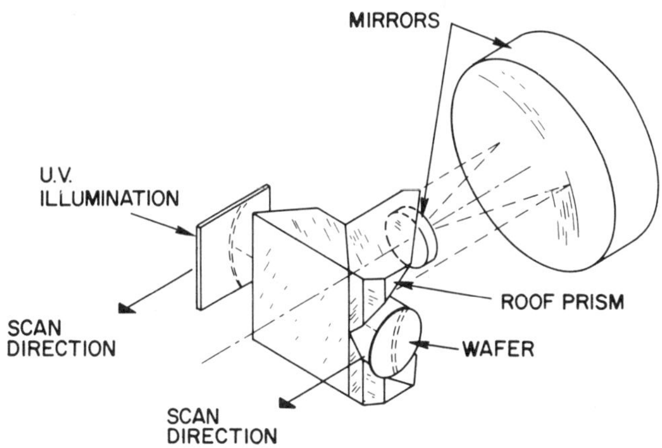


**Figure 3.** Optical system for 1:1 scanning projection cameras built by Perkin Elmer (100, 200 and 300 series Micraligns).

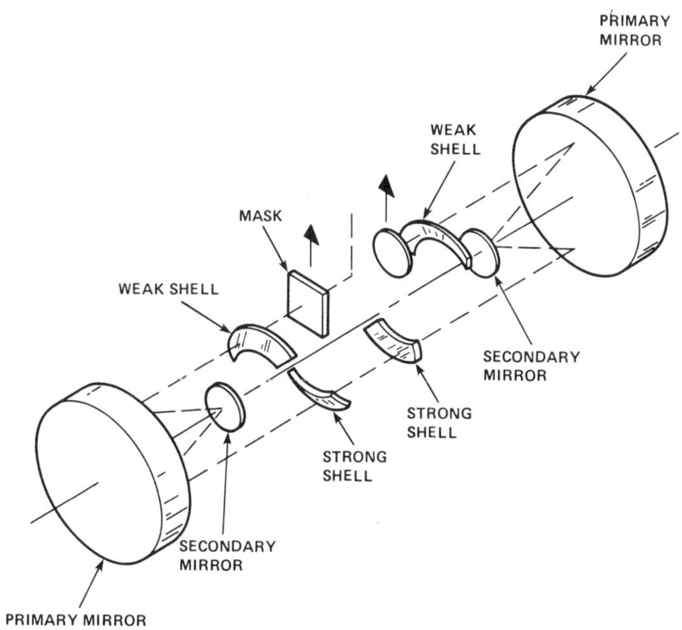


**Figure 4.** Optical system for the Perkin Elmer Micralign 500. Prism in the earlier Micraligns is replaced by a second set of spherical mirrors.

Overlay accuracy of the 100 and 200 models is mainly limited by nonflatness of the surfaces of the roof-prism. This nonflatness leads to distortions of up to ~2$\mu$. However, overlay of about $\pm 1\mu$ ($3\sigma$) can be obtained by choosing cameras with similar distortions. In the model 300, the prism surfaces have been improved and distortions reduced to less than $\pm 0.25\mu$ allowing overlay between different cameras to be well under $\pm 1\mu$. Distortion in the 500 is similarly improved. The 500 also has a better alignment system that images the alignment marks directly onto photo-diode arrays rather than viewing them with a television camera.

The major advances with the 500 are the ability to expose larger wafers (up to 150 mm, compared to 100 mm for the earlier cameras), and to adjust magnification. The latter makes it possible to compensate for the inevitable microscopic changes in wafer size that arise during device processing. Image size in the direction of the slit is changed by small axial motions of the 'stronger' pair of refractive elements. Correction in the opposite direction is made by micro-scanning the mask during exposure in order to slightly increase or decrease the scan length for the mask compared to that for the wafer.

If distortions of the sample are truly isotropic, the only errors that remain after magnification correction are residual distortions in the optics, and mask errors. The easiest way to reduce mask errors it to go to 5X or 10X step and repeat systems. Ultimately, though, it should be possible with electron beams to reduce mask errors to insignificant levels. Distortion in the optics is already below $0.1\mu$.

As already mentioned, resolution with mirror lenses can be improved by operating at shorter wavelengths, the only difficulty being that of finding resist/lamp combinations that give satisfactory exposure times. Several resists have already been found that are satisfactory for 300 nm exposure and there are a number of others with potential for operation at 250 nm. The major difficulty with resists for the deep UV regime is not sensitivity but the need to find resins that are transparent enough to allow uniform exposure to the bottom of the resist.

If sensitivity proves to be a problem then exposure times could be reduced by using an excimer laser rather than an arc lamp as the source.[10] An excimer laser combined with an efficient condenser can deliver considerably more power than a mercury arc source. Excimer lasers are pulsed lasers that operate at discrete frequencies throughout the UV spectrum. They can deliver 500 mJ in a single 20 ns pulse and their pulse rate can be as high as 1000 Hz but is generally closer to 50 or 100 Hz. This is still adequate for scanning exposure. Illumination systems for laser illumination are complex because they have to increase the divergent angle of the light as well as to uniformly illuminate the mask. The spherical bandwidth (~15A) and multi-model characteristics of excimer lasers reduce speckle, the major problem with general application of lasers to lithography, to negligible levels. Operation at a single wavelength simplifies the correction of aberrations in refractive lenses, which will be of importance if they are applied to S/R cameras, but increases the chance that standing wave interference effects will be a problem.

Present full wafer mirror scanning systems are limited to a numerical aperture of about 0.16 by the size of the optical components, and by the need to cover a 125 mm wafer in a single scan and yet avoid having the mask or wafer vignette the optical system. If the full wafer capability is sacrificed, and a system designed to only expose a portion of the wafer in a single scan, or a partial scan, then it should be possible to increase the numerical aperture to 0.35 or higher. Eventually it may be possible to build a system with a numerical aperture of 0.4 which, at a wavelength of 250 nm, would produce 60% contrast for linewidths of 0.5$\mu$. Depth of field would be very small ($\pm 0.8\mu$ for incoherent case), but methods such as capacitive sensing are available to monitor the wafer surface to $<\pm 0.1\mu$ and piezo-electric manipulators could be used to provide means for continuously correcting focus.

Step and Repeat Cameras

Figure 5 shows the general layout of a step-and-repeat (S/R) camera with a refractive lens. A large number of commercial cameras of this type are available. The best refractive lenses built to date have numerical apertures of about 0.3 and cover fields of about 2 cm$\times$2 cm. Higher numerical aperture is available for smaller field-size as shown in Figure 6. Most lenses are designed to operate at a single wavelength that corresponds to a strong line in the mercury spectrum (365 nm, 405 nm or 436 nm), but lenses have also been corrected for two wavelengths (405 nm and 436 nm) in order to reduce the deleterious effects of standing wave interference in the resist. However, with two layer resists this is not really necessary.

As dimensions get smaller it is of increasing importance to improve overlay and this places stringent requirements on the distortion of the lenses. The distortion of the best lenses remains at about 0.1$\mu$ and this error contributes significantly to final overlay when different cameras are used for different layers.

Because the field size of refractive lenses is very much smaller than a silicon wafer, they must always be used in the step and repeat mode. The sample position in a S/R camera is either tracked by a laser interferometer after an initial reticle to wafer alignment, or the reticle and sample are aligned with respect to each other at every chip site.[11-16] Alignment at every chip

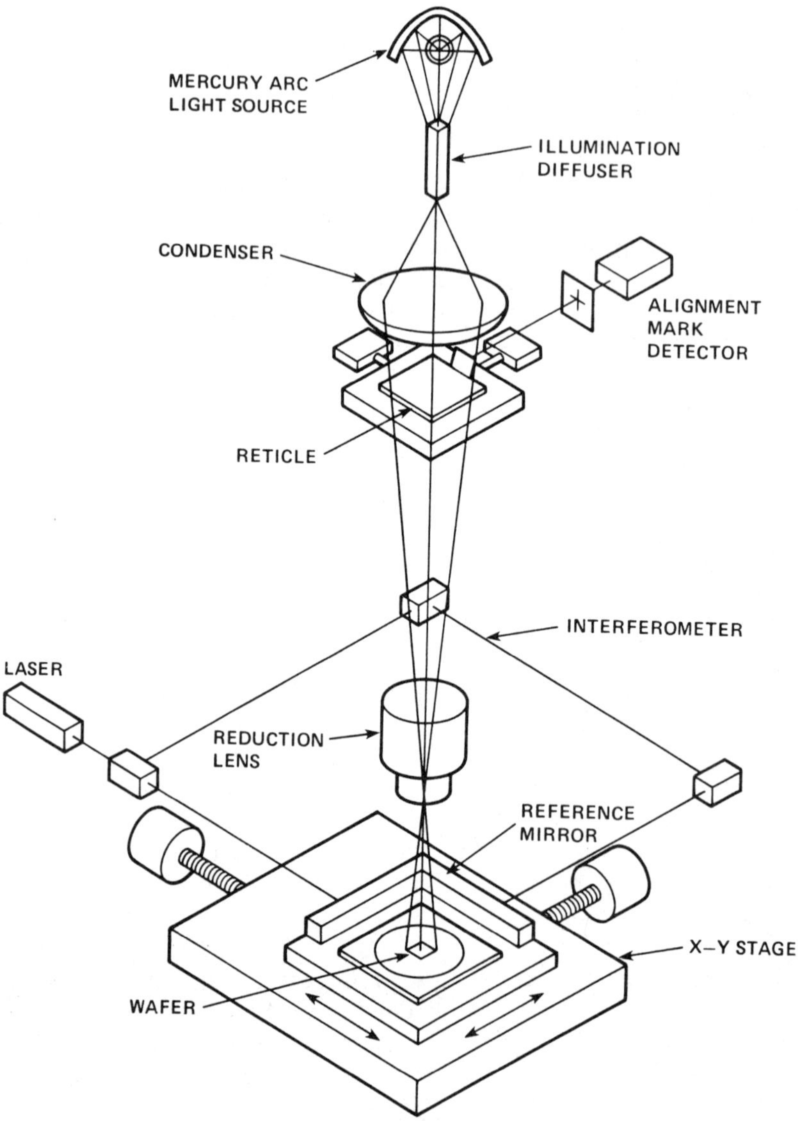


**Figure 5.** Generalized step-and-repeat camera. Ideal systems have both direct wafer to reticle alignment, and a stage capable of maintaining accuracy through dead-reckoning.

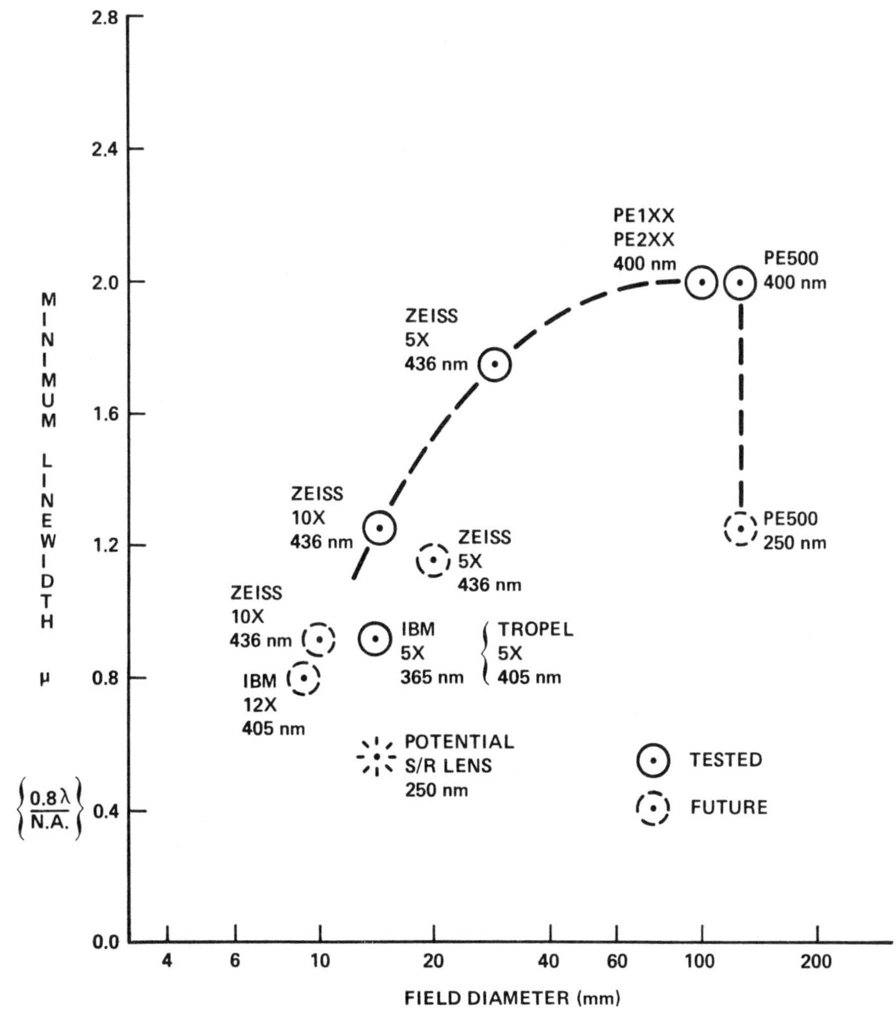


**Figure 6.** Field size and minimum linewidth for microcircuit lenses. IBM lenses were designed by Wilczynski and Tibbetts.[51]

avoids errors due to drifts between the wafer and the interferometer reference point, but has proven very difficult. Almost all automatic alignment schemes have suffered from the difficulty that for certain resist thickness and alignment mark type, light is reflected equally from the mark and from the background surface, and the mark 'disappears'. To avoid this difficulty it is necessary to carefully control resist thickness or to remove the resist from the marks. Both solutions complicate the process. Best results have been outlined with dark-field illumination.

In the absence of the above mentioned problems it should be possible to align to an accuracy better than the resolution of the optical microscope; that is better than $0.1\mu$. The diffraction limit of the optical sensing system can readily be exceeded because the signal to noise is generally adequate to allow precise threshold detection on the image of the edges of the alignment marks.

In principle, resolution of optical lithography can be similar to that of optical microscopy. A lens with a numerical aperture of 0.95 would produce linewidths $<0.4\mu$ over an area of about 0.2 mm×0.2 mm. The small field would not be an inherent limitation because the image could be stepped or scanned, however, the large numerical aperture would reduce depth of field to about $\pm 0.2\mu$ restricting resist thickness to $<0.2\mu$. Improvement could be obtained by operating with deep-UV light, but refractive lenses are difficult to build at this wavelength because of the lack of materials that are adequately transparent and yet have relatively high refractive indices. Lens elements made from low refractive index glass must have smaller radii of curvature and this increases higher order aberrations.

## X-RAY LITHOGRAPHY

X-ray lithography is proximity printing with soft X-rays of wavelength between 0.4 nm and 5 nm.[17,18] The wavelength is so short that diffraction effects are negligible down to linewidths of $0.1\mu$, even for a mask to wafer spacing of $50\mu$. The mask cannot be made on a quartz plate as is used for optical lithography, because the plate would absorb the X-rays, so it is formed on a thin ($<10\mu$) membrane. The membrane is generally made from a combination of two materials with one chosen for dimensional stability (silicon, boron nitride, *etc.*) and the other for ruggedness (*e.g.*, polyimide).

X-ray lithography offers one outstanding advantage over other methods; the exposing radiation is neither scattered in the resist nor is it diffracted. This results in high aspect ratio resist patterns (see Figure 7) and there is no proximity effect as there is with electrons.

**Figure 7.** High aspect ratio $1\mu$ resist pattern obtained with X-ray lithography (Spiller *et al.*[18]).

On the other hand, the mask is fragile and may not be dimensionally stable, and the large divergence of conventional electron bombardment X-ray sources gives rise to image distortions if mask and wafer are not perfectly flat (see Figure 8). In order to keep pattern distortion below $0.1\mu$ for a 125 nm diameter wafer and a source to wafer distance of 50 cm, the mask and wafer must be held flat to better than $1\mu$. Wafer buckling that arises during many standard integrated circuit processes make this very difficult to achieve and it is unlikely that full wafer exposure can only be accomplished for wafers larger than about 7.5 cm.

The run-out error could be avoided if resist sensitivity was improved to the point that exposure could be made at a point further removed from the source. This requires a sensitivity approaching 1 $mJ/cm^2$, however, and while there are some resists with sensitivity at this level,[19] they do not yet produce high aspect ratios and therefore lack the major advantage of X-ray lithography over optical lithography. High aspect ratio can be obtained by using a tri-layer resist process, but this option is also open to optical lithography.

The run-out error could also be avoided by going to a S/R approach and limiting mask size to a few centimeters. This would also alleviate problems due to mask and/or wafer distortions but exposure times with presently available resists and conventional sources are too long for this to be economical. Exposure times of a few seconds would be needed for a source to wafer distance of about 50 cm which is similar to the requirement for the full-wafer case.

One source that produces an X-ray beam with all the properties needed for S/R exposure is the electron storage-ring (Figure 9).[20] The X-ray beams emitted from a storage ring are so intense and have such small divergence that the problems of run-out error and long exposure time are removed. For several years physicists have been using the electron storage ring as the brightest known short wavelength light source. A storage ring consists of a circular stainless-steel vacuum tube in which electrons circulate at speeds close to the speed of light. The electrons are injected into the ring from an accelerator, and their energy increased and subsequently maintained by accelerating them through a microwave cavity each time they are deflected by one of these magnets they emit radiation with a spectrum that extends from visible light to hard X-rays. By choosing the appropriate electron energy (1 GeV), copious soft X-rays suitable for X-ray lithography are emitted.

The radiation from a storage ring is emitted in a broad sweep with a very narrow vertical spread. At a distance of about 10m from the ring the beam is typically 1 cm high and 30 cm wide. An oscillating, grazing incidence mirror is used to broaden it to the desired height.

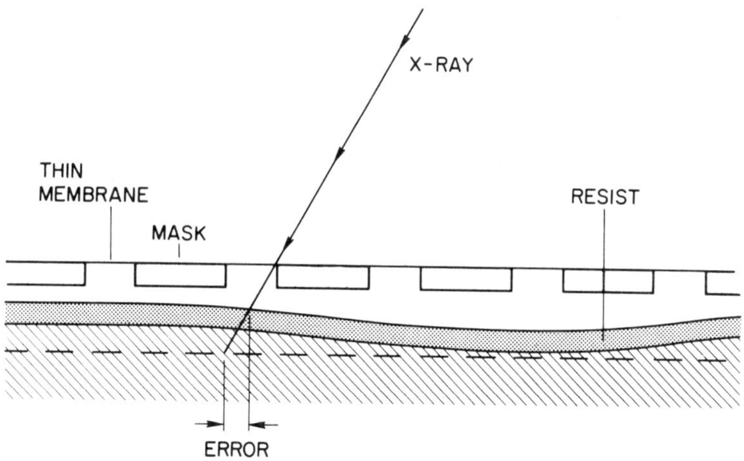


**Figure 8.** Run-out error encountered with conventional point source X-ray lithography.

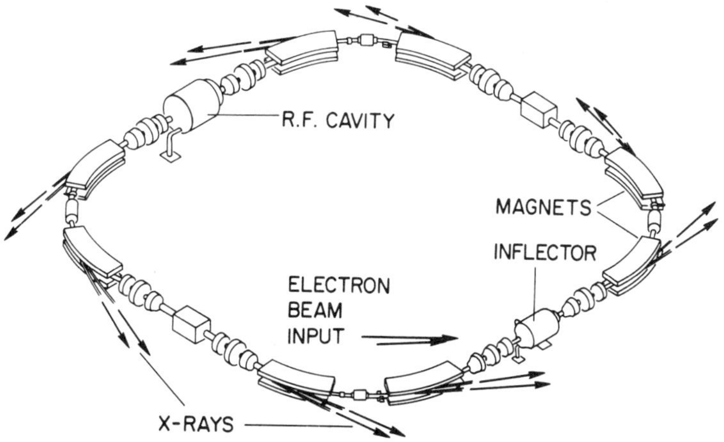


**Figure 9.** Electron storage ring is the highest intensity, lowest divergence X-ray source known, and is capable of providing adequate flux to expose hundreds of 125 mm diameter wafers per hour. A number of exposure stations would be placed around the ring.

An alternative method for generating the X-rays is to "wiggle" the electrons while they pass down the straights in the ring between the bending magnets. Wiggling is accomplished with multi-pole magnets that produces a sequence of alternating magnetic fields. The bending radius can be smaller in this case and the X-ray energy at each deflection adds up to yield increased total output for a given electron energy. Lower energy reduces the cost of the ring.

Even with this reduction in cost it might at first seem difficult to understand how a ~$10M storage ring could be economical for the manufacture of integrated circuits. The answer lies in the vast total X-ray flux that emanates from the ring. Enough radiation is emitted at each of the bending magnets, or wigglers, to expose 40-50 125 mm diameter wafers per hour,[20] a rate similar to that of present optical S/R cameras. Therefore, fully utilized a storage ring could produce the equivalent of 6 to 12 optical cameras. A S/R camera costs $500K to $1M, so fully loaded, a storage ring becomes competitive with optics, and, of course, provides higher resolution and better aspect ratio in the developed resist pattern. Several laboratories are working on designs for 'compact' storage rings that will be more practicable for production environments.[21,22]

X-ray lithography with a storage ring offers high throughput and a S/R approach that is relatively tolerant to distortions of mask or wafer, but the problem of overlay remains. Alignment accuracy with X-ray lithography is presently similar to that with optical lithography because the alignment methods are optical. Two basic methods have been tried; the first uses simultaneous imaging of mask and wafer with an optical system operating either with two different wavelengths or with two different optical path lengths, and the second examines the superposition of diffracted beams from gratings or zone plates placed on the mask and wafer. Both methods are theoretically capable of achieving accuracies of $<0.1\mu$, and this has been demonstrated in the laboratory, but is yet to be demonstrated in a device process.

In addition to alignment errors, overlay accuracy depends on mask errors and mask and wafer distortions. Errors and distortions are likely to be relatively serious because the mask is fragile and has the same dimensions as the wafer. With conventional sources, this problem is somewhat alleviated because isotropic changes in wafer or mask size can be corrected by changing the mask-wafer gap, but adequate overlay for sub-micron devices has yet to be demonstrated.

Resolution of X-Ray Lithography

Two factors set the resolution for X-ray proximity printing: (1) diffraction between mask and wafer, and (2) the range of the photoelectrons formed when the X-ray protons are absorbed in the resist. The same diffraction criterion used above for optical proximity printing can be applied for X-rays to give the relationships shown in Figure 1. Photoelectron range has been measured experimentally[23] and estimated from the Gruen range.[24] In all cases, ultimate resolution is about 0.02$\mu$ which coincidentally is close to the ultimate resolution of electron beam lithography.

For integrated circuit applications, the significant factor to be observed from Figure 1 is that a gap in excess of 100$\mu$ can be tolerated for 1$\mu$ resolution. In most semiconductor applications, penumbral blurring due to the finite source size is more likely to limit resolution.

X-ray lithography is perhaps the most promising method for large production of sub-0.75$\mu$ devices because it should produce wider process margins than optics and lower cost than scanning electron beams. Many new materials and methods are required, however, and the margin of improvement will have to be considerable before the change will be worthwhile.

## ELECTRON BEAM PROXIMITY PRINTING

Beyond optics, the most likely competitor to X-rays seems to be electron beam proximity printing (see detailed description by W. Kulcke elsewhere in this text). In the electron beam proximity printer, a flood beam of electrons passes through a stencil mask to expose the resist. The mask is placed far enough from the wafer to prevent damage, but close enough to allow the beam to have acceptable divergence. The current density in the beam is proportional to the square of the divergence angle. The mask must be a freely suspended stencil because there is no practicable substrate that does not scatter electrons. The problem of unsupported areas falling out is overcome by making a double exposure with two complementary masks. Exposure is made in a S/R manner so that the mask can be kept relatively small (2 cm-5 cm), and the problems of fragility and dimensional instability minimized.

Bohlen *et al.*[25] have developed a stencil mask technology that is capable of linewidths well below 1$\mu$. The pattern is first formed in a gold layer on a boron doped (to a depth of 3$\mu$) silicon wafer. The silicon is then reactively ion etched to a depth in excess of 3$\mu$ using the gold as a mask. The reactive ion etching is anisotropic and produces vertical walled slots in the wafer. After reactive ion etching the wafer is wet etched from the back with an etchant that stops at the boron doped layer. This leaves a silicon membrane about $3\mu$ thick with holes through which the electrons pass.

The advantages of electron beam proximity printing over X-ray lithography are: (1) The beam can be electronically tilted to shift and/or scan the image (scanning of the shadow image produces the alignment signal). (2) It is relatively easy to get short exposure times.

The disadvantages are: (1) Double exposure is needed to overcome the stencil problem, thus adding to the number of errors that contribute to overlay. (2) Variations in exposure that arise for different pattern shapes and densities due to the electron beam proximity effect can only be corrected by changing the size of the shapes in the mask.

## SCANNING ELECTRON BEAMS

Scanning electron beam lithography has found an important place in integrated circuit production because it does not need a mask. In research and development it has the additional advantage of producing higher resolution than any other method. Many early micron and sub-micron integrated circuit devices were made first with electron beams, and structures are small as a few tens of nanometers have been made for electrical characterization.[26,27]

The sub-micron electron beams used for lithography cannot be scanned over a whole wafer because of deflection aberrations and deflection system noise, so electronic scanning is combined with mechanical movement of the sample. Several methods are used to maintain the position of the beam with respect to the sample. In the 'vectorscan' type of system shown in Figure 10 a direct beam to sample is made before each complete section (chip) of the pattern is exposed. The measurement is made by collecting electrons scattered from four marks located at the corner of each field. Magnification, rotation, position and orthogonality are checked. After registration, the chip is scanned with a two-dimensional electronic scan.

In the 'EBES' electron beam mask-maker shown in Figure 11 the beam is electronically scanned in one direction only, and the sample moved mechanically in the other direction.[28] Continuous mechanical movement is used, chips being written strip by strip, the same strip on every chip being written before proceeding to the next strip. The position of the beam is checked initially with a direct beam to sample measurement, but after this a laser interferometer keeps track of the sample. Alternatively, the beam position can be etched as frequently as needed, but this decreases throughput.

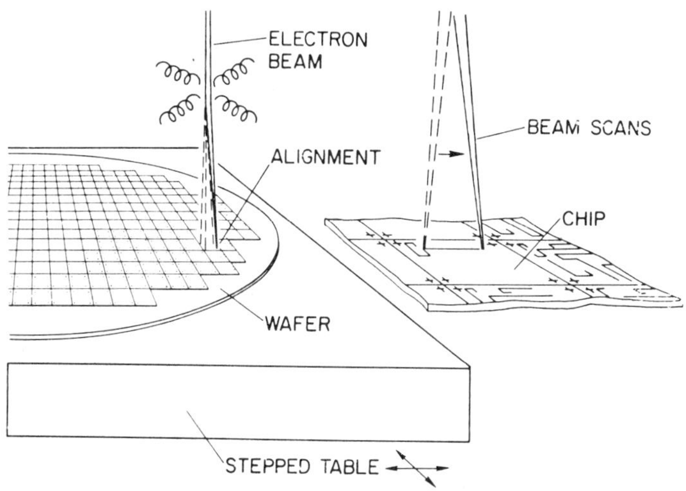


**Figure 10.** Writing method used in vectorscan electron beam systems. Sample is mechanically stepped between chip sites. The beam is first registered to the sample by scanning alignment marks, and then the chip written with two-dimensional electronic deflection.

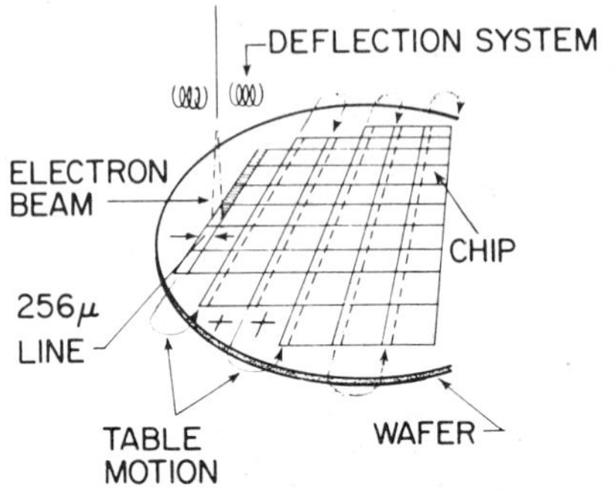


- I - DIMENSIONAL LINE SCAN - (RASTER)
- CONTINUOUSLY MOVING TABLE
- FULL WAFER ALIGN
- LASER INTERFEROMETER FEEDS TABLE POSITION BACK TO E-BEAM

**Figure 11.** Operating mode for EBES electron beam mask-maker.[28] The beam is scanned along a 256$\mu$ line while the sample moves in a continuous serpentine motion. A laser interferometer tracks the position of the sample and errors in position area corrected by feeding signals to the beam deflection unit.

Errors in position are corrected by feeding signals to the electron beam deflection coils. EBES type systems are used extensively for mask making. They can make masks with dimensions as small as $0.5\mu$, and $3\sigma$ overlay accuracy better than $0.25\mu$. Throughput of EBES is relatively low as discussed later.

Most EBES-type systems use a small round beam, although shaped beam versions have also been described.[29] It is important that the focusing/deflection system be telecentric in order to avoid magnification errors when the sample surface is not flat.

### Throughput of Scanning Electron Beam Systems

As already discussed, throughput is the major problem that prevents broader application of electron beam lithography. Throughput used to be limited mainly by electron optical performance, that is by the need to deliver an adequate number of electrons to the resist. Now it depends on a combination of this limit with other factors including the noise bandwidth characteristics of the deflection system, digital to analog conversion rates, switching speed for digital pattern generation hardware, alignment time, mechanical movement of the sample and the time taken to load samples into the vacuum. The relative importance of these different factors depends on the system configuration. The electron optical limit was alleviated by a series of developments which included resists with improved sensitivity, brighter electron guns that use $LaB_6$ of field emission cathodes, shaped beams and deflection/focusing systems that operate at larger beam angles.

### Shaped Bean Versus Round Beam

Two basic approaches are used to maximize the electron beam current and minimize writing time. The first is to use a rectangular beam the size of which can be dynamically varied to fill the pattern features.[30-32] This method minimizes the number of beam addresses while maximizing the current. The second approach is to use a high brightness cathode to focus the maximum current into a round spot that is about four times smaller than the minimum feature size.[33-35] Full pattern flexibility is retained with this approach (*i.e.*, the ability to write curved and angled lines) but the current cannot be as high as it is with a shaped beam, and the number of beam addresses is very much greater.

With shaped beams the source for the electron optical column is an image of a square aperture that is formed in the plane of a second square aperture (nonsquare apertures have also been used to produce angled shapes).

The final beam takes on a broad selection of rectangular shapes as the position of the image is varied with respect to the second aperture.

Provided the electron gun produces adequate current at a given brightness, the brightness in the final shaped beam is the same as that in the round beam, in fact the average brightness is higher. This means that for the same beam aperture and beam brightness, the beam current in the shaped beam can be higher by at least the ratio of the spot areas.

The larger the spot, the larger the beam current, and the smaller the number of beam addresses. This approach cannot be taken too far, however, because as the current increases, electron-electron interactions become excessive and edge definition degrades. For example, in EL3, which has a beam current density of 50 $A/cm^2$ and a maximum spot size of $4\mu \times 4\mu$ ($16\mu^2$), spot edge definition degrades beyond $0.25\mu$ for spots larger than about $6\mu^2$.[31] Larger spots could be used if the current density was reduced, but this would increase the exposure time for the majority of shapes, which are generally smaller than $6\mu^2$, and reduce throughput. The optimum spot size is that which maximizes the time average beam current. In order to optimize exposure time, relatively high current is used, and the problem of poor edge definition on large shapes overcome by outlining these shapes with small narrow rectangles.

Round Beam Versus Shaped Beam

For the current in a round beam to equal that in a shaped beam, the spot current density must exceed that in the shaped beam by the average number of resolution elements projected in the shaped beam, assuming the same deflection/focusing system and pattern generation concept (raster/vector, *etc.*) are used. This number is typically 30-100. However, as already mentioned, the current in the shaped beam is limited by electron-electron interactions and, for example in EL3, this limit is reached at a gun brightness of about $3 \times 10^5$ $A/cm^2$ steradian (25 kV). So a brightness of $1\text{-}3 \times 10^7$ $A/cm^2$ steradian would be adequate to produce the same net current in a round beam.

Such brightness is marginally available from field emitters, and workers are using them in an attempt to reach high throughput,[35] but the reliability of field emitters at high total currents needed to produce a final beam current of several microamps is not proven for production environments.

Lanthanum hexaboride cathodes fall short of the required brightness for a round beam to compete with a shaped beam only producing $1\text{-}3 \times 10^6$ $A/cm^2$

steradian (25 kV) with adequate reliability for unattended production applications.[36]

In summary, it is not possible with presently available sources to obtain the same throughput with a round beam as with a shaped beam.

Throughput of Vectorscan Systems

In vectorscan systems the pattern is written with a round beam, the diameter of which is typically one quarter of the minimum linewidth. The beam is incremented to sixteen positions to write a minimum image square. The beam is only directed to those areas of the chip that require exposure. Chip writing time can be estimated from the beam incrementing rate, and the average fraction of the chip area than needs to be exposed. For example, with a $1\mu$ image, a 5 mm$\times$ 5 mm chip contains a total of $25\times10^6$ minimum images, or $16\times25\times10^6$ beam positions. If, on average, 25% of the chip must be exposed, then $10^8$ beam positions must be written per chip. At 10 MHz, which is the maximum rate used so far to successfully make integrated circuits, this will take 10 seconds. Time taken for table stepping, registration, wafer loading and data transfer must be added to the writing time to come up with the overall time per wafer. In the most advanced systems, overhead times are below 1 second per chip and a 125 mm wafer containing 500 chips can be exposed in about 2 hours.

It should be possible in the near future to reach incrementing rates of 50-100 MHz, provided adequate current density can be delivered to a practical resist, a throughput of about 5 125 mm wafers per hour will be obtained.

Throughput of EBES Systems

EBES systems also use a small round beam with a diameter about one quarter of the minimum linewidth, but unlike the vectorscan systems the pattern is written in a raster manner. This means that the beam has to be addressed to every element on the wafer whether or not exposure is required (advanced EBES systems skip kerf areas and areas where no exposure is required but still write the rest of the areas in a raster manner). If the grid on which the pattern is written is finer than the beam diameter, as is typically the case for masks where the ratio of minimum image to design grid can be as high as 20:1, then the number of beam addresses becomes greater than would be expected from the beam diameter. This reduces throughput. In a vectorscan system each rectangle can be placed with a precision that is smaller than the beam diameter without impacting throughput, only the precision of the pattern control electronics is affected.

Throughput in EBES type systems scales inversely with the square of the image design grid. For example, it takes about 3 hours to write a $1\mu$ minimum linewidth mask for a 125 mm diameter wafer using a design grid of $0.1\mu$ and an incrementing rate of 80 MHz (the maximum rate considered possible today). This is more than an order of magnitude faster than the opto-mechanical mask-makers that preceded electron beam systems, but is an order of magnitude slower than the speed available with variable-shaped electron beam systems, where a large number of resolution elements is exposed at each beam position.

## Throughput of EL3, a Variable-Shaped Beam Vectorscan System

The key specifications of EL3 are shown in Table 1. EL3 combines a high current variable shaped beam column with a dual deflection system.[37] Maximum beam size is $4\mu \times 4\mu$ and shapes smaller than this are available in $0.1\mu$ increments. Spot edge definition is better than $0.2\mu$. A highly accurate but relatively slow magnetic deflection coil deflects the beam in a raster-like sequence to the center of an array of subfields located on $75\mu$ centers. After arriving at the center of each subfield, the variable shaped beam is vector-addressed over the subfield with electrostatic deflection plates. The subfields overlap each other so that there are no discontinuities, and are so small that super-precision deflection is not needed. The array of subfields can be as large as 10 mm×10 mm. Errors in both magnetic and electrostatic deflection are 'learn-corrected' by scanning a reference grid, storing errors in the control computer, and feeding back corrections during writing.

Throughput of EL3 (see Table 1) is limited almost equally by current density (and/or resist sensitivity), mechanical stepping and control system speed. Further improvement will require advances in all areas.

## Optimum Electron Beam System

The most obvious way to improve throughput in scanning electron beam systems is to combine the best features of the individual systems into a single system. The optimum combination appears to be a variable shaped beam column, and a continuously moving table preferably with a continuous, or semi-continuous, alignment system. The shaped beam ensures maximum beam current, and continuous table movement potentially eliminates many overhead times. Registration will have to be accomplished without stopping the table either by means of a laser interferometer, or through direct beam to sample reference. Such a system could perhaps be 50% to 100% faster than present systems.

Resolution for Electron Beam Exposure

Ultimately, resolution in electron beam lithography is set by the range over which the primary electrons interact with the resist. That is by the distance over which the low energy secondary electrons are created (the resist is exposed mainly by subsidiaries) and by the range of the secondaries in the resist. For thin resists, and thin substrates (thin compared with the primary electron penetration) this resolution limit falls at about 12 nm.[38]

For thick substrates, backscattered electrons from the substrate decrease contrast and the minimum dimension increases to about 40 nm. For thick resists, and samples thick compared to the primary electron penetration range, electron scattering in the resist (forwardscattering) and backscattering of electrons from the substrate, become more important than the electron interaction range. In these cases, in order to produce an acceptable exposure distribution, exposure dose must be altered according to the local pattern density to compensate for variations in the backscattered dose (proximity effect).[39] Compensation for this effect is valuable even at dimensions of $1\mu$-$2\mu$, particularly for thick (>500 nm) resist layers.

Figure 12 shows the contrast for electron beam exposure versus linewidth (assuming an infinite array of lines and spaces) for different resist thicknesses. It can be seen that with a thick substrate and a thin resist, the contrast is the same for $0.05\mu$ lines as it is for $1\mu$ lines, provided the resist is thin ($<0.1\mu$). This is because the backscattered distribution is much wider than either the secondary distribution or the forward-scattered distribution. For the thicker resist cases, it is assumed that the lateral scattering in the resist can be approximated by a Gaussian distribution.[39]

The fraction of the exposure due to backscattered electrons is reduced for multi-layer resists because the underlying resist generally has a lower average molecular weight than substrates such as silicon. The contrast for such a sample would be higher than is shown in Figure 12. When the total pattern area is small compared to the area from which the backscattered electrons emerge, resolution on bulk substrates should be as good as on thin membranes. For example, the backscattered contribution to an isolated 25 nm line on a silicon substrate is only 0.4% of the incident exposure.

For scientific studies, where ultimate resolution is required, and throughput is not a concern, scanning transmission electron microscopes with beam sizes below 1 nm have been used to fabricate what have been called nanostructures.[40,41] Although it is possible from the fabrication point of view to fabricate equally small structures on solid substrates, these nanostructures have generally been made on thin ($<0.1\mu$) $Si_3N_4$ membranes so that they can

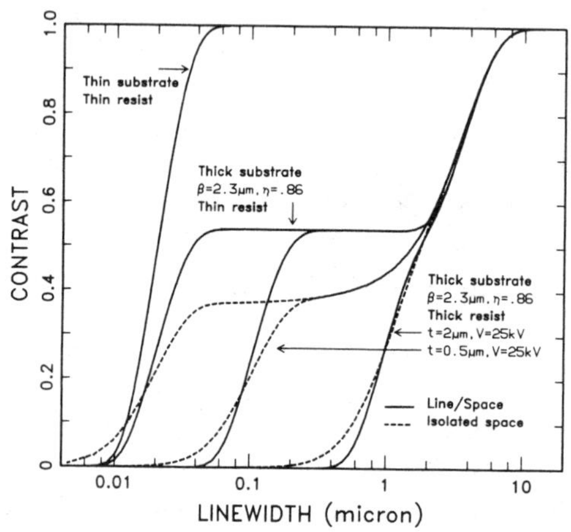


**Figure 12.** Contrast for line/space patterns and for isolated spaces written with 25 kV electrons.[38] Thin substrate improves contrast rather than resolution.

be examined with high resolution scanning transmission electron microscopy. It is almost impossible to examine such samples in a conventional secondary electron SEM because the same straggling of secondary electrons that limits resolution in the resist pattern, also limits resolution of the SEM. Linewidths of 10 nm are possible in PMMA and center to center spacing of 40 nm for raster-like patterns of equal lines and spaces.[38] Because fundamental electron interaction phenomena limit resolution to approximately 10 nm for practical fabrication processes invented to date, the resolution of a STEM does not limit the process.

Figure 13 shows a nanostructure, used to explore conduction in fine wires.

## ION BEAM LITHOGRAPHY

Focused ion beams can be used to expose resist, to write directly diffusion patterns into semiconductor substrates and to repair masks. These techniques can potentially simplify semiconductor device production and perhaps reduce cost. Many of the technological challenges with ion beams are similar to thosc encountered with electron beams, but the development of ion sources and focusing/deflection systems are at a much earlier stage of development so application to manufacturing is several years away.

### Ion Beam Resist Exposure

High energy ions can be used to expose resist materials (>100 keV ions are needed to penetrate typical resist layers) with the promise of less image blurring due to lateral scattering (proximity effect) than electrons.

Experiments indicate that ions are almost two orders of magnitude more efficient than 20 keV electrons.[42] This is because the energy of the ions is more completely absorbed in the resist layer. The high sensitivity can be a problem, however, for very high resolution because of inherent noise fluctuations created by the small number of ions needed to expose an image element.[43] At an exposure dose of $10^{-8}$ C/cm$^2$, only 6 ions will expose a $0.1\mu \times 0.1\mu$ element resulting in an exposure uncertainty of 40%. This is not satisfactory. Typically a hundred times this exposure is needed to reduce the uncertainty to an unacceptable level of <5%. In other words the minimum usable sensitivity for ion exposure is $10^{-6}$ C/cm$^2$. This in effect removes the sensitivity advantage of ions over electrons because resists with a sensitivity of $10^{-6}$ C/cm$^2$ are available for electrons.

**Figure 13.** Transmission scanning electron micrograph of two nanobridge SQUID's (Superconducting Quantum Interference Devices). A SQUID consists of a superconducting ring containing two 'weak-links'. In this instance, the weak links are niobium wires 25 nm wide fabricated by electron beam.

Even without the problem of statistical fluctuations the performance of ion beam systems is marginal for economical exposure of resist because of constraints similar to those of electron beams. In the electron beam case, as discussed above, a combination of shaped-beam optics, large-angle highly corrected lens/deflection systems, and high brightness electron guns only make it possible to reach an exposure speed that is economically satisfactory for special applications such as mask making and gate-array personalization. The brightness of ion sources approaches that of electron sources, although with poorer reliability, but the lens/deflection systems are not yet able to operate at comparable aperture. Electron beam systems have been corrected to the point that beam half-angles of $10^{-2}$ radian can be maintained for field sizes up to 10 mm×10 mm.

Two types of ion sources produce high enough brightness ($\geq 10^6$ A/cm$^2$ steradian 20 keV) for them to be considered for semiconductor fabrication applications; the field ion source[44] and the liquid metal source.[45,46] The field ion source produces relatively small energy spread ($\leq 3$ eV) and when combined with a short focal length ($\leq 1$ cm) electrostatic focusing system, should be able to produce beam sizes as small as 10 nm with adequate current ($10^{-11}$ amp) for laboratory microfabrication experiments. As with field emission electron sources, however, the field ion source only produces a limited total current and the maximum final beam current is limited to about $10^{-10}$ amp.

The liquid metal source produces higher currents (~1 nA into ~0.25$\mu$) making it the leading contender for fabrication applications, but the energy spread is higher (10-12 eV), and the minimum beam diameter is limited to about 100 nm. Forty nm wide slots have been machined in thin gold films with a 100 nm diameter ion beam.[47] The higher beam current and brightness have been produced with liquid metals, mainly gallium. When other species are needed, for example implantation, or for greater penetration into resists, eutectic alloys are used but the brightness is reduced.[48] Boron, arsenic and silicon ion beams have been formed in this way.

As already mentioned, the electrostatic focusing and deflection systems needed for ions are in a much earlier stage of development than those for electrons, in addition, their design is more difficult because focusing and deflecting fields cannot be overlapped as they can with the magnetic field employed with electrons. Significant correction of aberrations can be accomplished in this way. Multi-pole lenses (quadrupoles and octopoles) are used in focusing high energy ion beams, and some combination of these may offer a solution to the difficulty in the future, but this is yet to be accomplished. The easiest way to cover a large field will be to move the sample mechanically and join together smaller fields. Laser interferometry can be used to track the sample position and errors in position can be fed back to the ion deflection in the same way used in electron beam mask makers. This approach, in principle, yields unlimited field size.

Ion Beam Proximity Printing

Ions can be used for proximity just as UV light, electrons and X-rays, however, the mask is more difficult than for any of the other alternatives. A bright ion source is not needed, only one that produces adequate current. If necessary, the beam can be scanned across the mask to complete the exposure. Masks have been made from silicon membranes which have adequate transparency to protons, and it is possible in principle to use the same stencil masks developed for electron beam proximity printing. With silicon membranes the ions are scattered and the resolution depends on the mask to wafer separation. Image blurring is estimated to be $0.1\mu$ for a mask to resist separation of $15\text{-}25\mu$.[49] This separation is already marginal from the point of view to wafer damage. Problems might be encountered because of sputtering with the metal stencil masks.

These mask problems would seem to limit the chances of this particular method succeeding in competition with UV, X-ray or electron alternatives.

Resolution of Ion Beam Processes

Ultimately, resolution in a sputtering process is limited by ion penetration into the substrate and/or by the range over which momentum can be transferred effectively enough to remove atoms from the sample surface. The diameter from which atoms can be sputtered has been reported to be about 10 nm for incident ion energies up to 12 keV.[50] This means that it should be possible to fabricate structures of about 10 nm which also happens to be the size of the smallest ion beam that can be produced with present ion sources.

For resist exposure, the resolution limit will be set by the range over which the ions interact with the resist. As with electron beam exposure, ions will create secondary electrons up to several nanometers away from the beam, and these electrons can travel further before their energy is absorbed. Ultimate resolution will probably be about 10-20 nm, as it is with electrons. At present this limit is beyond the capabilities of the ion optical systems.

## REFERENCES

1. P. Tischer, "Electronics in Microelectronics," W. A. Kaiser and W. E. Proebster, eds., North Holland, 1980.

2. B. J. Lin, "Microcircuit Engineering 81," Swiss Federal Instit. Technol., Lausanne, Switzerland, 47 1981.

3. R. Hershel, SPIE Conference, Santa Clara, California, 1982.

4. M. C. King, "Principles of Optical Lithography," Chapt. in VLSI Electronics Microstructure Sci., Vols. 1 and 2, ed. Einsbruch, Academic Press, 1981.

5. B. J. Lin, *SPIE Proc.* **174**, 179.

6. K. L. Tai *et al., J. Vac. Sci. Technol.* **16**, 1977 (1979).

7. D. A. Markle, *Solid State Tech.* **17**, 50 (1974).

8. J. J. Greed and D. A. Markle, SPIE Conference, Santa Clara, California, 1982.

9. A. Offner, *Photogr. Sci. Eng.* **23**, 374 (1979).

10. K. Jain, C. G. Willson and B. J. Lin, *IEEE Electron Device Lett.* **EDL-3**, 53 (1982).

11. S. Wittekoek, Microcircuit Engineering 80, R. P. Kramer, ed., Delft Univ. Press., Holland, 155, 1980.

12. J. Lauria *et al.,* Microcircuit Engineering 80, R. P. Kramer, ed., Delft Univ. Press, Holland, 181, 1980.

13. G. Dubreoueq *et al.,* Microcircuit Engineering 80, R. P. Kramer, ed., Delft Univ. Press, Holland, 191, 1980.

14. H. E. Mayer *et al.* Microcircuit Engineering 80, R. P. Kramer, ed., Delft Univ. Press, Holland, 191, 1980.

15. J. Dey *et al.,* Microcircuit Engineering 80, R. P. Kramer, ed., Delft Univ. Press, Holland, 211, 1980.

16. J. S. Wilczynski, *J. Vac. Sci. Technol.* **16**, 1929 (1979).

17. D. L. Spears and H. I. Smith, *Electron. Lett.* **8**, 102 (1972).

18. E. Spiller and R. Feder, *Sci. Am.,* Nov. 1978.

19. G. N. Taylor and T. M. Wolf, Microcircuit Engineering 81, Swiss Federal Instit. Technol., Lausanne, Switzerland 381, 1981.

20. W. D. Grobman, IEDM Digest, IEEE, New York, p. 415, 1980.

21. E. Gronid *et al.,* Microcircuit Engineering 81, Swiss Federal Instit. Technol., Lausanne, Switzerland, 122, 1981.

22. J. LeDuff *et al.,* Microcircuit Engineering 81, Swiss Federal Instit. Technol., Lausanne Switzerland, 130, 1981.

23. E. Spiller and R. Feder, "X-Ray Lithography," in H. J. Queisser, "X-Ray Optics," Springer-Verlag, Berlin, 1977.

24. A. E. Gruen, *Z. Naturforsch*, **12a**, 89-95 (1957).

25. H. Bohlen *et al., J. Vac. Sci. Technol.* **16**, 1889 (1979).

26. R. Laibowitz *et al., Appl. Phys. Lett.* **35**, 891 (1979).

27. R. Voss *et al., Appl. Phys. Lett.* **37**, 656 (1980).

28. D. R. Herriott *et al., IEEE Trans. Electron. Devices* **ED-22**, 385 (1975).

29. S. Sano *et al.,* Microcircuit Engineering 80, R. P. Kramer, ed., Delft Univ. Press, 950, 1981.

30. H. C. Pfeiffer, *J. Vac. Sci. Technol.* **15**, 887 (1978).

31. E. Goto, T. Soma and M. Idesawa, *J. Vac. Sci. Technol.* **15**, 883 (1978).

32. M. G. R. Thomson, R. J. Collier and D. R. Herriott, *J. Vac. Sci. Technol.* **15**, 891 (1978).

33. T. H. P. Chang *et al.,* Proc. 75h Int. Conf. Electron and Ion Beam Sci. and Technol., R. Bakish, ed., Electrochem. Soc. Inc., Princeton, NJ, 97 (1974).

34. G. Varnell *et al., J. Vac. Sci. Technol.* **16**, 1787 (1979).

35. J. C. Eidson *et al., J. Vac. Sci. Technol.* **19**, 932 (1981).

36. F. H. Hohn *et al., J. Appl. Phys.* **53**, 1283 (1982).

37. R. D. Moore, *J. Vac. Sci. Technol.* **19**, 950 (1981).

38. A. N. Broers, *J. Electrochem. Soc.* **128**, 166 (1981).

39. T. H. P. Chang, *J. Vac. Sci. Technol.* **12**, 127 (1975).

40. A. N. Broers *et al., Appl. Phys. Lett.* **29**, 596 (1976).

41. M. Isaacson and A. Muray, *J. Vac. Sci. Technol.* **19**, 1117 (1981).

42. T. M. Hall *et al., J. Vac. Sci. Technol.* **37**, 656 (1980).

43. A. Wagner, Proc. SPIE Conf. Santa Barbara, California, 167 (1983).

44. G. R. Hanson and B. M. Siegel, *J. Vac. Sci. Technol.* **16**, 1875 (1979).

45. R. Clampitt *et al., J. Vac. Sci. Technol.* **12**, 1208 (1975).

46. V. E. Krohn and G. R. Ringo, *Appl. Phys. Lett.* **27**, 479 (1975).

47. R. L. Seliger *et al., J. Vac. Sci. Technol.* **16**, 1610 (1979).

48. K. Gamo *et al.,* Microcircuit Engineering 80, R. P. Kramer, ed., Delft Univ. Press, Holland, 283 (1980).

49. D. B. Rensch *et al. J. Vac. Sci. Technol.* **16**, 1897 (1979).

50. J. A. McHugh, "Secondary Ion Spectrometry," Capt. in "Methods of Surface Analysis," S. Wolsky and A. Czanderna, ed., Elsevier Pub. Co., Amsterdam, 1975.

51. R. E. Tibbetts and J. S. Wilczynski, Proc. 1980 Int. Lens Design Conf., SPIE Vol. 237, R. E. Fischer, ed., SPIE Bellingham, Washington, DC, 321 (1980).

# LITHOGRAPHIC MATERIALS

Joachim Bargon*

IBM Research Laboratory
5600 Cottle Road
San Jose, California 95193

## ABSTRACT

Photosensitive coatings, the workhorses of modern microelectronic technology, were originally developed for application in planar printing technology. From these both positive and negative photoresists have found their way into semiconductor manufacturing lines, originally because of their high resolution capability. From a chemical point of view, many different resist systems exist, but very few have found practical application. The dominating negative resists are based upon cyclized rubber as a host matrix containing a photosensitive crosslinking catalyst. The chief positive photoresists consist of a phenolic host resin which is photosensitized via an admixed dissolution inhibitor of the naphthoquinone diazide type. Their chemistry is closely related to the old fashioned blueprint copying process. Photoresists of this type have contributed significantly to the development of novel semiconductor processing concepts. They have survived the change over from wet chemistry etchants to dry plasma etching, and from visible light exposure to UV or electron beam lithography. More specialized resist systems have since been identified but the traditional performers impress by their universal applicability. Lithographic techniques have become very sophisticated, and accordingly many new lithographic materials such as multilayer resists, dual exposure resists, photosensitive insulators (polyimides), inorganic resists and dry developable systems have been added to the rather impressive menu of the modern microelectronic technology.

*New address: Institut für Physikalische Chemie, Universität Bonn, Wegelerstrasse 12, D5300 BONN-1, West Germany

# I. INTRODUCTION

Lithography originally refers to the art of printing using planarized stone into which patterns have been engraved or etched. Aloys Senefelder,[1] who invented this process towards the end of the 18th century, already used wet chemistry for the etching and called his procedure "chemical printing." Later the patterning of the printing form was achieved via light induced chemical reactions. This process, called "photolithography," has considerable tradition as well and has been outlined elsewhere.[2]

Photolithography takes advantage of a light induced change in solubility of a photosensitive coating. Accordingly upon development with an appropriate solvent (developer) either the exposed or the unexposed area is washed out.

In a *positive* planographic printing plate the image area which remains on the substrate *corresponds directly to the mask* through which the exposure was carried out, *i.e.,* the remaining, chemically unchanged coating provides a *positive replica of the mask.*

A *negative* plate, however, leaves behind a *complementary replica of the mask, i.e.,* a negative copy. The expressions positive or negative photoresists have subsequently been carried over from the application of photosensitive lacquers in printing technology to modern microelectronics. Many interesting resist systems have been described in the literature,[3] but only very few different types have found acceptance.

The commercially successful positive resists are almost all based upon the so-called "naphthoquinone diazides," which are typically used as composite systems in combination with phenolic resins pioneered by Kalle in Germany.

These same diazides or diazo compounds had also been a key compound in the now somewhat old-fashioned "blueprint process," of which for example the "OZALID" brand name reflected the name of the chemical that made it work. This claim becomes more obvious when spelling OZALID backwards, which yields DI(L)AZO. The "L" had been inserted to prevent a troublesome tongue twister.

The most prominent positive photoresists of modern microelectronic technology, for example the Shipley or AZ-Photoproducts resist family of the 1300 series (AZ1350, AZ1370, *etc.*) are all based upon chemistry closely related to the "blueprint process." The chemistry of the naphthoquinone diazide resist systems has been originally worked out by Süs[4] in Germany during and shortly after the Second World War.

## II. RESISTS FOR FINE LINE LITHOGRAPHY

*Resists* used in the manufacture of microelectronic devices serve to protect underlying substrate during etching, *i.e.,* they *resist etching.* Typically they are organic polymers, and depending on their response to the exposing radiation they can either be classified as positive or negative. *Positive* resists have become *more soluble in the irradiated* regions relative to the unexposed areas, whereas *negative* resists become *less soluble in the irradiated* regions relative to the unexposed ones.

Positive behavior can come about in two different ways:
(1) *Main Chain Scission.* The main chain of the polymer can break, resulting in an overall reduction of its molecular weight. Since low molecular weight polymer dissolves in the developer, whereas the normal high molecular weight form does not, the exposed areas thus become washed out. This main chain scission process requires, however, that rather stable chemical bonds have to be broken by the exposing radiation. This requirement renders the *degradation* concept attractive only for the most energetic deep UV or X-ray regime; but less so for conventional photoresists. Efforts to lower the bond strengths of the main chain polymer in order to make the polymer more sensitive to light or ionizing radiation, are frustrated by a concomitant lowering of its thermal stability, typically an unacceptable consequence.
(2) *Dissolution Inhibition.* The other alternative is exemplified by the previously mentioned naphthoquinone diazide system, which acts as a *dissolution inhibitor* for a phenolic resin. Originally it was thought that the unexposed naphthoquinone diazide, *i.e.,* the photoactive compound (PAC), would couple with the resin to yield an azo-dye upon development in the alkaline developer. This coupling would certainly render the resin *less soluble than before in the unexposed area.* A current view, however, believes that photolysis of the PAC yields an acid, as shown by Süs,[4] which is very soluble in base, thus causing the resist to become *more soluble in the exposed areas.* The destruction of the PAC, therefore, effectively removes a dissolution inhibitor.[5] This dissolution inhibitor concept among other things has subsequently led to the successful design of a positive electron beam resist.[6] The latter is made up of the same phenolic resin as found in the naphthoquinone diazide type positive photoresist but instead of the above PAC it contains an e-beam sensitive polysulfone as the dissolution inhibitor.

Negative resists work on two entirely different principles:
(1) *Cross-Linking.* One possibility which causes decreased solubility in the irradiated areas is radiation induced cross-linking. Again, in this case chemical bonds have to be broken first by radiation, which favors this process for high energy radiation, such as deep UV and X-ray radiation. However, bifunctional compounds with suitable chromophores have extended this process to

conventional photolithography as well, and the Kodak Thin Film Resist (KTFR) is a characteristic example.
(2) *Radiation Induced Polymerization.* Another possibility is polymerization of a suitable monomer which can be initiated by a suitable photoactive compound. In practical resists this alternative is typically found combined with a crosslinking concept.

Inorganic resists in turn also lead to a change in solubility, but achieve this goal by inducing diffusion of atoms into specific zones.

The advent of plasma etching and advanced processing techniques has led to still another lithographic approach, namely the deposition of etch-resistant patterns on surfaces from the vapor phase. Simultaneous or subsequent irradiation induces curing of the etch resistant barrier. This novel concept will be discussed in more detail later.

## III. RESIST PARAMETERS

Resists for microelectronics are photosensitive materials, and thereby they are related to photographic emulsions. The requirements, however, that resists have to satisfy, are rather different from materials for taking pictures. For photographic applications, the emphasis is primarily on sensitivity at the expense of resolution. This scenario favors the silver halide process in photography virtually at the exclusion of any other competing approach. For applications in microelectronics, however, the resolution required translates into vanishing "grain sizes." A comparison of light sensitive systems[2] shows that photoresists are at the lower end in sensitivity relative to other imaging systems.

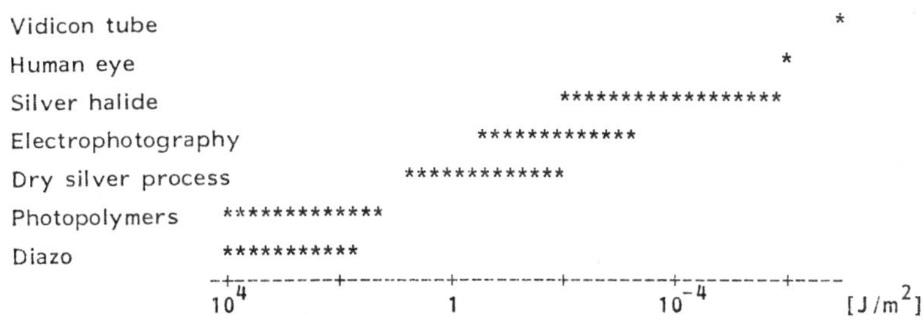


**Fig. 1.** Energy Requirements of Light-Sensitive Systems

Resolution requirements, stemming from the smallest line dimensions to be reliably reproduced by lithographic techniques used in microelectronics, have been stepped up continuously during the past twenty year period. Thus between 1965 and 1980 linewidths have shrunk from initially 6 $\mu$m to about 2 $\mu$m. Commensurate with this decrease of the actual line dimensions the overlay has been reduced from 2.5 $\mu$m to about 0.8 $\mu$m. By about 1985 the line dimensions are expected to be at the 1 $\mu$m level with overlays at 0.5 $\mu$m.[7] Design trends in microelectronics have been and remain towards increasing the scale of integration and thereby of device complexity. This thrust has made it possible to reduce fabrication costs and performance of components and whole systems alike in a rather impressive fashion. The primary technology parameters, which have made the increased complexity and performance possible are essentially related to three basic areas: (1) circuit innovation, (2) dimensions and (3) total chip size. In the decade 1970-1980 about 60% of the increased productivity came from circuit innovation, about 30% from decreased lithographic dimensions and about 10% from increased chip size. To illustrate this breakthrough, it is worthwhile appreciating that for example an individual memory cell, which used to require six transistors, can now be realized with a single transistor in a modern dynamic RAM.[8] In the future design cleverness will decrease in significance and correspondingly improved performance of electronic circuitry will have to rely more on reduced line dimensions, *i.e.,* on lithographic resolution.

1. Resolution

The resists used in micron or submicron lithography must be able to replicate the desired line dimensions accurately and reliably. Resist properties that affect resolution are *contrast* (which translates into differential solubility rates during development), *swelling* and *flow* of the developed pattern during removal of the developer in the so-called "baking" step.

For dimensions around 1 μm and less, swelling is perhaps the most important resist parameter limiting resolution.[9] The factors affecting the dissolution rate of polymeric resists have been studied in detail by Ouano[10] who has shown that two stages can be differentiated. At first solvent diffuses into the polymer matrix, a slow process which is greatly assisted by the irradiation induced formation of gaseous or low molecular weight products such as $CO_2$, $H_2$, $CH_4$, $CH_3OH$, *etc.*, especially in electron beam resists. This process causes swelling, the degree of which depends upon such factors as molecular weight, polymer structure, glass transition temperature and molecular size of the solvent. Swelling is even more a concern for negative than for positive resists. In the second stage true conversion of the swollen polymer occurs, a process which is also dependent upon the molecular weight and structure of the resist polymer.

Since all these different factors contribute to the dissolution characteristics of the resists, they also determine (among other things) another important resist parameter, namely sensitivity.

2. Sensitivity

Whereas initially the key parameter of resists was resolution, more and more emphasis is being placed on sensitivity. At the onset of microelectronic technology exposure tools operating in the visible region of the spectrum had no problem providing relatively high levels of exposure brightness. As the resolution requirements are increasing, however, the wavelength of the primary irradiation is shifting into the UV regime. There the brightness of the light source is much more of a problem, and also the efficiency of optical elements is considerably lower. Since the total exposure energy deposited in a resist film is the product of energy density, time, and absorption efficiency, a loss in the first and last coefficient must be compensated by longer exposure times. Time, however, translates into throughput of wafers per hour, a consequence which affects the number of exposure tools required and thus costs. Accordingly, over the years resist requirements have been stepped up to satisfy higher sensitivity demands as exposure tools were pushed to shorter wavelengths. Such considerations are especially valid for e-beam, ion-beam and X-ray lithography.

Increased sensitivity, however, frequently results in other trade-offs such as decreased temperature stability or etch resistance. An ideal resist of course would combine a string of individually optimized parameters satisfying all requirements simultaneously in an ideal though most likely incompatible fashion.

3. Etch Resistance

Originally, semiconductor manufacturing processes used wet etching processes extensively. Etching solutions typically consisted of aqueous solutions of HF and $NH_4F$, and etch resistance was required with respect to such liquid media.[11] With the advent of plasma etching,[12] etch resistance towards dry etching conditions replaced the original requirements. It is difficult to unite the demand for high sensitivity and plasma-etch resistance at the same time. Fortunately, however, the conventional workhorse of fine line lithography, namely the phenolic resin/quinone diazide family of resists, shows a reasonable resistance to either wet etching or dry etching conditions. This lucky condition may have contributed significantly to the success of plasma etching.

The wet etching stability of various linear condensation polymers to acids and bases is typically dominated by the nature of the linkages in the polymeric chain.[13] Plasma etching stability, on the other hand, seems dominated by the reactivity of the polymeric chain with the reactive intermediates like fluorine atoms, and is thus more dependent upon the nature of the monomer units which make up the polymeric backbone.[14] It is typically found that aromatic units render a polymer rather resistant to plasma etching, but unfortunately at the price of reduced radiation sensitivity. Other approaches, such as using fluorinated polymers to increase resistance to $CF_4$ derived plasmas or polysiloxanes to increase their resistance to oxygen plasmas, do not suffer from the reduced sensitivity drawback.

4. Adhesion

Liquid etching techniques in particular demand a high level of adhesion of the resist films to the underlying substrates. Adhesion promoters are frequently used and have actually been found to sensitize resists, such as the sensitivity of phenolic formaldehyde resin/quinone diazide to electron beams in particular.[15] For dry etching techniques the adhesion requirements are relaxed.[16] The area of adhesion is little understood and can thus be regarded as a "black art."

5. Gas Permeability

Since resists are typically applied by spin-coating of solutions, the solvent has to be removed subsequently by heating in an oven. During this so-called "baking" step, the solvent has to penetrate the resist film, since otherwise the formation of bubbles results. The details of this process are insufficiently understood.

6. Mechanical Properties of Thin Films

Resist layers tend to crack if their thermal or mechanical properties differ substantially from those of the substrate. This is a particular problem for inorganic materials, as organics are typically more "forgiving," if there is a mismatch of thermal expansion coefficients. Although the mechanical properties of bulk polymers are typically well known and easily measured, thin polymeric resist films pose more of a problem, and data are difficult to determine since conventional measurements fail.

7. Defect Density

Defect density is a consequence of a variety of reasons, typically due to voids, contaminants and the likes. To avoid pinholes resist thickness had to be in the order of $1\mu$ and more. Shrinking line dimensions may require thinner coatings, and furthermore they demand more homogeneous resist composition. Particulate contamination has to be minimized which can be achieved by careful filtration techniques and by tight controls of resist formulations during their storage. This eliminates consequences of coagulation and gel formation. Negative resists frequently represent more of a problem than positive ones in this respect.

8. Development Anisotropy

Although it is now possible to control the etch profile via the operating conditions of plasma etching systems, the frequently used lift-off process requires an undercut profile of the developed resist structures. Where line profile and linewidth tolerances are crucial, computer based modeling of the exposure and development stage may have to be considered together, as has been outlined in various studies.[17]

IV. PHOTORESISTS

Optical photoresists, the workhorses of integrated circuit manufacturing, have a long history. Consequently there is a wide range of different resist systems known and available, which will be discussed in two convenient categories of positive and negative working materials. The chemistry of many systems has been extensively reviewed[18] and only some essential features will be discussed here.

1. Positive Photoresists

Positive resists are typically made up of combinations of two compounds: a *polymeric resin* and a *photoactive compound* as a sensitizer. The principle of this type of resist approach is that the photoactive compound (PAC) acts as a dissolution inhibitor of the resin in aqueous base which functions as the developer. The polymeric resin is either weakly or strongly acidic, and in the absence of inhibitor it dissolves in aqueous base. Typical resins are derived from condensation of phenol or cresol with formaldehyde, *i.e.,* so-called novolacs (for example in resins manufactured by Shipley, Azoplate, Hunt and Kodak) but can also be derived from polyacrylic acid or poly-p-hydroxystyrene. The most widely used dissolution inhibitors are members of the o-quinone diazide family, typically derived from either 1-amino-2-hydroxynaphthalene-5-sulfonic acid or its isomeric equivalent, namely 1-hydroxy-2-amino-5-naphthalene sulfonic acid.

1.1 Photoactive Compounds

Quinone diazides are typically yellow to red colored crystalline compounds. A review of their chemistry and properties have been compiled by Kazitsyna *et al.*[19] The presence of benzene rings decreases the solubility of these substances in water but increases their solubility in organic solvents. Thus the 1,2- or 2,1-naphthoquinone diazides can be easily recrystallized from water, but they are also readily soluble in organic solvents.[20] The introduction of negative substituents into quinone diazides reduces their solubility in both polar and nonpolar solvents.

Quinone diazides are sensitive to both heat and light. The thermal stability of substituted quinone diazides is higher for compounds with electro-negative substituents. Whereas aliphatic diazides derived from 1,2-diones are known to occur as a mixture of a cis and a trans form in equilibrium due to restricted rotation around the central carbon-carbon bond,[21] no such isomerism is possible in the cyclic naphthoquinone diazides. The partial double-bond character derived from the barrier to isomerization suggests that both the carbonyl as well as the diazide moiety are coplanar with the rings. Properly positioned substituents shift the optical absorption spectrum of these photoactive compounds, and an appropriate choice of different isomeric variations has led to a vast number of patented combinations.

1.2 Thermal Stability of Positive Photoresists

Heating of the resists in the so-called "prebake" and "postbake" steps removes residual solvent before exposure, eliminate the consequences of standing waves resulting from the exposure, and improves adhesion to the substrate. Various authors[22,23] have used physical measurements to quantitatively characterize the changes in performance of the Shipley photoresist AZ1350J as a consequence of heating. They concluded from mass spectrometric studies of this positive photoresist that the thermal decomposition mechanism is essentially similar to the photolytic decomposition mechanism, as worked out earlier by Süs and his coworkers[4,24] (Scheme 1). Up to 170°C, which is significantly higher than typical baking temperatures, only nitrogen but no CO or $CO_2$ were detected as decomposition products. Speculations about the types of thermal decomposition products to be expected have been forwarded by Levine[25] and by Yates and Roble.[26] Furthermore, Paramonov *et al.*[27] showed using infrared spectrometry that during heat treatment of unexposed photoresist films ester formation occurred between the phenol units of the resin and the indenecarboxylic acid derived from the PAC, probably via a ketene intermediate and subsequent Wolff rearrangement.[28]

1.3 Photoreactions of Naphthoquinone Diazides

The photochemistry of the naphthoquinone diazides has initially been investigated by Süs[4,7] and his coworkers and shown to be:

Scheme 1:

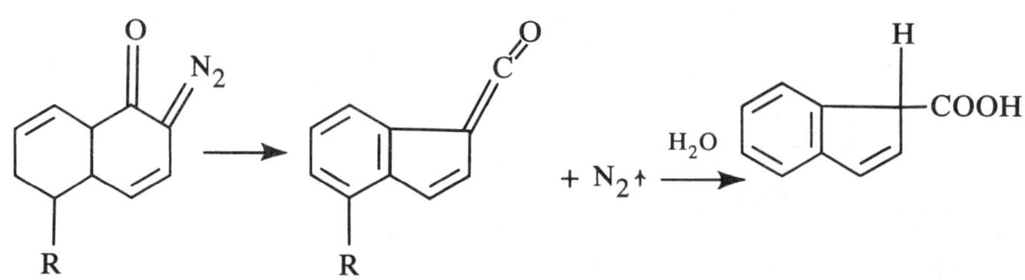


The decomposition product of this sequence, an indenecarboxylic acid, is believed to be formed from a ketene intermediate, which in turn is expected to be a rearrangement product according to the Wolff[28] mechanism of an intermediate ketocarbene. This ketene intermediate has later been trapped and identified by means of IR spectroscopy by Pacansky and Johnson.[29] Subsequent infrared and carbon-13 NMR studies of Pacansky and Lyerla[30] have substantiated and further elucidated the above mechanism. They were able to show that the end decomposition product is not the 1-indenecarboxylic acid as originally proposed by Süs,[4,24] rather it has the isomeric 3-indenecarboxylic acid form. Otherwise the Süs mechanism is essentially correct.

Another key part thereof, namely the isomerization of the above postulated intermediate ketocarbene into a ketene, *i.e.*, the so-called Wolff rearrangement[28] (WR), has recently been studied in more detail with quantum chemical means.[32] This arrangement was discovered by Wolff[28,33] and independently by Schroeder.[34] The theoretical studies revealed that the photochemically induced WR proceeds smoothly on the singlet energy surface but is impeded on the triplet energy surface by a rather high barrier, which could lead to other products. This theoretical insight had been experimentally supported before by Trozzolo and Fahrenholtz[35] who had found that if a ketocarbene is generated photochemically in its triplet state at low temperature it does not undergo the Wolff rearrangement to more than 3%. That the photoinduced Wolff rearrangement does indeed occur from the singlet excited state has further been demonstrated in experiments by Padwa and Layton.[36]

Additional support for the theoretical results[32] comes from mass spectrometric studies of Zeller,[37] who showed that the photoinduced WR of o-naphthoquinone diazide does not lead to a scrambling of a carbon-13 label introduced at the ketocarbon. This finding demonstrates that no oxirene intermediate occurs, a sometimes postulated intermediate of academic interest.

According to the results of both the theoretical and experimental studies conducted on simpler model compounds of the resist PAC, the photochemical decomposition of naphthoquinone diazide-type resist should preferentially occur from the singlet excited state to maximize the resist sensitivity. It has to be emphasized, however, that studies on simple models might not necessarily reflect the facts which govern the decomposition of the more complex resist sensitizers accurately enough to render the following postulates safe beyond challenge.

Looking at the sensitizers of real photoresists of the o-naphthoquinone diazide type, one finds that the recommendations which one would derive from the above theoretical considerations are *not* followed. Thus the PAC of the Shipley resist AZ1350J, the triester depicted in Fig. 2, is a benzophenone derivative. Benzophenone, however, is known to undergo very efficient intersystem crossing, thereby converting any absorbed light into triplet energy very efficiently. In other words, benzophenone is known in photochemistry as a very powerful triplet sensitizer, which can thus be expected to short-circuit some of the photoenergy absorbed by the PAC of AZ1350J. The intersystem crossing occurring in the benzophenone moiety is unlikely to contribute to the

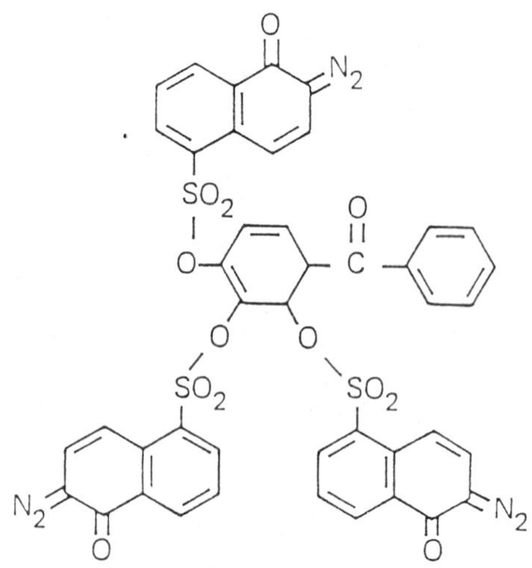


**Fig. 2.** Photoactive Compound of the Shipley 1350J Positive Photoresist

WR of the PAC and thus should not lead to any removal of dissolution inhibitor from the photoresist. To avoid such potential inefficiency, moieties which are known to act as triplet sensitizers such as benzophenones should better be omitted. Indeed PACs containing o-naphthoquinone diazides attached to aromatic hydrocarbons (*i.e.,* pyrene) rather than to benzophenone have been used in some commercial resists and have also been reported in the patent literature. Benzophenone derived PACs are rather widespread and work satisfactorily in many commercial systems, however.

Other concerns about a PAC as found in AZ350J center around the fact that a triester causes a relatively high level of reaction of the intermediate ketene with other fragments of the PAC, as discussed in more detail by Levine.[48] These side reactions reduce the formation of a base soluble acid equally well. Such concerns are even more important, if the o-naphthoquinone diazide-containing photoresists are used as electron beam resists. One significant difference to the standard photoresist mode stems from the fact that e-beam exposure occurs in a vacuum. This renders the moisture content in the resist lower and thus eliminates a key reactant from the Süs mechanism, namely the water which is required to convert the intermediate ketone into the acid. Since the possible reactions of the ketene obey the conditions of competitive kinetics, the reduction of moisture renders other reactions more likely. Thus reactions of the ketene with the phenolic resin to yield ester gain in importance, and a more detailed discussion has been provided by Pacansky and Lyerla.[30]

Various techniques to follow the decomposition of the PAC in a positive photoresist have been applied[38] and line profile studies[39] as well as modeling efforts thereof[40] have been published. The kinetics and linearity of the exposure and development process has also been extensively studied.[41]

## 1.4 Phenolic-Formaldehyde Resins

Phenols react with formaldehyde under both acidic and alkaline conditions to yield a wide array of condensation products. Details of the reactions under different conditions have been given by Keutgen.[42] For application in photoresists acid catalyzed products which are typically low molecular weight and thus base-soluble, are being preferred. These so-called novolacs are commercially available resins, which consist of a mixture of oligomers. Koebner[43] has synthesized the individual low-molecular weight members of the p-cresol-derived series, of the type

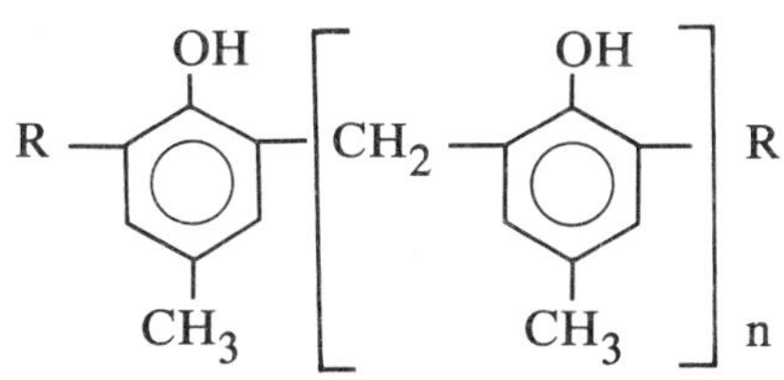


for n=0 to 4

Kamurera *et al.*[44] synthesized the members of this series up to n=9. According to these authors and Hunter *et al.*,[45] the oligomer containing a total of 5 phenol units (n=4) has a melting point around 112-120°C. Accordingly, resin oligomers for positive resist applications should have molecular weights at least that high in order not to flow during the baking steps. The molecular weight of the phenolic resins can in principle be approximated form their IR spectra[44] but modern chromatographic techniques yield far superior results.[46,48]

Commercial resins contain a wide distribution of different oligomers and isomers. Thus Varcum resins contain as many as 15-20 organic compounds such as phenol and o-, m- and p-cresol derived oligomers of various isomeric nature, molecular weight and distribution.[47] A typical resin, Varcum 6000-1, has been analyzed to have a weight-average molecular weight $M_w$=11000 and a number-average molecular weight $M_n \approx 1000$.[46] NMR and IR studies have also been conducted to elucidate[48,49] the molecular structure. Recently, different synthetic approaches have been worked out that allow the fabrication of much better defined novolacs.[47] The p-cresol derived resins have received attention as potential resist resins because of their favorable

UV-absorption spectra, which renders them attractive for deep UV lithography.[46]

**Table 1**

| Characteristics of Standard Positive Photoresist Systems | | | | |
|---|---|---|---|---|
| Product Code | Application | Coating Thickness [μm] | Solids Content [%] | Viscosity [cSt at 25°C] |
| 1350B | Photomasks | 0.3-0.5 | 16.5 | 4.6 |
| 1370 | Low Speed | 1.0-1.5 | 26.5 | 17 |
| 1350J | Spray-on | 1.4-2.2 | 30 | 30 |
| 1375 | High Step Coverage | 2.1-3.3 | 37 | 80 |

Increased use of optical step-and-repeat exposure systems has created a greater need for more uniform image resolution, linewidth control and reduced exposure time. Significant progress in this regard has been achieved with the so-called striation free formulations of the 1300 resist systems, which are available with suffix-SF from both AZ-Photoproducts and KTI or as the 1400 series from both Shipley and KTI. These recent versions provide for better coating uniformity and thus better line control down to the 1 μm resolution range.

The Shipley 2400 series of resists as well as the AZ4000 family of resists from American Hoechst (AZ-Photoproducts) are designed for use with UV exposure systems. The AZ4000 resists respond to radiation in the extended range of 310 to 450 nm. They are microfiltered to 0.2 μm and specially packaged for ultra cleanliness. The striation free version for photomask fabrication (AZ4040) has a solids content of 17%; the striation-free versions for spray-on spin coating have either 25% (AZ4110) or 31% (AZ4210) as solids content and correspondingly higher viscosities. Two standard versions with solids content of 36% and 42% are intended for high step coverage (AZ4330) or for thick film applications (AZ4620).

All systems described so far use 2-ethoxyethyl acetate a the principal solvent. They tend to show poor adhesion to oxides, a problem which can be reduced significantly upon applications of surface primers such as hexamethyldisilazane (HMDS). HMDS is typically applied in either pure liquid or diluted solution form immediately prior to resist application.[50] These resists just like all other related positive photoresists of the novolac/diazide type can be hardened after development using UV-flood exposure, as discussed by Hiraoka and Pacansky.[51]

Related resists to those discussed so far are HPR-200 striation-free resists of Philip A Hunt, Lincoln, Rhode Island. Hunt's Waycoat WX-118 resist, on the other hand, even though a novolac/quinone diazide positive resist system, has a considerably increased useful exposure range from 250 to 450 nm.

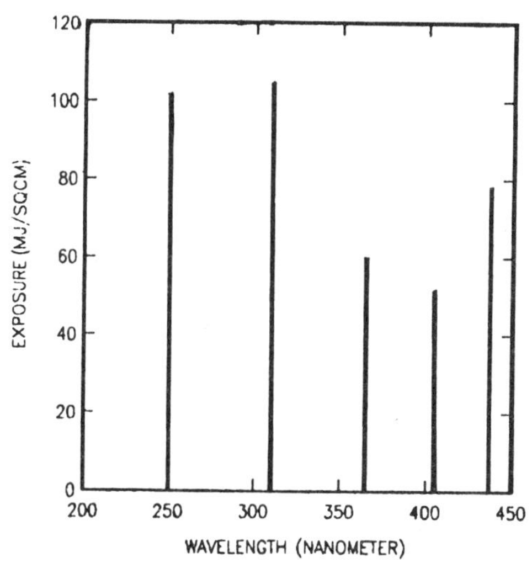


Figure 3 shows the exposure requirements of WX-118 (to match 3.0 micron mask images) at different exposure wavelengths.[52] Photobleaching of the dissolution inhibitor upon exposure is only evident at wavelength longer than 300 nm.

Eastman Kodak's Micro Positive Resist 809 resolved one of the major weaknesses of most positive photoresist systems by providing excellent adhesion to oxides. The newer Micro Positive Resist 820 permits wide process latitude and allows post-bake temperatures up to 150°C. The energy required to expose 1 $\mu$m thick resist films is shown as a function of the exposure wavelength.

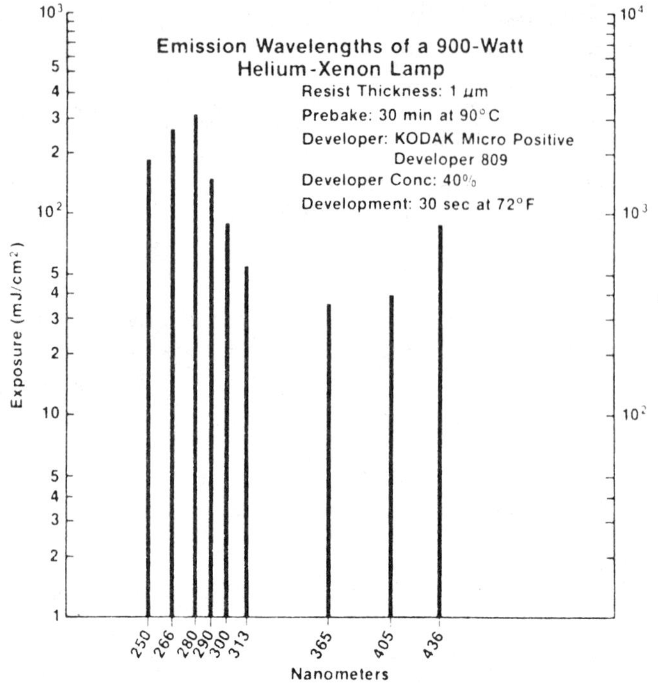


**Fig. 4.** Energy Required to Expose Kodak 820 Resist

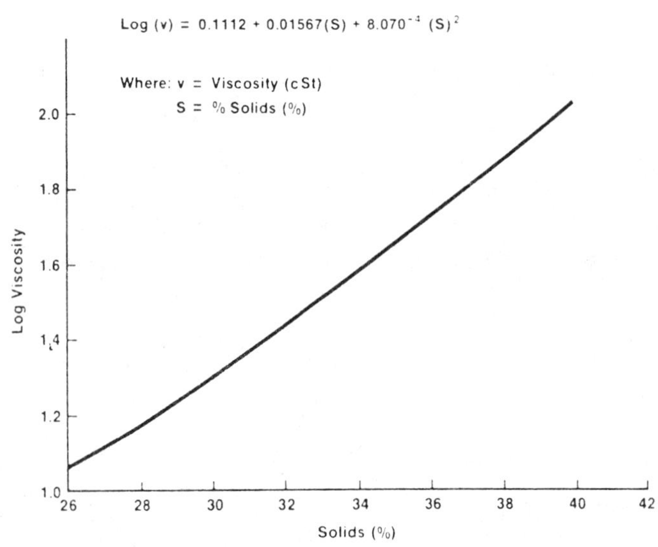


**Fig. 5.** Solids *versus* Viscosity Dependence of Kodak 820 Resist

$$\text{Thickness} = \frac{\left[23950.3 + 244.96\,(v) - 0.3612\,(v^2)\right] \cdot \left[-0.01567 + (-1.135 \times 10^{-4} + 3.228 \times 10^{-3}\,\mathrm{Log}\,(v))^{0.5}\right]^2}{\sqrt{RPM}}$$

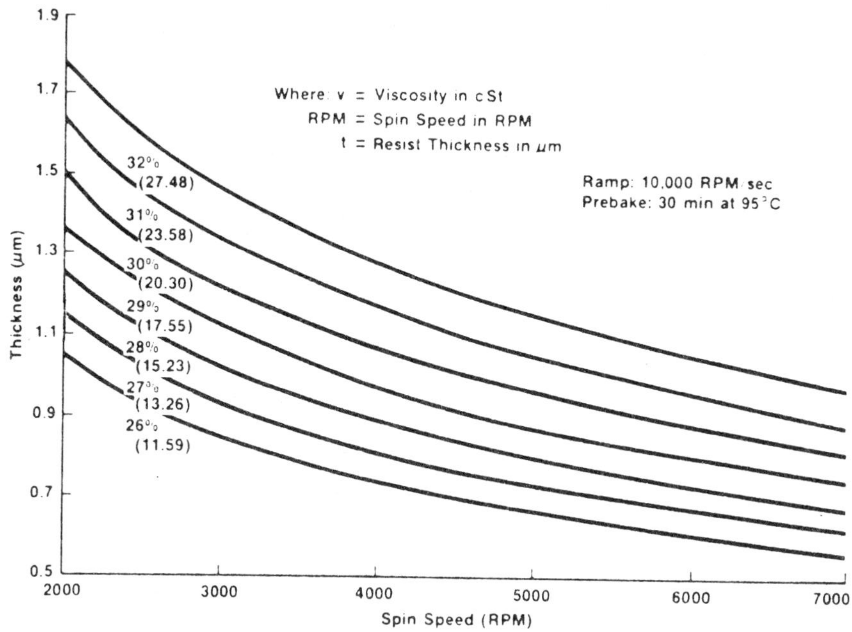


**Fig. 6.** Thickness *versus* Spin Speed of Kodak 820 Resist

## 1.5 Economical Aspects

According to estimates published in Chemical and Engineering News[53] it is believed that positive resists for integrated circuits account for 40 to 45% of the market in 1983, but are growing at 25% or 30% per year. Negative resists still account for the lion's share (55% to 60%) but are only growing 10% a year. Kodak, Hunt and Shipley are believed to supply 90% of the US market's needs.

The advantage of positive resists is seen in better resolution, which should become more important as dimensions are reduced to the 1 μm range. Negative resists swell in solvents during development and are otherwise sensitive to oxygen; therefore, they must be used under nitrogen or vacuum. Negative resists are more resistant to etching, adhere well to substrates, are more sensitive to radiation and less expensive. With a big push towards dry processing they might well remain a dominating if not the dominating material in the future.

**Table 2**

**Major Supplies of Photoresists**

| Company | Liquid Photoresists Positive | Negative |
|---|---|---|
| 1. Allied | x | |
| 2. American Hoechst (AZ-Photoresist Products) | x | |
| 3. J. T. Baker | x | x |
| 4. Ciba Geigy | | x |
| 5. Dynachem (Norton Thiokol) | x | x |
| 6. Kodak | x | x |
| 7. EM Chemicals, U.S.A. (E. Merck, Darmstadt, W. Germany) | | x |
| 8. W. R. Grace | | x |
| 9. Philip A. Hunt | x | x |
| 10. KTI (Union Carbide) | x | x |
| 11. Mac Dermid | x | |
| 12. Mead | x | x |
| 13. Norland Products | | x |
| 14. Shipley | x | |

2. Image Reversal

Image reversal describes a process whereby a positive tone resist is treated in a nonstandard way to yield a negative tone image, *i.e.,* upon this treatment a normally positive resist behaves as a negative one. This process was first described in the patent literature and seems to have been discovered by chance. Thus, it has been found that using imidazoles or triazole derivatives, *i.e.,* compounds which are typically used as adhesion promoter, image reversal can be achieved in the following way: At first the positive diazo-naphthoquinone type resist is being "doped" with about 1% of imidazole. Then this doped resist is exposed to UV light, and without development subjected to a baking step at about 100°C. Subsequent flood exposure of the entire resist area reveals upon development in the typical base developer a *negative* tone image. On the other hand, if this same doped resist is conventionally UV exposed and directly developed, the usual *positive* tone image is obtained. Accordingly this identical resist formulation can be developed to yield either positive or negative pictures, depending on the nature and sequence of the processing steps.

The chemistry of this unusual dual behavior has recently been elucidated by Willson and coworkers.[54] According to their results the fate of the photoactive compound leads either to a base soluble or base insoluble photoproduct, as outlined in the following: Normally, the photoinduced decomposition of the diazonaphthoquinone follows the Süs mechanism,[4,24] which proceeds via an initial ketocarbene to an intermediate ketene, which typically reacts with omnipresent water to an indenecarboxylic acid. For the following it is of significance to note that according to Pacansky and Lyerla,[30] and thus deviating from the Süs postulate, this acid is a 3-indenecarboxylic acid and not an isomeric 1-indenecarboxylic acid.

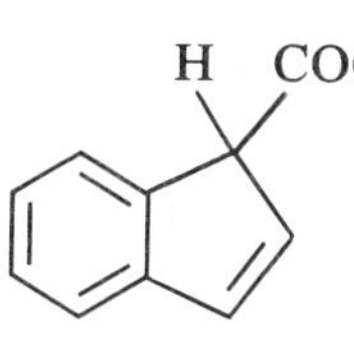


1-indenecarboxylic acid

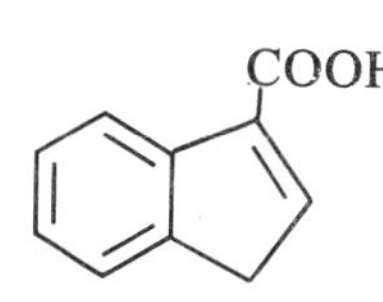


3-indenecarboxylic acid

The 3-isomer is an $\alpha,\beta$-unsaturated acid, which is known to undergo base-catalyzed decarboxylation[54] resulting in an indene-type hydrocarbon. This process requires only temperatures around 75°C and thus occurs during the intermediate baking cycle following the initial exposure. The hydrocarbon so obtained is no longer soluble in base, and thus does not support the solubility of the resist in the base developer any more.

Scheme 2: Image Reversal

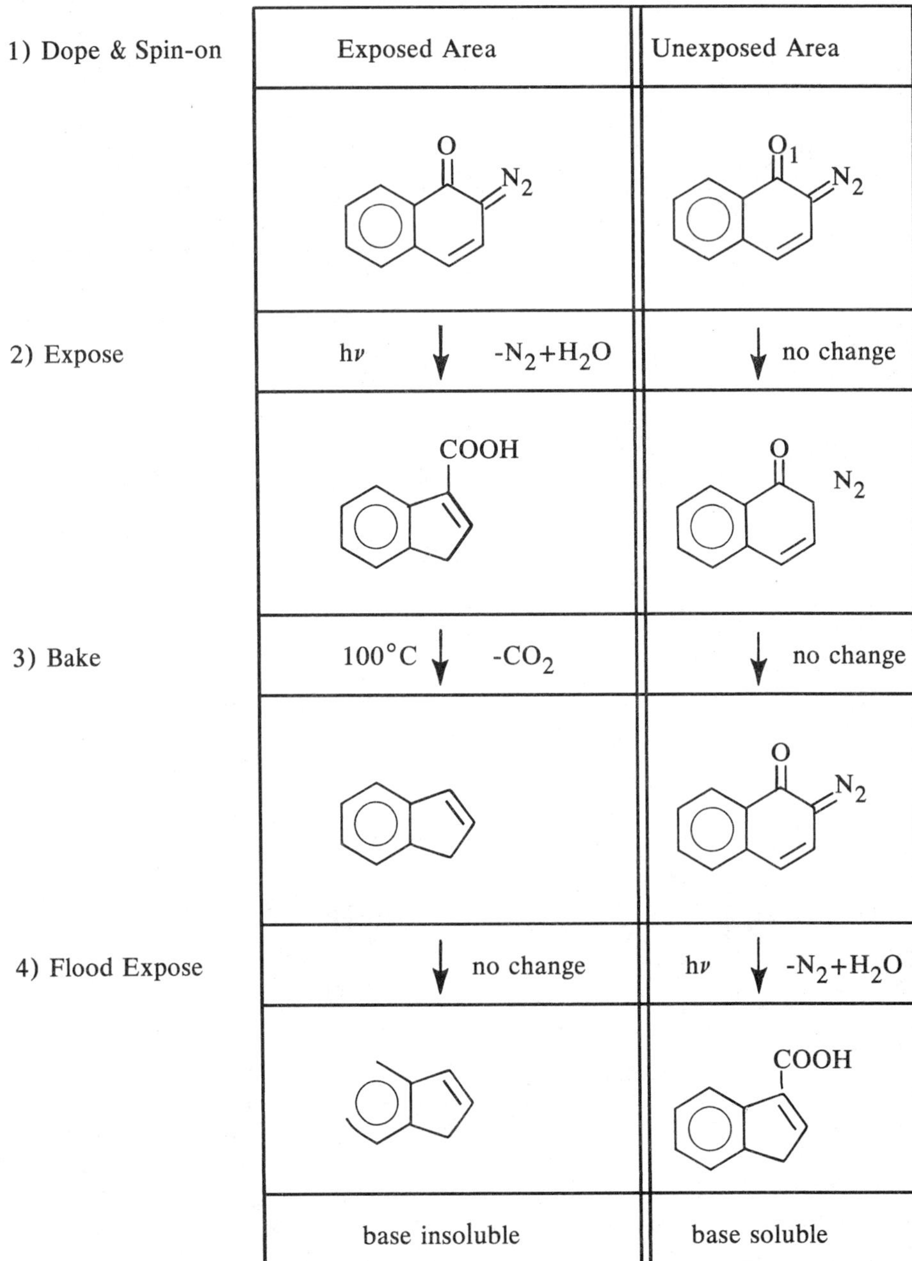

3. Negative Photoresists

One of the first synthetic negative photoresists used for microelectronic applications was the Kodak Photosensitive Resist (KPR), which is a solution of intrinsically photosensitive polyvinyl cinnamate[55] of average molecular weight 200,000

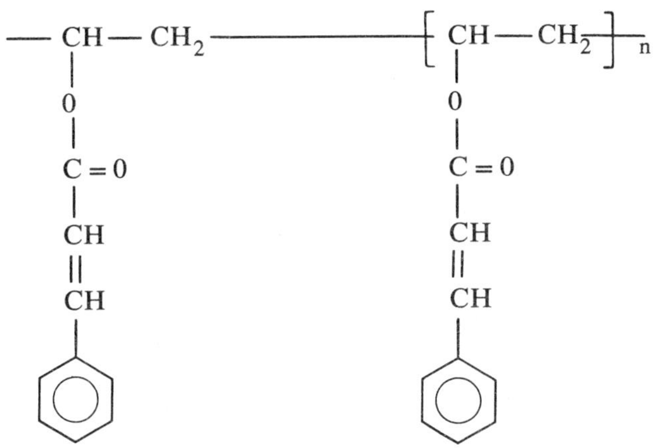

poly(vinyl cinnamate)

The cinnamic acid sidegroups of this polymer are known to undergo intermolecular crosslinkages upon exposure, which resembles the well known dimerization of cinnamic acid to truxillic or truxinic acid

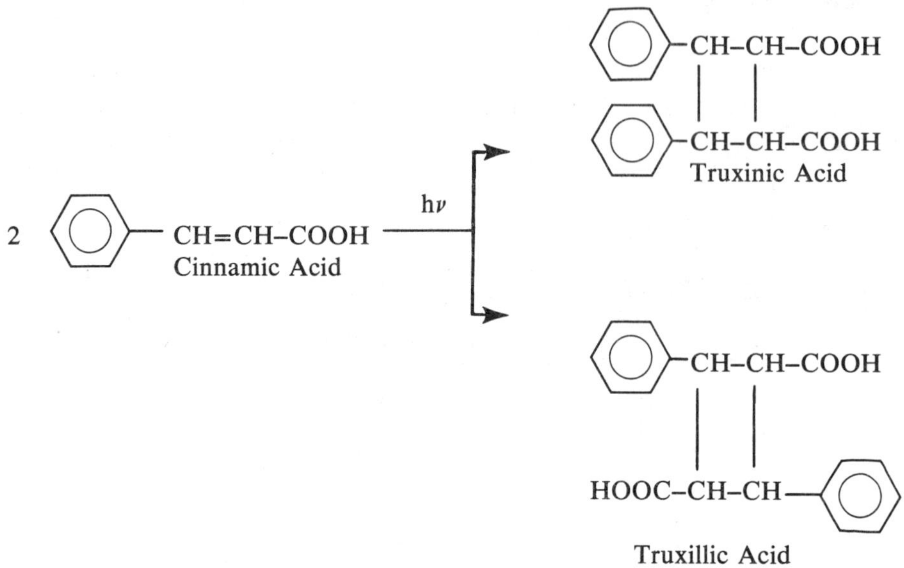


The KPR resist system is sensitized to the appropriate wavelength region using 3-methyl-1,3-diaza-1,9-benzanthrone. This resist system gave rise to a relatively high incidence of pinholes and also caused undercutting; therefore, it has been replaced by the most common negative photoresist system described in the following, namely those based upon cyclized poly(isoprene).

### 3.1 Cyclized Poly(isoprene) Negative Photoresists

This family of resists involves the crosslinking of a polymeric matrix via a photosensitive additive. They are thus composite resist systems. Typically the polymer is cyclized cis-1,4-isoprene[56] even though more recently cyclized cis-1,4-butadiene[57] has also been proposed

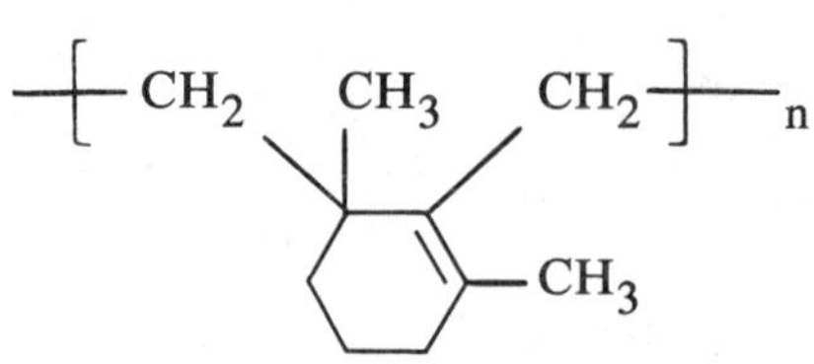


cyclized cis poly(isoprene)

The cyclization renders the poly(isoprene) both adherent to the substrate and gives rise to better films, which are resistant to both acidic and alkaline wet etchants. Typically the polymer is about 85% cyclized, has a molecular weight in the range 50,000 to 150,000, a softening temperature around 65°C, and no appreciable self absorption in the wavelength region where the photoactive compounds absorbs.

The photosensitizer is a bis-aryldiazide,[58] namely

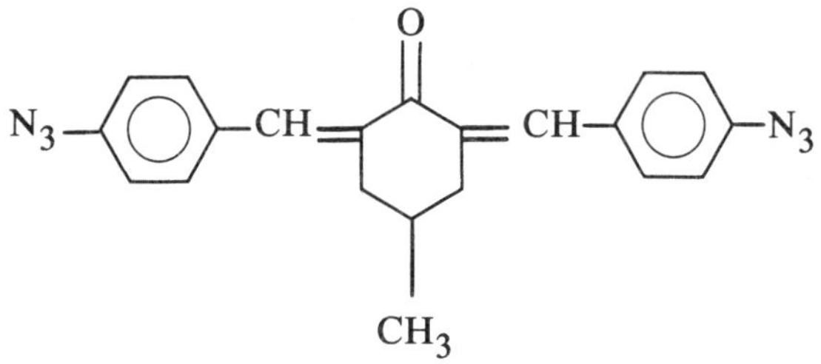

2,6-di(4'azidobenzal)-4-methyl cyclohexanone

The bis-aryldiazide is believed to photodecompose under elimination of nitrogen into a highly reactive bis-aryldinitrene via sequential absorption of two photons.[60] The nitrenes are known to react preferably via insertion into carbon-hydrogen bonds[61] or to some extent via addition to carbon-carbon double bonds. They may also abstract allylic hydrogen or dimerize to form azo-dyes. A major drawback is their tendency to react with molecular oxygen to form highly undesirable nitro compounds; therefore, oxygen must be excluded during exposure to achieve full crosslinking of the resist.

The Kodak Metal Etch Resist (KMER) was the first of the highly successful polyisoprene-type negative photoresists to be introduced commercially. Its high content of inorganic impurities limited the minimum line width that could be achieved consistently to about 4 μm.

The Kodak Thin Film Resist (KTFR) was a close relative to KMER with low levels of inorganic contaminants.[59] Using vacuum contact printing it allowed the fabrication of structures with 2.5 μm lines. The Kodak Micro-Neg 747 Resist is a high purity poly(isoprene) negative photoresist as well.

A sufficient concentration of the photoactive compound would be one mole of diazide per two moles of polymer, since only one crosslinkage per polymer is required to render the resulting network insoluble. Commercial resists use higher concentrations of the photosensitizer to increase sensitivity. A typical formulation is 2 parts of diazide per 100 parts of rubber by weight dissolved in mixed xylene isomers. An added adhesion promoter assures both good adhesion, chemical resistance and high sensitivity simultaneously.

This photoresist system is sensitive in the wavelength region 300-515 nm with peak absorption at 410 nm. Thus this photoresist is ideally suited for projection exposure at 405 and 436 nm. With a viscosity around 40 cP and a solids content of 9.65%, dimensions down to 2.5 μm can be obtained.

Negative photoresists suffer from a thickness loss between coating and development, a problem, which for the polyisoprene systems is reduced when the sensitivity of the resist is increased.[62]

3.2 High Temperature Negative Resists: Photosensitive Polyimide Precursor Resists

Polymers have found application in semiconductor processing steps not only as photoresists but also as organic insulators. The driving force behind this development has been their low dielectric constants, their mechanical properties, their tendency to flow and to planarize topography among other things. The most successful polymers along this line are the high temperature stable polyimides.[63]

Another concern namely to save processing steps, leads to the concept of "permanent resist." This is a photosensitive polymer that will first serve as a photoresist but can subsequently be left behind as for example an organic insulator.

Furthermore, the evolution of semiconductor processing has led to manufacturing steps which require higher processing temperatures up to 450°C, which conventional photoresists will not tolerate. Rubner[42] has recently described a system, which achieves crosslinkages in the precursor to polyimides, and thus combines the concept of a high temperature negative photoresist with that of the permanent resist. Systems based upon these ideas have recently become commercially available as Selectilux HTR-2 from EM Chemicals, Hawthorne, New York, a division of E. Merck, Darmstadt, West Germany.[65] This polyimide precursor photoresist has a sensitivity of 70 to 80 $mJ/cm^2$ μm and is applied like a common negative photoresist. After curing in a postbake mode it gives rise to a structured polyimide layer with a resolution in the 2.5 micron range. The resolution depends upon the thickness of the polymer layer. The polyimide precursor negative photoresist is essentially a solution of a polyamic acid ester in a solvent mixture of N-methyl-pyrrolidone and cyclopentanone. Polyimide is typically formed via an imidization reaction of polyamic acid formed from pyromellithic anhydride and p,p'-diamino diphenylether (Scheme 3a)

Scheme 3a: Polyimide Formation

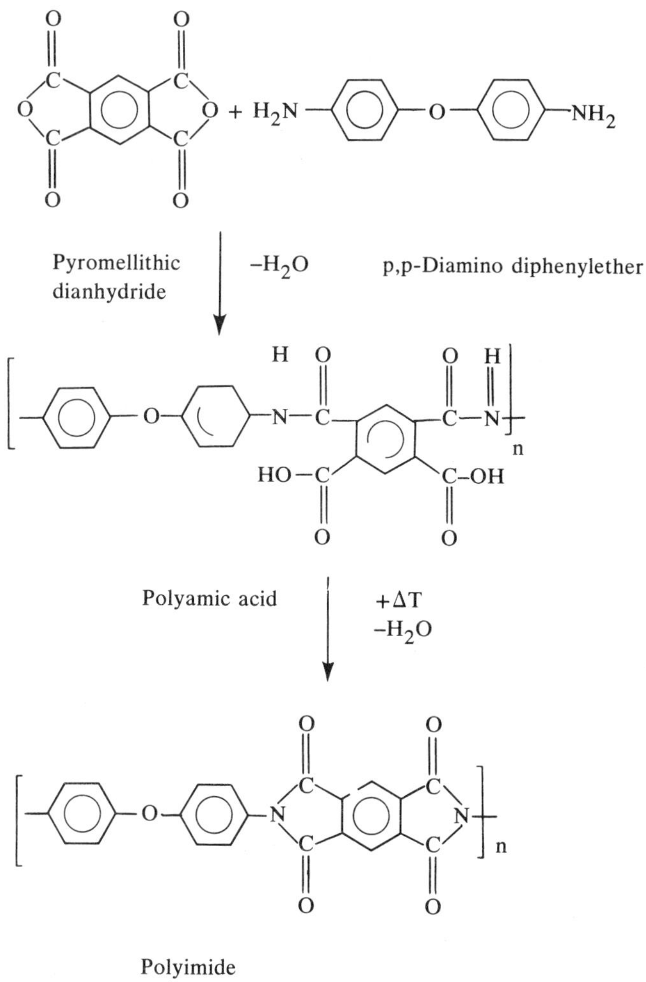


A light sensitive polyamic acid is obtained from a half ester of the pyromellithic acid with allyl alcohol (R-OH) or hydroxyethylmethacrylate. In this case, the intermediate polyamic acid is obtained according to Scheme 3b. Here the acid chloride is formed from the half ester of pyromellithic acid.

Scheme 3b: Polyimide Precursor Negative Photoresist

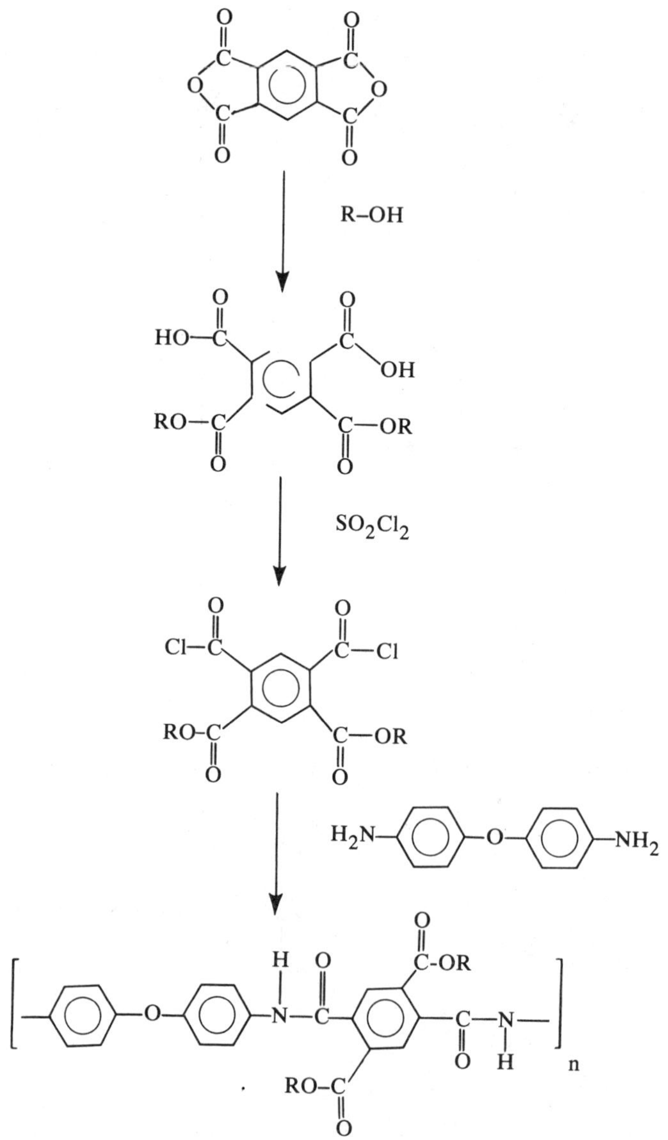

Imidization of the exposed and developed polyamic acid yields polyimide. At 275°C more than 99% of the polyimide precursor is converted to polyimide and most of the photo-crosslinked groups are volatilized. This process results in a weight and thickness loss of 40%.

The polyimide precursor resist is applied in the same way as any common negative photoresist. Adhesion promoters for polyimide, such as silanes or aluminum chelates, can be used and promote adhesion to the substrate. Vinylsilane based adhesion promoters are already added to the photoresist formulations. The resist is applied by common spin coating techniques. 50g resin in 60g solvent yield a 5000 cP viscosity solution which yields 13 $\mu$m thick layers at spinning speeds of 5000 rpm. Solutions with a viscosity of 500 cP yield 5 $\mu$m thick layers under the same conditions, a 130 cP solution 2.5 $\mu$m thick layers.

The finally cured film is essentially equivalent in its properties to nonphotosensitive polyimide. Complete removal of the photo-crosslinked groups is achieved at temperatures of 400°C or more, resulting in a final weight and thickness loss of 55%. The polyimide pattern definition will not deteriorate. Accordingly, cured films derived from polyimide precursor negative photoresists have resistivities around $10^{16}$ ohms•cm, a dielectric constant of $\epsilon = 3.3$ up to 100 Hz, a dielectric loss factor of $\tan\delta 2 \times 10^{-3}$ at 1 kHz, a glass transition point between 305 and 350°C and long term stability up to 300°C, but they are stable for several hours at 400°C.

The polyimide patterns can be removed either by oxygen plasma stripping or by using ethylene-dianine for 6 min at 80°C.

### 3.3 Other Negative Photoresists

Copolymers of methyl methacrylate and 2,3-epithiopropyl methacrylate have been used as crosslinking systems using aryl diazomium salts as Lewis photosensitizers[66] and acid generators. Additional systems have been outlined by Feit.[67] However, it does not appear that these alternate systems will displace the polyisoprene systems in the conventional wavelength regime.

For many years negative resists seemed to dominate positive resists. However, pressure for 1 micron and sub-micron resolution favored positive resists for mask fabrication and has also caused entire fabrication lines to switch to positive optical resists. A key reason for the limited resolution of negative photoresists is the developer induced swelling. A negative resist with the resolution properties of current positive photoresists would be highly desirable.

## V. DEEP UV RESISTS

Among the high resolution lithography techniques which are capable of achieving submicrometer line dimensions, UV lithography has, because of its convenience, increased significantly the importance ever since the early impressive results of Lin.[68] He termed lithography using radiation in the wavelength range 200 to 260 nm "deep UV lithography." Currently employed (*i.e.,* conventional) "near UV lithography" operates in the wavelength region 350 to 450 nm, leaving the range in between for "mid UV lithography" (260 to 350 nm). For both the mid and the deep UV range, exposure systems and resists have only become available very recently.

### 1. UV Light Sources

Part of the obstacles to deep UV techniques stem from the inefficiency of conventional light resources. Lamp sources can be divided into high pressure arc lamps, which are usually Hg or Hg-Xe lamps in the 0.5 to 1.5 kW range of electrical input energy, deuterium lamps, and microwave excited Hg discharge lamps.[69]

Of these, the first type are currently used in commercial projection equipment, whereas the latter are only attractive for blanket exposure, *e.g.,* for multi layer systems. Deuterium light sources are too inefficient in the deep UV (<0.01%) and at input power levels >200W, they suffer from plasma instability and short lifetime. The mercury line dominating the deep UV is the well known 254 nm line, whereas in the mid UV a high intensity mercury emission line is located at 313 nm, and a lower intensity one at 334 nm.

The advent of excimer lasers has provided highly efficient UV light sources, notably the ArF laser at 195 nm, KrCl at 222 nm, KrF at 249 nm and XeCl at 308 nm. They can provide the basis for an ultrafast, high resolution lithography[70] with exposure times down by two orders of magnitude relative to the conventional lamp light sources. So far, however, exposure systems based upon lasers have remained limited to applications in development laboratories, and their entry into the mass production environment does not yet seem imminent.

### 2. UV Resists

In turn the availability of commercial deep UV exposure systems, such as the Perkin Elmer Micralign Model 500 projection printer, necessitated the development of more sensitive UV resists. The ideal high resolution deep UV resist should possess good sensitivity in the 200 to 280 nm range with little or no absorption at longer wavelength to eliminate the need for inefficient spectral filtration. Furthermore, it should be compatible with the conventional processing requirements for electron microcircuitry.

Moreau and Schmidt[71] had demonstrated in 1970 that PMMA could function as a high resolution deep UV resist. Unfortunately, its sensitivity at 254 nm is relatively low.

## 2.1 Positive Deep UV Resists

Several positive resist families have since been investigated, some derived from successful e-beam resists like PMMA, some from conventional "near UV' photoresists, and in addition some totally new systems have been explored, which have uniquely been designed for deep UV application from basic principles.

### 2.11 Methacrylate Type Resists

Poly(methyl methacrylate), PMMA, homopolymers have an absorption spectrum which peaks around 215 nm with a maximum absorption coefficient of about 0.45 $\mu m^{-1}$. At 230 nm the absorption coefficient is 0.40 $\mu m^{-1}$ and from 260 nm to the visible PMMA ceases to be sensitive as a photoresist.[68] This low absorption coefficient (for comparison[72] Shipley AZ1350J photoresist has an absorption coefficient of 0.86 $\mu m^{-1}$ at 404 nm) combined with the inefficiency of conventional lamp light sources results in long exposure times (in the order of minutes). Together with the poor performance under plasma etching conditions[14] PMMA and even some of its copolymers do not satisfy the criteria of an ideal deep UV resist.

One concept for increasing the performance of methacrylate type resists proceeds via introducing UV chromophores into the polymer. Thus Reichmanis and Wilkins[73] have increased the deep UV sensitivity of methacryaltes by incorporating $\alpha$-keto-oxime chromophores in form of a copolymer. A small (<10%) incorporation of 3-oximino-2-butanone methacrylate, $CH_2{=}C(CH_3){-}CO{-}O{-}N{=}C(CH_3)CO{-}CH_3$, in PMMA resulted in a four fold increase in photosensitivity of the copolymer in the range 230-260 nm. Additional incorporation of methacrylonitrile into the above copolymer further enhances the deep UV sensitivity of the resulting terpolymer up to 85 times above that of PMMA. Unfortunately, and perhaps surprisingly, the plasma etch rates of such terpolymers are not significantly different from PMMA, namely in a standard $CF_4$+8% $O_2$ plasma the etch rate of the terpolymer is only half that of polysilicon.[73] Typically, poly(methacrylonitrile), PMAN, and methacrylonitrile containing copolymers show a rather remarkable performance under plasma etching conditions[74] (Figure 7).

The photoinduced decomposition mechanism of the copolymer is considered to be initiated by side chain cleavage followed by main chain scission, a concept that is rather universal for PMMA type systems.[74] Another approach to increased UV sensitivity is based upon the addition of sensitizers. This concept can also be combined with the addition of chromophores. Accordingly, the sensitivity of the above terpolymer can further be increased about three fold upon sensitization using either naphthalene-2-acetic acid, N-acetyl carbazole or t-butylbenzoic acid.[73]

An alternative approach to introduce UV chromophores into polymethacrylates has been chosen by Hiraoka and his coworkers.[74] They prepared copolymers of methacrylonitrile and methylmethacrylate or methacrylic acid, which have no significant UV absorption initially. Upon heat treatment in air, *i.e.,* after a 15 min prebake step in air at about 130°C, a broad UV absorption at 246 nm builds up, which causes the formation of cyclic structures[76] (Scheme 4). These structures act as chromophores similar to the $\alpha$-keto-oximers utilized by Reichmanis *et al.*,[73] and they are responsible for a rather interesting behavior of the prebake copolymers.

Exposure at 254 nm, *i.e.,* in the deep UV, degrades the polymer weight (initially about 250,000 of the 1:1 copolymer) and results in a positive deep UV resist with a sensitivity of 50 $mJ/cm^2$. With sensitizers like p-t-butyl benzoic acid their sensitivity can be increased to about 10-20 $mJ/cm^2$.

Exposure at 300 nm, however, *i.e.,* in the mid UV range, introduces cross linkages and renders the prebake copolymers to behave as negative mid UV resists with a sensitivity of 100 $mJ/cm^2$. Postbake of the exposed and developed pattern increases the plasma etching resistance significantly, as is outlined[74] in Figure 7.

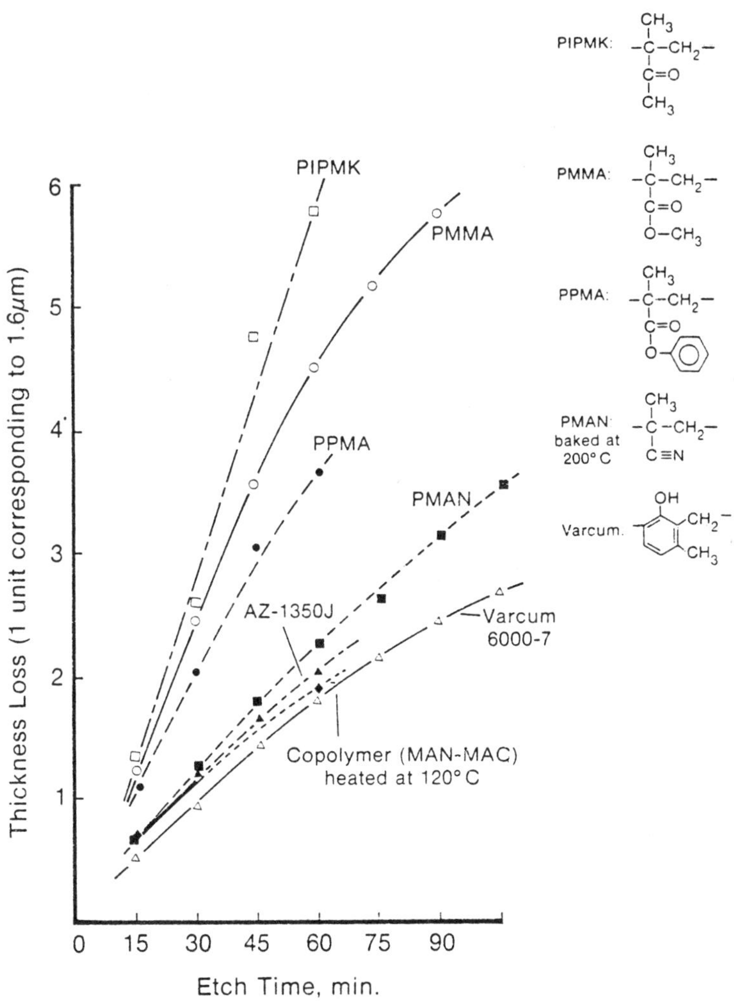


FIGURE 7. Comparison of RIE of resistances of copolymer (methacrylonitrilemethacrylic acid), heated at 120°C, with other resist materials.

Scheme 4: Heat Treatment and Photochemistry of Co Polymers

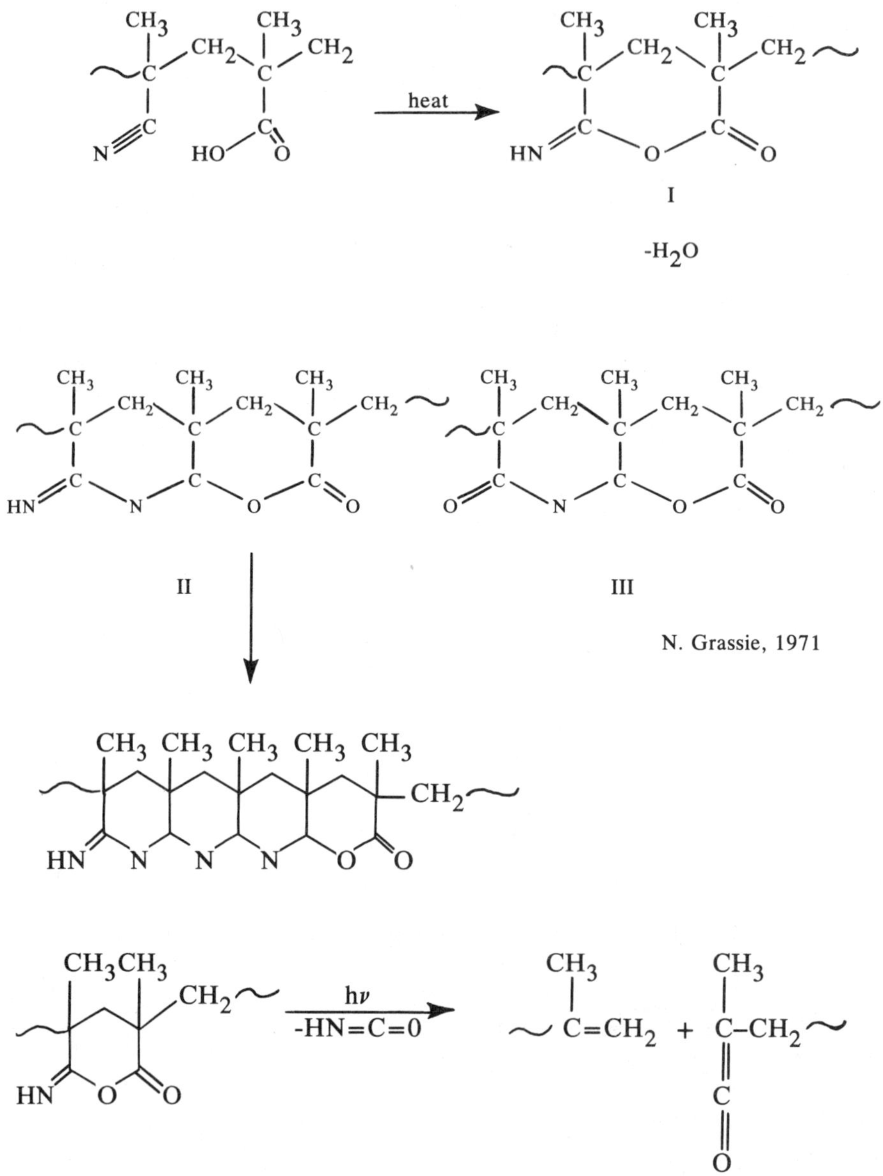


The above is the major reaction of the prebaked films at 254 nm exposure. It has been demonstrated that poly(dimethacrylimide) underwent elimination of isocyanic acid upon UV-light irradiation, accompanied by main chain scission, as follows:[8]

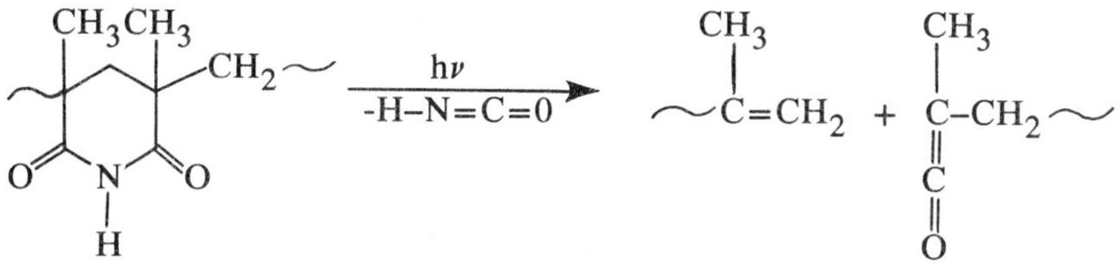

2.12 Poly(methyl isopropenyl ketone)

Poly(methyl isopropenyl ketone), PMIPK, has been used as an electron beam resist as outlined by Levine.[77] It absorbs light shorter than 340 nm and its UV sensitivity[78] is about 5 times that of PMMA. Photoirradiation transfers PMIPK into an excited state and causes main chain scission associated with about a five-fold decrease in the number average molecular weight.[78]

A further increase in sensitivity can be achieved by using sensitizers such as p-t-butyl benzoic acid, which increases the sensitivity by a factor of three.[78] A positive UV-resist based upon PMIPK has become commercially available by Tokyo Ohka Kogyo Co. with the trade name ODUR-1011 and ODUR-1013.

2.13 Poly(olefin sulfones)

Another family of polymers, which had originally been developed as sensitive electron beam resists,[79] are copolymers of sulfur dioxide and appropriate olefins. Normally, these copolymers absorb light below 200 nm, but arylated olefins, such as styrene, extend their absorption range into the 200-300 nm region with a peak around 265 nm. More extended aromatic systems in the side chain of the poly(vinyl arene sulfones) allow to shift the absorption peak closer to the visible. Terpolymers such as poly(styrene-co-acenaphthalene sulfone), for example, which has an absorption peak at 300 nm, is approximately 50 times more sensitive than the simple poly(styrene sulfone) copolymer under identical exposure conditions.[5] Aromatic units incorporated into the main chain of the polysulfones lower their radiation sensitivity instead.

The radiation induced decomposition mechanism of poly(olefin sulfones) has been studied by a number of groups.[6,80-82] Significant seems the observation by Pacansky and coworkers that upon addition of novolac type resins to poly(olefin sulfones), energy transfer occurs between the two components, which for example in the e-beam mode, causes a decrease of the cross section of the sulfone upon addition of novolac.[82] In the photochemical mode, *i.e.,* when used as UV resists, this energy transfer concept might potentially point to a way for increasing the UV sensitivity of poly(vinyl aryl sulfones) via sensitization.

## 2.14 Novolac Type Systems

Classical positive photoresists based upon novolac-type matrix resists are unsatisfactory in the deep UV range, in part because the typical cresol-formaldehyde novolacs absorb strongly at 254 nm, whereas they are essentially transparent at 404 nm. The traditional matrix resins, such as the Varcum resins, are complex mixtures of as many as 15 to 20 organic components derived from condensation products of formaldehyde with ortho-, meta- and p-cresols with a broad molecular weight distribution. Efforts have been made to utilize resins prepared from the pure isomers of ortho-, meta- or para-cresols. Studies by Gipstein and coworkers[83] showed that the resin derived from the para isomer is most transparent at 254 nm (Figure 8). A second drawback of conventional novolac-type positive photoresists, such as Shipley AZ1350J, when used in the deep UV, stems from the fact that the photoactive compound (PAC) is not ideal for the 254 nm region either. The PAC namely is used as a dissolution inhibitor, and it is desirable that in the exposed regions the photoproducts of the PAC do not absorb. Otherwise, the radiation may be absorbed in the photoproducts formed at the top, thus effectively shielding the PAC in the bottom layers from exposure. The near UV resists (AZ1350J) do not satisfy this requirement, since the photoproducts of the PAC absorb strongly at 254 nm. The combined absorption of the matrix resin and the PAC renders films of 1 $\mu$m or more essentially opaque at 254 nm; therefore, the permissible thickness of the traditional near UV photoresists for usage in the deep UV is not suitable for most applications.

The optimum deep UV sensitizer should have a strong but bleachable absorbance at 254 nm, a high quantum efficiency, a thermal stability that is consistent with wafer processing, it should be soluble in suitable coating solvents for spin coatings, and its photochemistry should result in a high solubility difference between exposed and unexposed areas to assure a high imaging contrast. Ideally, the sensitizer should only absorb in the deep UV range and not at all above 300 nm, to assure high resolution without requirements for filtration of the exposure light. Grant and coworkers[84] identified 5-diazo-Meldrum's acid (2,2-dimethyl-4,6-dioxo-5-diazo-1,3-ioxolane) as a suitable deep UV sensitizer and formulated a novolac-type positive resist based upon this PAC. Interestingly, however, it remained unclear whether the PAC of the Meldrum's diazo type acts as a dissolution inhibitor. Thus DeForest[18b] has suggested that during the development step of novolac-type resists containing substituted diazo-naphthoquinones as the PAC, base induced diazo coupling occurs between the diazo sensitizer and the phenolic resin. Grant *et al.*[84] found no evidence for coupling between their

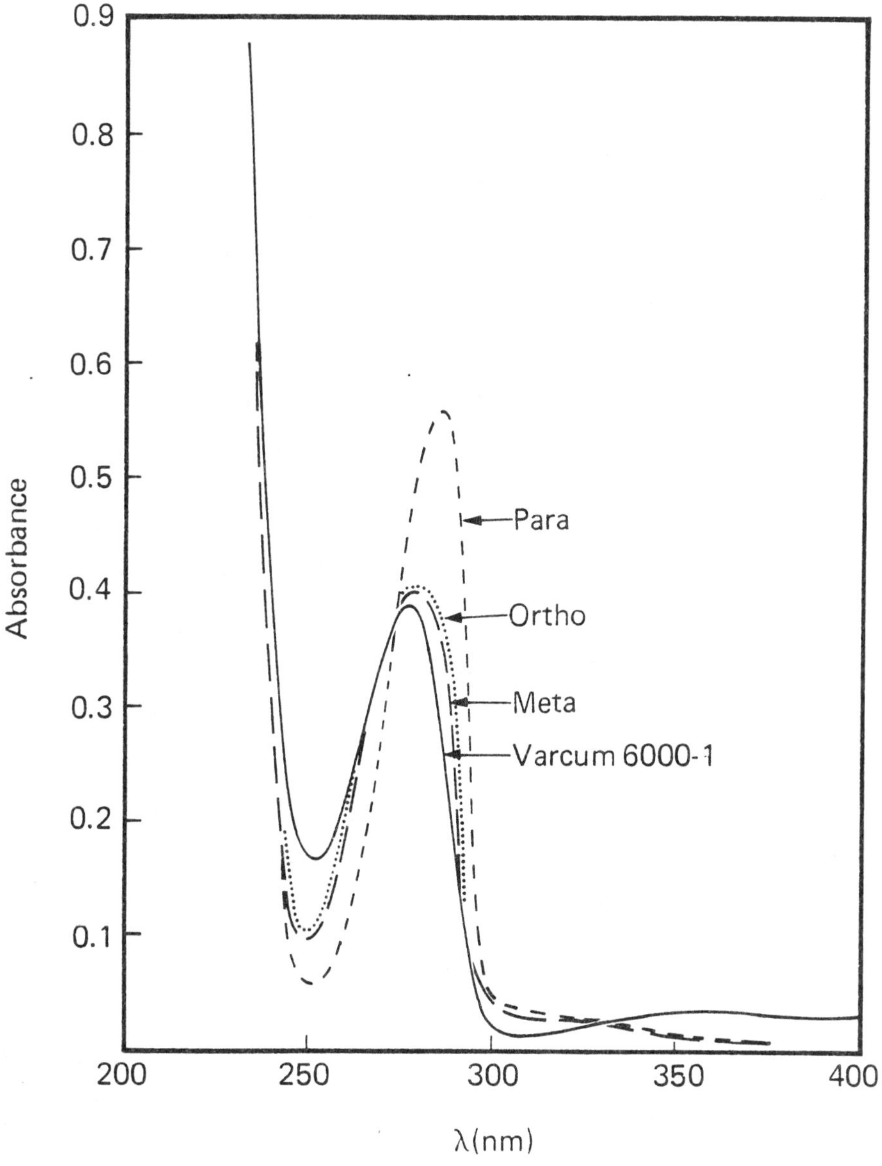


**FIGURE 8.** UV spectra of $0.2\mu$ novolac resin films on quartz.

diazo sensitizers and sodium phenolate, however. Instead they assign more significance to a hypothesis by Ouano,[85] who argued that the PAC acts essentially as a "blowing agent," giving off volatile products upon exposure, which increase the free volume of the polymer film and render it more soluble because of facilitated diffusion of the developer into the polymeric film.

Indeed, Bowden and coworkers[6] have used a nondiazo type dissolution inhibitor, namely a degrading polymer, as an electron beam sensitizer for novolac type resists. E-beam exposure of the polymer, namely poly(2-methyl-1-pentene sulfone) causes spontaneous depolymerization into olefin and $SO_2$. Even though Bowden *et al.* attribute the role of the dissolution inhibitor to the molecular weight reduction and thus increased solubility of the polysulfone, it is also possible that the generated $SO_2$ gas again acts as a "blowing agent," by merely making the exposed areas more easily accessible to the aqueous base developer. The two competing hypotheses for explaining the mechanism of positive resists, *i.e.,* dissolution inhibitor removal *versus* increased developer penetration via formation of volatiles as suggested by Ouano[85] deserves more attention and additional investigation. Indeed, the latter concept could easily point to a large variety of thermally stable but photochemically liable "blowing" agents, which could provide convenient deep UV photoactive compounds to sensitize a variety of matrix resins.

The advantage of highly aromatic matrix resins such as novolacs over polyvinyl or polyvinylidene type resists is in part due to their superior resistance to plasma etching conditions. In addition there is a tendency to use one and the same resist or at least matrix resin for all types of lithography, especially photo-, deep UV- and e-beam lithography, and the novolac derived systems comprise such potential. Since many processing data have been accumulated for these widely used photoresist variety, appropriately sensitized resists for deep UV and e-beam applications represent very promising formulations and convenient choices.

3. Negative Deep UV Resists

Phenolic resins, in particular poly(p-vinyl phenol) can be used as matrix resins for negative tone deep UV resists. A negative resist based upon poly(p-vinylphenol) and a bis azide (3,3'-diazidodiphenylsulfone) has been developed by Hitachi Ltd and is known as MRS-1 (Micro Resist for Shorter wavelengths.)[86] In this resist, absorption of deep UV light is strong. Therefore, reactions occur only on the top layer. This behavior avoids standing wave problems due to reflections. The advantages of this top layer exclusive reaction are related to those of multilayer resist systems using different resists. Steep wall profiles and 1 $\mu$m lines can be obtained, and the resistance of the MRS resist formulation to plasma etching conditions has been reported to be superior to those of novolac type positive resists.

A few conventional negative e-beam rėsist, such as poly(glycidyl methacrylate) and its copolymers, as well as epoxidized poly(butadiene), have been evaluated as deep UV resists by Japanese workers,[78,87] who found these materials to behave as *positive* deep UV resists. This behavior was attributed in the case of the methacrylate type resin to main chain scission, believed to result from absorption of the UV light by the ester carbonyl group. In the case of the epoxidized poly(butadiene), the epoxide ring did not open as a consequence of the irradiation.

Deep UV initiated cross linkages of polymer chains is not only of interest in negative resists but has also become important in the area of UV hardening of developed resist patterns in positive resist. Thus deep UV light exposure of AZ-type positive resists in air have been shown by Hiraoka and Pacansky[89,90] to result in UV hardening. Hiraoka and Welsh[88] have investigated the photoinduced reactions of substituted poly(styrenes) and of novolac resins, and they found efficient UV induced halogen removal from para-halogen substituted poly(styrenes) as well as photooxidation, which resulted in cross linkages and hardening of resist patterns. Similarly, UV exposure in air of both poly(p-hydroxystyrene) and novolac resin were shown to result in photooxidation, leading to the formation of cross linkages via intermediate aldehyde and carboxylic groups. Poly(p-nitrostyrene), however, was found to undergo photo-rearrangement as the major reaction, with photooxidation playing an insignificant role.

## VI. DEEP UV HARDENING OF RESISTS

For a variety of processing steps thermal stability of the developed photoresist pattern is essential. Novolac type positive resists typically flow around 100°C which limits their range of application. Hiraoka and Pacansky[89,90,91] have shown that upon flood irradiation of the exposed photoresist with deep UV irradiation (254 nm), the flow resistance can be pushed to about 200°C. Similarly Van Pelt[92] and later Allen and coworkers[93] have shown that other positive working photoresists can be hardened via deep UV flood exposure, notably the resists AZ1350J, AZ1470, AZ2400, Hunt WX-118, Hunt-204 and Kodak 809. All of these resists can thus be rendered to withstand a 30 minute bake cycle at 180°C. The flood irradiation is believed to increase the molecular weight of the resin at the surface, most likely via cross linking. Interestingly, flood irradiation with near UV light (>300 nm) does not result in the formation of a surface hardened skin, and with the exception of AZ2400, was actually found to decrease the thermal flow resistance of the resists.[93] Special microwave powered deep UV sources have since been developed for resist hardening with an accumulated irradiance of 27 $mW/cm^2$ in the range 200-245 nm.

## VII. ELECTRON BEAM RESISTS

Electron-beam resists typically consist of polymers, which after exposure to an electron beam degrade (positive mode) or crosslink (negative mode). Thus irradiation changes their solubilities, and subsequent treatment with a suited solvent system allows the resists to be developed into a pattern with resolution in the sub micron regime. Electron-beam resists can be, but do not have to be, sensitive to light, and either type is known.

### 1. "Classic" Resists

Indeed early attempts[95] to use electron beams for lithographic purposes used conventional photoresists such as the Kodak Thin Film Resist KTFR, a standard negative photoresist based upon cyclized rubber as a matrix and a diazide as a crosslinking sensitizer as outlined in detail above. The early results were disappointing, however, yielding typically 10 micron wide lines with a smoothly sloping edge of about 2 microns on either side (using 10 kV beam). Radiation induced changes in polymers have been investigated long before the advent of e-beam lithography. Impirical rules have been formulated which correlate the polymer structure and the changes to be expected upon irradiation. These concepts have been expanded to allow for a prediction of the radiation sensitivities of polymers.[96] It has been found that typically:

A) Polymers with quarternary carbon atoms in the backbone chain, *i.e.*, vinylidene polymers

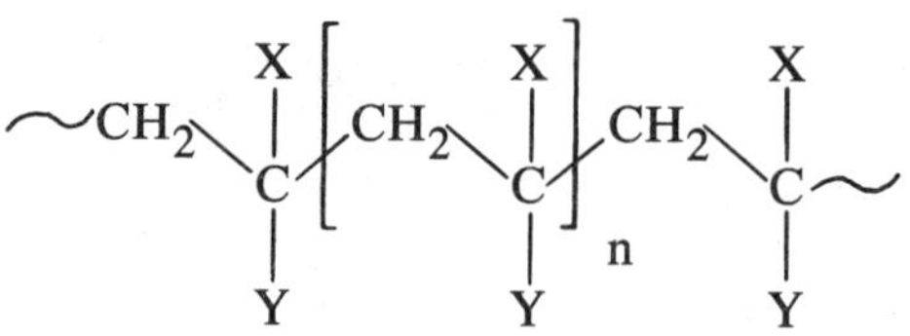


undergo main-chain cleavage, and thus suffer a molecular weight reduction upon irradiation.

whereas:

B) Polymers with teriary carbon atoms in the backbone, provided they contain no quarternary carbon atoms, *i.e.*, vinyl polymers

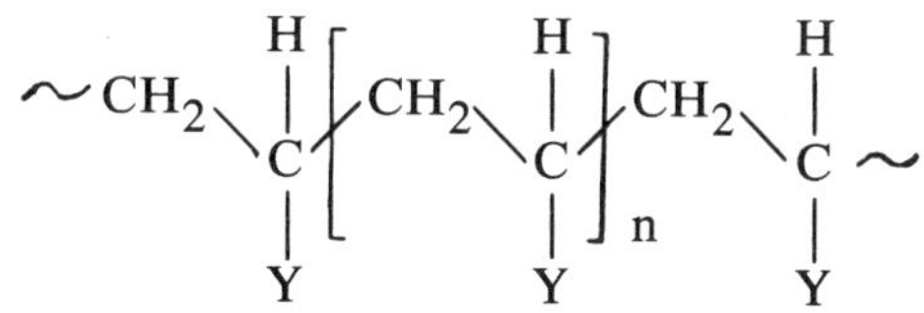

undergo crosslinking reactions, which increase their molecular weight.

Accordingly, degrading polymers, *i.e., vinylidene polymers*, represent *positive* e-beam resists, whereas crosslinking polymers, *i.e., vinyl polymers* act as negative resists.

Vinylidene polymers were first employed as positive e-beam resists by Haller and coworkers,[97] who compared the e-beam induced degradation of poly(isobutylene), poly($\alpha$-methyl styrene) and poly(methyl methacrylate). These authors found no major differences in sensitivity among the three materials, which typically required a minimum exposure at 10 kV of $5 \cdot 10^{-5}$ to $10^{-4}$ Coulomb/cm$^2$. At that time poly(methyl methacrylate) was found to have the best overall properties and thus became a "classic" e-beam resist. Poly(isobutylene) gave rise to inferior resolution due to its rubber-like properties.

Early semiconductor processing conditions required that e-beam resists should be resistant to chemical etchants, for example those made up of aqueous solutions of HF or $NH_4F$. Even though dry etching techniques were known during the early evaluations, for example ion etching as outlined by Davidse,[99] it was not until the advent of plasma etching,[100] that the requirements for an ideal e-beam resists changed from resistance to wet etchants to plasma etching resistance.

Composite positive e-beam resists, which work according to the dissolution inhibitor principle, are, for example, the conventional positive photoresists, such as the standard naphthoquinone diazide systems[102] AZ1350J, AZ2400 (both of Shipley), as well as Polychrome PC-129 and Kodak 809. However, some other naphthoquinone diazide derived positive photoresists yield negative tone pictures upon e-beam exposure,[102c] for example, the Microline resist PR435 of GAF, New York.

To elucidate this seemingly inconsistent behavior, Pacansky and Coufal[103] have studied the e-beam induced reactions of the parent o-naphthoquinone diazide and found that like photodecomposition, the e-beam induced degradation of the diazide follows the Süs[4] mechanism. It has to be noted, however, that e-beam exposure is typically conducted in a vacuum system; therefore, a key ingredient for the formation of indenecarboxylic acid, namely water, is only scarcely available, which increases the probability of reactions between the intermediate ketene and the resin. Hiraoka and Gutierrez[102c] showed that plain (*i.e.,* unsensitized) phenolic or cresylic formaldehyde resins undergo cross-linking reactions when exposed to electron beams at a dosage higher than $10^{-3}$ $C/cm^2$. They also found that added photoactive compounds tend to sensitize this cross-linking reaction. The radiation induced decomposition of the photoactive compounds, however, is typically the dominating reaction. Phenol- or 4-hydroxy-benzophenone derivatives of o-naphthoquinone-diazide-5-sulfonylchloride derivatives rendered the resins positive e-beam resists, whereas derivatives of aniline or 4-aminobenzophenone (such as the photoactive compounds in the Microline PR-435 *positive* photoresist) rendered the resins *negative* e-beam resists.[102c] Bowden and coworkers[6] have pointed out that a radiation sensitive dissolution inhibitor need not be photosensitive. They used a second polymer as a dissolution inhibitor, namely poly(2-methyl-1-pentene sulfone), itself a very sensitive positive e-beam resist.[79]

Formation of composites such as the polysulfone above and phenolic resins as outlined here, provides a convenient and promising method to custom tailor the properties of resists, such as the reactive ion etching resistance in this case. The polysulfone/novolac resist system namely unites the high radiation sensitivity of the poly(olefin sulfones) with the superior plasma resistance properties of the novolac resin family.

A third approach to positive e-beam resists is also based upon a composite material, made up of a polymeric host resin and a radiation sensitive additive. Here, however, the radiation sensitive compound does not only act as a dissolution inhibitor, rather its irradiation generated composition product, namely an acid, also induces the acid catalyzed cleavage of the host resin. A system of this type has recently been disclosed by Itoh and coworkers[13] whereby the matrix resin is a polydialdehyde, for example a poly(phthalaldehyde) with acetate end groups. The radiation sensitive compound is typically an oxonium or sulfonium salt, such as GE 1014, but could in principle be any precursor, which upon irradiation yields a strong Lewis or Bronstedt acid. The obvious disadvantage of these systems is their pronounced sensitivity to acids which disqualifies them for most wet etching techniques and thus limits their use to plasma etching conditions. A major factor in the quest for a more sensitive e-beam resist is derived from the fact that e-beam lithography exposes only one spot at a time, whereas in photolithography all spots are simultaneously illuminated in a flood exposure approach. Therefore, for e-beam lithography (with its higher costs for exposure systems) to be competitive with photolithography, the exposure time of any given spot has to be very short, which of course translates into high sensitivity of the e-beam resists. The desire for more sensitive e-beam resists has initiated a variety of development efforts, which have followed rather different pathways. Of these a few characteristic ones will be discussed in more detail in the following.

2. Modified Poly(methyl methacrylate)

Efforts to improve the adhesion properties of PMMA to metal substrates led to the evaluation of methyl methacrylate/methacrylic acid copolymers as e-beam resists. During these investigations it was accidentally observed that the e-beam sensitivity of these copolymers increased upon heat treatment prior to exposure.[105] This finding could subsequently be correlated with the formation of anhydride structures in the infrared spectra.[106] The anhydride formation was shown to give rise to six-membered rings from two acid groups of two adjacent methacrylic acid monomer units. These glutaric anhydride (GA) type structures have a high cross section for attachment of slow secondary electrons. In this context, it has to be remembered that roughly 50% of the secondary electrons generated from a 25 keV electron beam have energies of less than 4 eV.[106] The GA model compound has been previously investigated[107] and has been shown to decompose according to Scheme 5.

Scheme 5: Decomposition of Glutaric Anhydride Structures

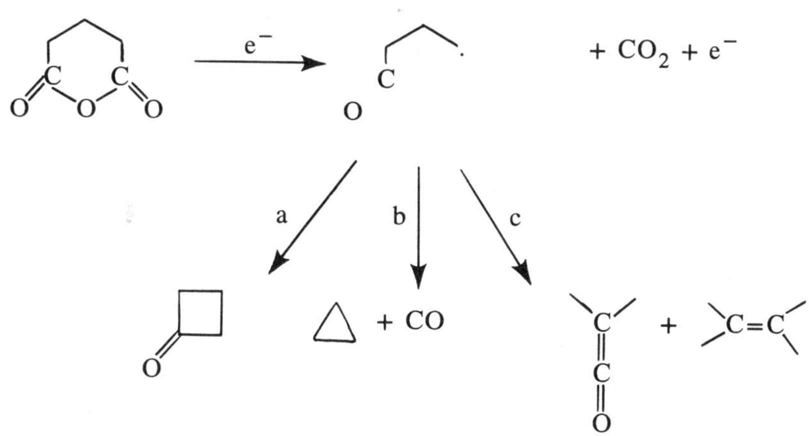


| | | | |
|---|---|---|---|
| GA: | 54% | 45% | <1% |
| Polymer: | — | — | all |

Thus GA yields mainly cyclobutanone by closure of the 1,4-biradical intermediate.

When incorporated into a polymer backbone, however, the six-membered anhydride ring yields the 1,4-diradical intermediate as well, but its subsequent reactions differ remarkably from those of the model compound GA. Thus only cleavage occurs and no cyclobutanone or cyclopropane structures can be found in the irradiated polymers. It appears that the strain characteristic for polymeric chains is the reason for the different reaction pathways followed by the 1,4-biradicals as a function of their substituent.

The high cross section of the anhydride ring for dissociative electron attachment is attributed to a very favorable match between the geometry of the intermediate $CO_2^-$ ion and the O–C–O unit in the six-membered GA units. Similarly the isoelectronic ions $OCS^-$, $CS_2^-$, as well as $H{-}NCO^-$ have also been predicted by molecular orbital calculations[108] to be bent, and accordingly cyclic structures of the type

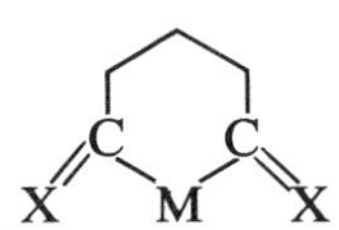

where X=0 or S and M=0, S, N-H or N-alkyl might all show a favorable cross section for dissociative electron attachment just like the case X=M=O outlined above.

We have succeeded in obtaining the correspondingly modified poly(methacrylic anhydrides) by heating the anhydride containing polymers in the presence of $MH_2$ with M=S, N-H or $NCH_3$. Thereby water is eliminated (preferably under vacuum) and the anhydride oxygen is replaced by M. Whereas the polymers with M=S might qualify as X-ray resists, those with M=NH or $NCH_3$ show an improved glass-transition temperature (Tg) and hence a higher temperature stability. These materials have been evaluated as high-temperature e-beam resists.[109]

Even poly(acrylic acid), (PAA), which is a negative, *i.e.*, a crosslinking resist, becomes a positive material[110] upon heating because of the formation of six-membered anhydride structures. The imide derived from PAA tends to crosslink, whereas the N-$CH_3$ modified material degrades upon e-beam exposure. Additional insights into the events and changes occurring during the heating step of methyl methacrylate, methacrylic acid copolymers was provided by Roberts.[111] He found that baking of acid containing copolymers gives rise to interchain crosslinks, which render the polymers completely insoluble in all organic solvents:

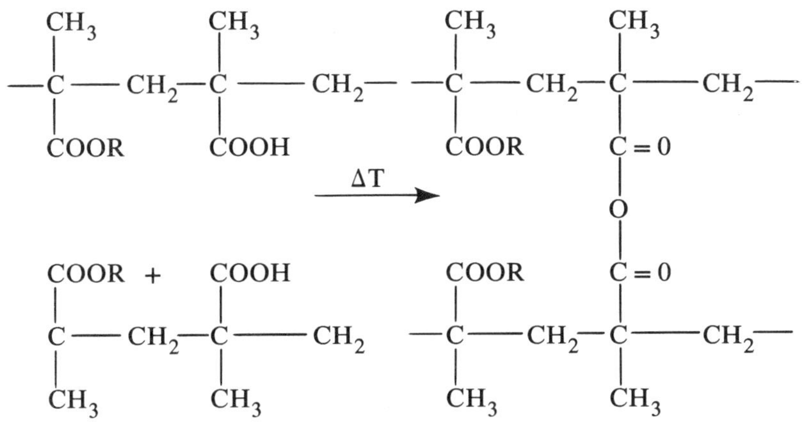

However, the chemical bonds joining the anhydride group, to the main chain are particularly susceptible to rupture by ionizing radiation. Consequently e-beam exposure brakes the crosslinkages and restores the straight chain structure and solubility. Therefore, the exposed parts of the polymer become soluble again in developers such as methyl isobutyl ketone. Mixtures of two copolymers, namely poly(methyl methacrylate-co-methacrylic acid) and poly(methyl methacrylate-co-methacryloyl chloride) give superior results and prove the correctness of the above concept. The sensitivity of the resulting resist depends upon the concentration of the anhydride groups and thus upon the composition of the copolymer mix as well as on the prebake conditions. Sensitivities between 2 to 20 $\mu C/cm^2$ have been reported.[111]

Another approach to more sensitive methacrylate type resists assumed that the e-beam induced decomposition of PMMA follows Scheme 1.[112]

Scheme 6: Mechanism of radiation induced scissioning of PMMA.[112]

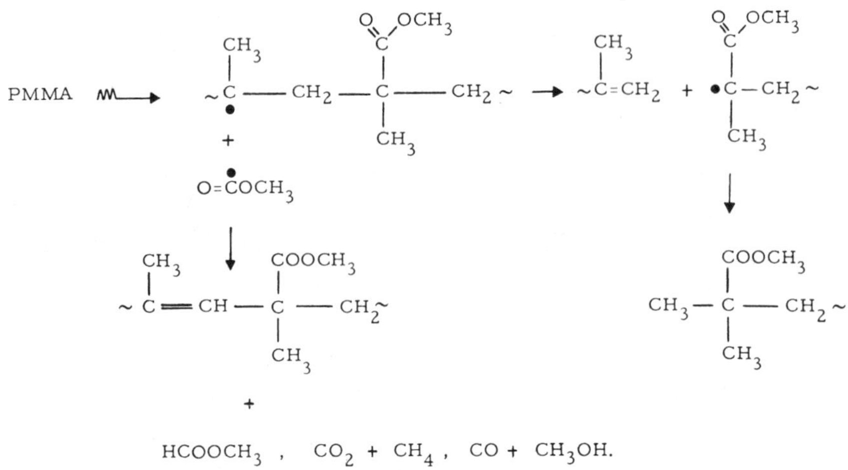


Extrapolation to more sensitive materials led to an investigation of poly(methyl methacrylates) with the following structure:

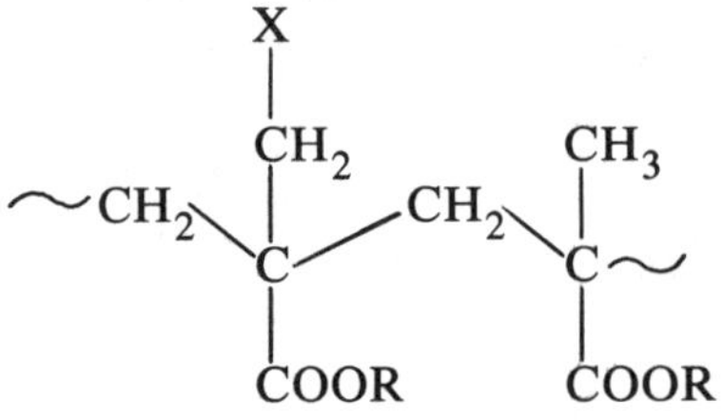


A variety of these were synthesized, where $X^-$ was a better leaving group than $H^-$. As a guideline for finding more suited groups X served the appearance potential for $X^-$ from compounds of the type $CH_3$-$CH_2$-X, which had previously been determined by others.[113] Table 3 shows the appearance potential (AP) and the observed increase relative sensitivities partially modified PMMA as well as those of preheated copolymers from methacrylic acid (25%) and methyl methacrylate (75%).

**Table 3**

**Relative Speed of Modified Homo and Copolymers**

| X | AP[eV] | PMMA | Copolymer |
|---|---|---|---|
| H | 9.5 | 1 | 1 |
| Cl | 3.5 | 3 | 4 |
| F | 2.0 | 5 | 4 |
| $NO_2$ | 0.5 | 10 | 10 |

Accordingly, partially chlorinated PMMA is about 3 times more sensitive than PMMA itself. Other poly(methacrylates) appear to follow Scheme 1 as well, more of less independently upon the nature of their side groups.

3. Poly(olefin sulfones)

Poly(olefin sulfones) are essentially alternating copolymers between sulfur dioxide and an olefin. They have been found to undergo degradation upon $\gamma$-irradiation,[114] with a G (scission) of 10-12, a value which is far higher than that of any other known positive resist. Some of the poly(olefin sulfones) give rise to poor film properties, but others, notably poly(butene-1-sulfone) and poly(2-methyl pentene-1-sulfone) make excellent positive e-beam resists[115] with sensitivities of 1 to 3 $\mu C/cm^2$.

In part the extraordinary sensitivity of these materials has been attributed to the tendency of a broken polymer backbone to spontaneously split off monomer units. This unzipping of the polymeric chain thus enhances the effect of the radiation. In the extreme case the depolymerization process volatilizes the exposed resists, thus causing an auto-development process, which has been named "vapor development" by Thompson and Bowden.[115a]

When poly(olefin sulfones) are imbedded into novolac resins to serve as e-beam sensitive dissolution inhibitors, they seem to transfer part of their absorbed energy to the novolac resin. This process could lower their effectiveness and thus decrease their sensitivity. The details of this mechanism are not fully understood.

4. Other Systems

A multitude of other polymers has been used as e-beam resists, among them poly(itaconic anhydride),[116] poly(pivalolactone),[117] substituted polystyrenes,[118] poly(methacrylonitriles),[119] *etc.* The dominating positive systems, however, seem to be PMMA and derivatives, polysulfones and novolac based photoresists. Polysiloxanes are gaining rapidly in significance as negative e-beam resists.

## VIII. DUAL EXPOSURE RESIST SYSTEMS

In order to take advantage of the high resolution patterning capability of e-beam lithography without sacrificing the higher throughput of photolithography, a combination of photo- and e-beam exposure has been proposed. This mixed lithography approach can be achieved using a multilayer resist system with specialized resist layer for photo- and e-beam lithography, respectively. The added complexity resulting from different resists, however, favors the Same-Level Mixed Lithography (SLML) process proposed by Berker and Bernacki.[120] The SLMS uses the same resist for both e-beam and photoexposure in one layer and requires only one development step. Typically the small geometries on any given level are exposed using an electron beam, while the large geometries of the *same* level are defined via optical lithography. Using diazoketone based resist systems both positive (Shipley AZ3000) or negative (AZ2415) e-beam images can be obtained, which allows a high degree of flexibility in the choice of the exposure mode of a complex pattern. Gate lengths as small as 0.6 $\mu$m have been transferred into doped polysilicon via anisotropic plasma etching. The virtues of the SLML technique are that it combines the high resolution capability of e-beam lithography with the high throughput of optical lithography while maintaining a simple process with only one coating, one development and one transfer step. The drawbacks, however, are the same that have limited the widespread use of e-beam lithography in general. Due to the recent progress especially in the area of resolution via deep UV photolithography and because of improved linewidth control with optical aligners in multiple resist structure, the SLML process is not likely to find widespread application in the immediate future in areas other than mask fabrication.

## IX. MULTILAYER RESIST

Frequently the opposing demands which are being put on a lithographic resist cannot be simultaneously met. This dilemma has led to the concept of a multilayer resist structure, whereby any given layer has a specialized role. Multilayer resist systems (MRS)[121] offer the possibility of increasing the resolution of almost all lithographic tools since only a *thin* top layer of pattern defining resist is needed. This top layer is put onto one or more relatively thick base layers, which may be optimized to meet the other requirements of microelectronics. Thick resist layers are typically required since a decrease of the lithographic dimensions within the wafer plane have to be compensated by thicker metal conductors to maintain an acceptable resistance. For example, in logic circuits, as Hatzakis[121] has pointed out, as the width of the first metal interconnections approaches one micrometer, the metal thickness must be around one micrometer as well in order to keep the current density in the conductor at acceptable levels and to avoid electromigration effects. If these metal structures are obtained via lift-off techniques, the resist thickness must be higher than the metal thickness, which requires a resist thickness of >1 $\mu$m. In higher levels of logic and memory VLSI circuit topography generated below by, for example, double polysilicon in MOS technology or by crossing metal lines requires planarization, which in principle can be achieved by relatively thick (~2 $\mu$mm) resist layers. Very uneven resist layers, however, typically give rise to very poor resolution, since the low lying areas remain either underexposed or the top part is being overexposed. The aspect ratios of such thick resist patterns are further deteriorated by the long exposure times required. Similarly, in electron beam lithography, electron backscattering in the substrate and forward scattering in thick resist deteriorates resolution and gives rise to the so called proximity effect. If the thickness of the radiation sensitive layer is kept low, the consequences of back and forward scattering can be compensated much more effectively and, therefore, the resolution increases. Development times can be kept relatively short, and the uniform thickness of the top layer provides for better pattern definition as well.

The improvement of the contrast of optical tools using thin resist top layers has made it possible to extend the range of optical aligners and linewidth control to submicrometer dimensions (at 405 nm wavelength of the illuminating light). However, linewidth control at this level is only achievable with state-of-the-art optical aligners, provided that resist thickness variations across surface topography as well as reflections from buried metal lines or the wafer surface are narrowly controlled or eliminated.

All of these demands are much better met using multilayer resist structures. The existing multilevel systems can be divided into two categories, namely those where at least two layers are used as resists which are sequentially exposed and developed, or those which have a single resist layer.

1. The Portable Conformable Mask Scheme

The simple two-resist layered system has become known as the Portable Conformable Mask (PCM) scheme as proposed by Lin.[122] In this scheme the thin top layer is the pattern definition layer, which is optimized for resolution and convenience of handling. After exposure and development of the top layer its pattern is transferred to the thicker base layer via UV flood exposure. Early resist combinations were Shipley AZ1350 photoresist on top of polymethacrylate (PMMA), but other combinations such as Hunt MPR photoresist or Kodak 809 on top of PMMA have also been tried successfully[123] and used in production environments.[124] The advantage of the PCM process is that it can be done with conventional equipment. In addition to the handling of a simple resist system, UV-flood exposure is required, which can be achieved with either Xe-Hg short arc or microwave powered Hg sources. Incompatibilities of the different resist layers give rise to solubility and excessive cross linking in the pattern definition layer.

2. Single Resist Multilayer Systems

The success of the dry etching techniques, in particular the anisotropic etching capability of Reactive Ion Etching (RIE)[100] has made it possible to transfer the pattern from the thin top layer to lower thick layers which do not have to be either photo- or radiation-sensitive. This means that the bottom layer can be optimized for processing qualities such as high temperature stability, adhesion to the substrate, low plasma etch rate, long term stability, thermal expansion coefficient, *etc.*

The concept of pattern transfer from the thin top layer to thicker base layers requires anisotropic etching, which can be achieved in a parallel plate (diode) reactor[100] using an oxygen plasma.

Most organic polymers etch relatively fast in an oxygen plasma and etch rates of ~100Å/min are typical. Therefore, the top resist layer is also attacked and removed during the pattern transfer step in the oxygen plasma. To circumvent this difficulty, either an oxygen plasma resistant interface layer below the thin top resist layer has been used, *i.e.,* a three layer system, or the top layer has to be plasma resistant itself. The first approach introduces more complexity, the second requires new resist development.

The first resist system using reactive ion etching was proposed by Havas[125] in 1976. It was a trilevel system with an organosilicon layer as an interface between the thin top photoresist and the thick polymeric base layer, which had been baked to cause crosslinking. Upon exposure and wet development of the top layer the pattern was first transferred to the organosilicon layer via a $CF_4$ plasma. Subsequently the etching gas was changed to $O_2$ and the pattern transferred into the thick bottom layer. This step was accompanied by simultaneous removal of the top resist layer in the oxygen plasma, whereas the organosilicon layer acted as an etch stop and protective layer on top of the bottom layer. Since organosilicon polymers are virtually oxygen plasma resistant, the bottom layer can be over-etched to yield a profile suitable for lift-off. A related approach introduced by Morau and Maydau used plasma deposited silicon dioxide as an etch stop and interface layer, with a negative X-ray resist on top and 200°C baked Hunt photoresist

HPR-204 as the base layer. Pattern transfer from the top to the bottom layer was achieved again in two steps, namely first in a $CHF_3$ plasma, which transferred the pattern to the $SiO_2$ interface, and subsequently the final transfer to the thick base layer was achieved in an oxygen plasma.

### 3. Two Layer Single Resist Systems

If an organosilicon polymer can be used as the top pattern definition layer, two layers are sufficient for the RIE based pattern transfer. Organosilicon polymers are suitable e-beam resist, working in the negative mode, as described by Gazard *et al.*[126] and by Roberts.[127] As has been shown by Hatzakis *et al.*[121] both poly(vinyl methyl siloxane), PVMS, and poly(dimethyl siloxane), PDMS, function as very sensitive and high contrast negative e-beam resists on top of a thick layer of Shipley AZ1350J photoresist, into which the developed pattern can be transferred in a single oxygen plasma step. The same system can also be used in a deep-UV photolithographic mode, using 220 nm light for the exposure of the polysiloxane. In this mode 0.4 $\mu$m wide lines could be clearly defined and transferred into the Shipley photoresist base layer by RIE using an oxygen plasma. These results exemplify the possibilities of deep UV lithography very nicely, and accordingly, direct write e-beam processes may not become essential because of resolution reasons all too soon. E-beam lithography will retain its significance, however, for mask making purposes and where personalization of chips is required.[128]

The advantage of single resist two layer system has not yet been fully exploited, since the commercially available polysiloxanes such as PVMS and PDMS are not sensitive to light in the 300 to 400 nm range, and thus cannot yet be used in conventional optical exposure tools. More material development of photosensitive polysiloxanes is, therefore, required and corresponding efforts are underway.

## X. DRY DEVELOPING ORGANIC SYSTEMS

Resist systems which can be developed via heat treatment or oxygen plasma etching open up the possibility of a dry developed multilayer system.[129,130] In such cases typically the base and resist layer are still put on via wet steps by spinning on a solution of the different polymer systems. This spin coating is typically desirable primarily because the thick base layer especially is supposed to planarize underlying topography.

## XI. ALL DRY SYSTEMS

Certain dry developing systems and certain base layers, among them polyimides, can also be *deposited* from the vapor phase. In this case all the processing steps can be accommodated in a vacuum system without any need for either wet spin coating, development or stripping, thus allowing an all-dry processing procedure. Apart from having environmental advantage, an all dry process also avoids swelling of the polymeric layers or retention of solvent in permanent polymeric insulators. Planarization, however, cannot easily be achieved in an all-dry approach.

## XII. INORGANIC RESISTS

Among the resist which can relatively easily be deposited from the vapor phase are certain inorganic resists. Typically such systems consist of thin layers of $As_2S_3$ or multiple layers of $AgCl/As_2S_3$[131] or $Ag_2Se/GeSe$.[132] These layers are sensitive to UV, electron beams or X-rays and can either be developed dry by plasmas or wet by suitable developers.

## XIII. CONTRAST ENHANCEMENT LAYER

An alternative approach to increase the maximum resolution attainable with optical projection aligners uses a photobleachable top layer, also called contrast enhancement layer (CEL).[133] The photobleachable layer increases the ratio of a resist's threshold close to full exposure dose and thereby reduces the minimum contrast that will produce a usable pattern. The minimum contrast threshold required to form a usable pattern is a characteristic parameter for each resist/developer combination. The presence of a thin (1000 to 3000Å) photobleachable top layer increases the contrast of the illumination that effectively reaches the photoresist. Since exposure of the photoresist occurs only after bleaching of the top layer, the exposure times have to be increased two- to three-fold. The chemical composition of the CEL is currently still undisclosed. In principle this concept of a photobleachable top layer could be combined with a multilayer resist structure, but the resulting level of complexity will favor an approach with as few layers and processing steps as absolutely required to achieve the desired dimensions. Like in the case of organosilicon resist top layers these systems can be used without an interface layer and thus allow considerable processing convenience, especially since they represent an all-dry system.

However, multilayer systems solve many other problems, since they allow to obtain submicron dimensions with optical tools and photoresists.

They circumvent the need for a permanent resist, since polyimide can function superbly in this capacity when used as the base layer, thus no photosensitive polyimide is required; they planarize topography; they allow linewidth control, *etc.* Altogether the concept of multilayer structure effectively separates the variables and allows the utilization of previously known and well characterized materials as *composites*. The concept of a composite material has proven very powerful in other areas of technology where polymers are used, such as glass fiber reinforced epoxides or filled polymers, since in this way complex materials requirements can be met via superimposition of individual strengths of the known components, which make up the composite.

## XIII. MONOMOLECULAR RESISTS

Long chain fatty acids are known to form monolayers at liquid/gas interfaces.[134] From there they can conveniently be transferred, monolayer by monolayer, to yield films of controlled thickness. This so-called Langmuir-Blodgett technique[135] is the basis for commercially available Langmuir troughs, in which sample are dipped repeatedly into appropriate solutions to yield films of predetermined thickness in a fully automated fashion. Langmuir monomolecular films can be obtained as ultrathin compact films free of pinholes, which adhere well to a variety of substrates due to their polar end groups.[136]

Functionalized long-chain fatty acids, for example ω-tricosenoic acid, $CH_2{=}CH{-}(CH_2)_{20}COOH$, possess a polymerizable group at one end, which upon exposure to 4 kV electrons can be crosslinked, and thus they render these monomolecular layers the property of a negative e-beam resist[137] with a sensitivity of about 50 $\mu C/cm^2$. The advantage of such systems lies in the controlled thickness of the applied layers, and in their potential to achieve very high resolution in the e-beam mode. Resolution of better than 600Å has been demonstrated using 5 keV electrons.

The etch resistance of monomolecular resists to wet etchants, in particular to phosphoric acid, ammonium fluoride, and hydrofluoric acid, is very high. The etch rate in a $CCl_4$ plasma, used for aluminum etching, is only 60Å/min, where the etch rate of Al is 400Å/min.[137] Related systems containing olefinic and acetylenic bonds in the fatty acid chains have also been investigated by C. A. Brown and T. Wu and similar observations have been made.[138]

Monomolecular resists, though belonging to negative resist category, afford a very high degree of resolution in the e-beam mode, whose potential has not been fully exploited. Especially in combination with multilayer concepts they hold great promise for future applications.

## REFERENCES

1. A. Senefelder, "Vollständiges Lehrbuch der Steindruckerey," Karl Thienemann, Munich and Karl Gerold, Vienna (1818), p. 137.

2. H. W. Vollmann, *Angew. Chem., Int. Ed. Engl.* **19**, 99-110 (1980), and references therein.

3. a) J. Kosar, "Light Sensitive Systems," Wiley, New York (1965); b) K. I. Jacobensen and R. E. Jackson, in "Imaging Systems," Focal Press, London (1976), p. 173-222; c) P. Ziegler, *Polygraph* **32**, 1102 1600 (1979).

4. O. Süs, *Justus Liebigs Ann. Chem.* **556**, 65 (1944).

5. M. J. Bowden and L. F. Thompson, *Solid State Technol.* 72-82 (May 1979).

6. M. J. Bowden, L. F. Thompson, S. R. Fahrenholtz and E. M. Doerris, *J. Electrochem. Soc.* **128**, 1304-1313 (1981).

7. J. K. Hasson and H. G. Sarkary, *Solid State Technol.* 49-54 (May 1982).

8. D. Seitzer, this volume.

9. a) E. D. Feit, M. E. Wurtz and G. W. Kammerlott *J. Vac. Sci. Technol.* **15**, 944 (1978); b) R. D. Heidenreich and G. W. Kammerlott Proceedings of the 4th Photopolym. Conf., Soc. Plast. Eng, Ellenville, New York (1976), 94.

10. A. C. Ouano, *Polym. Eng. and Sci.* **18**, 306-313 (1978).

11. I. Haller, M. Hatzakis and R. Srinivasan, *IBM J. Res. Dev.* **12**, 251-256 (1968).

12. E. Kay, this volume.

13. D. G. Borden, Proc. Reg. Techn. Conf., Soc. Plast. Eng., Mid Hudson Sect., Ellenville, New York (1973), pp 10-19, and references therein.

14. L. A. Pederson, *J. Electrochem. Soc.* **129**, 205 (1982); G. N. Taylor and T. M. Wolf, *ibid.* **127**, 2665-2674 (1980).

15. B. Broyde, *J. Electrochem. Soc.* **117**, 1555-1556 (1970).

16. C. M. Melliar-Smith and J. Mogab, in: "Thin Film Processes," J. C. Vossens and W. Kern, eds., Chapter V-2, Academic Press, New York (1978).

17. a) T. S. Chang, D. F. Kyser and C. H. Ting, *IEEE Trans. Elec. Dev.* **ED28**, 1295-1300 (1981); b) W. Oldham, S. Handgaonkart, A. Neureuther and M. O'Toole, *ibid* **ED26**, 717 (1979).

18. a) M. S. Dinabury, "Photosensitive Diazo Compounds," Focal Press, New York (1964); b) W. S. DeForest, "Photoresist Materials and Processes," McGraw-Hill Book Co., Inc., New York (1975); c) L. F. Thompson and R. E. Kerwin, *Ann. Res. of Mat. Sci.* **6**, 267 (1976); d) E. D. Feit, "UV Curing: Science and Technology," ACS Symposium Series 151, S. P. Pappas, ed., Pub. Technology Marketing (1978); e) R. K. Watts, in "Very Large Scale Integration, VLSI," G. Ecker, W. Engl and L. B. Felsen, eds., Springer-Verlag, Berlin, Heidelberg, New York (1980), pp. 42-86.

19. L. A. Kazitsyna, B. S. Kikot and A. V. Upadysheva, *Russian Chem. Rev.* **35**, 388 (1966).

20. L. C. Anderson and M. J. Roedel, *J. Am. Chem. Soc.* **67**, 955 (1945).

21. G. Meloy, "The Structure of Diazo Ketones," Ph.D. Thesis, University of Cincinnati (1965).

22. F. H. Dill and J. M. Shaw, *IBM J. Res. Dev.* **21**, 210-218 (1977).

23. J. M. Shaw, MJ. A. Frisch and F. H. Dill, *IBM J. Res. Dev.* **21**, 219-224 (1977).

24. a) O. Süs, H. Stephen and R. Dietrich, *Liebigs Ann. Chem.* **617**, 20 (1958); b) O. Süs, J. Munder and H. Stephen, *Angew. Chem.* **74**, 985 (1962).

25. H. A. Levine, Polymer Preprints, Division of Polymer Chemistry, American Chemical Society, Washington, DC, Vol. 19, 1 (1969).

26. P. Yates and E. W. Robb, *J. Am. Chem. Soc.* **79**, 5760 (1957).

Applicaton," E. D. Feit and C. Wilkins, Jr., eds., ACS Symposium Series 184, 1-17 (1982).

112. a) A. R. Schultz, *J. Polym. Sci.* **35**, 369 (1959); b) H. Hiraoka, *IBM J. Res. Dev.* **1977**, 121, and references therein.

113. J. G. Dillard, *Chem. Rev.* **73**, 589 (1973).

114. J. R. Brown and J. H. O'Donnell, *Macromolecules* **5**, 109 (1972).

115. a) L. F. Thompson and M. J. Bowden, *J. Electrochem. Soc.* **120**, 1722-1726 (1973); b) L. Stillwagon, in "Polymer Materials for Electronic Applications," E. D. Feit and C. Wilkins, Jr., eds., ACS Symposium Series 184, pp. 19-27 (1982); c) H. Stillwagon, *Organic Coatings and Plastics Preprints* **43**, 236-242 (1980); d) T. Bowmer and J. O'Donnell, *J. Polym. Sci., Polym. Chem. Ed.* **19**, 45 (1981); e) T. Bowmer, J. O'Donnel and P. Wells, *Makromol. Chem. Rapid Comm.* **1**, 1 (1980); f) M. J. Bowden, L. F. Thompson, S. R. Fahrenholtz and F. M. Doerris, *J. Electrochem. Soc.* **128**, 1304 (1981).

116. J. Bargon and H. Hiraoka, *IBM Techn. Discl. Bull.* **19**, 3208 (1977).

117. J. Bargon, H. Hiraoka and L. W. Welsh, Jr., *ibid.* **19**, 3207 (1977).

118. a) E. D. Feit and L. E. Stillwagon, Soc. Plast. Eng., Mid Hudson Sect., Reg. Tech. Conf., Ellenville, New York, Oct. 1979, p. 68; b) H. Shiraishi, Y. Taniguchi, S. Horigone and S. Nonogaki, *ibid.* p. 56.

119. J. Bargon, H. Hiraoka and L. W. Welsh, Jr., U.S. Patent 4,389,482 (1983) to IBM.

120. T. D. Berker and S. E. Bernacki, *IEEE Trans. Electr. Dev. Lett.* **EDL-2**, 281 (1981).

121. a) M. Hatzakis, *Solid State Technol.* **24**, 74-80 (1981); b) M. Hatzakis, J. Paraszczak and J. Shaw, Proc. Int'l Conf. Microlithogr. (1981), p. 386.

122. a) B. J. Lin, "Development in Semiconductor Microlithography," SPIE (1979), pp. 114-121; b) B. J. Lin and T. H. P. Chiang, *J. Vac. Sci. Technol.* **16**, 1669-1671 (1979).

42. a) W. A. Keulgen, *Encyclopedia Polymer Science and Technology* **10**, 1-73 (1976); b) A. Knop and W. Scheib, "Chemistry and Application of Phenolic Resins," Springer-Verlag, Berlin, Heidelberg, New York (1979).

43. M. Koebner, *Angew. Chem.* **18**, 251-256 (1933).

44. a) H. Kämmer and W. Rausch, *Makromol. Chem.* **18/19**, 9 (1956), and **24**, 152 (1957); b) H. Kämmer and M. Gropmann, *Chem. Ber.* **86**, 1492 (1953); c) H. Kämmer, *Angew. Chem.* **70**, 316 (1958).

45. R. F. Hunter, R. A. Morton and A. T. Carpenter, *J. Chem. Soc.* 441 (1950).

46. E. Gipstein, A. C. Ouano and T. Tompkins, *J. Electrochem. Soc.* **129**, 201-205 (1982).

47. G. P. Gardini, Univ. of Parma, Italy, private communication.

48. P. N. King, R. H. Mitchell and A. R. Westwood, *J. Appl. Polym. Sci.* **18**, 1117 (1974).

49. a) J. C. Woodbery, H. P. Higginbottom and H. M. Aulbertson, *J. Polym. Sci.* **3**, 1079 (1975); b) L. Pospisil and M. Nauratil, *Int. Polym. Sci. and Technol.* **6**, 5 (1979), T/79 and **6**, 649 (1979), T42.

50. U.S. Patent 3549369, IBM Corporation (December 22, 1970).

51. H. Hiraoka and J. Pacansky, *J. Electrochem. Soc.* **128**, 2645-47 (1981), and *IBM Techn. Discl. Bull.* **24**, 1468 (1981).

52. W. F. Cordes, paper presented at the SPIE Microlithography Technology for the Mid-1980's Seminar, Santa Clara, California, March 31 and April 1, 1982.

53. S. C. Stinson, *Chem. and Eng. News* **61**, August 8, 7-12 (1983).

54. a) S. A. MacDonald, R. D. Miller, C. G. Willson, to be published; b) S. A. MacDonald, R. D. Miller, C. G. Willson, G. M. Feinberg, R. T. Gleason, R. M. Halverson, M. W. MacIntyre and M. W. Motsiff, Kodak Microelectronics Seminar, Interface 83, San Diego, California, October 1982.

55. K. G. Clarke, in Negative Photoresists, "Non-Silver Photographic Processes," R. J. Cox, ed., Academic Press, London-New York (1975), pp. 249-273.

56. W. Kornfeld, "Photoresist Technology Improvements," Contract No. AF33 (615)-1430, Project No. 4460, Task 446002, May (1966).

57. M. Harita, M. Ishikawa, K. Harada and T. Tsunoda, *Polym. Eng. and Sci.* **17**, 372 (1977).

58 J. J. Sagura and J. A. Van Allen, U.S. Patent 2,940,853 (1960).

59. R. F. M. Thornley and T. Sun, *J. Electrochem. Soc.* **112**, 11, 1151 (1965).

60. A. Reiser, H. M. Wagner, R. Marley and G. Gowes, *Trans. Faraday Soc.* **63**, 2403 (1967).

61. H. T. Thomas, *Proc. Kodak Microelectron Seminar* **89**, 89-92 (1974).

62. K. Neisius and H. J. Merrem, *J. Electronic Materials* **11**(4), 761 (1982).

63. H. Zügel, this volume.

64. a) R. Rubner and E. Kuhn, *ACS Div. Org. Coatings and Plast. Preprints* **37**(2), 118 (1977); b) R. Rubner, *Siemens Forsch. und Entwickl. Ber.* **5**, 92-97 (1976); c) R. Rubner, W. Bartel and G. Bald, *ibid.* **5**, 235-239 (1976); d) W. Kleeberg, R. Rubner and E. Kühn, German Patent 2,130,904 (1974); e) R. Rubner, H. Ahne, E. Kühn and G. Koldozieji, *Photograph. Sci. Eng.* **23**(5), 303 (1979).

65. a) Selectilux HTR-2, Preliminary Data Sheet, EM Chemicals, 5 Skyline Drive, Hawthorne, New York 10532; b) H. J. Merrem, R. King and H. Härtner, "New Developments in Photosensitive Polyimides," E. Merck, Res. and Dev. Dept., Frankfurter Str. 250, D61 Darmstadt-1, West Germany.

66. J. C. Dubois, A. Eranian and E. Datamandi, paper presented at 8th Intl. Conf. on Electron and Ion Beam Sci. and Technol., Electrochem. Soc. Meeting, May 21-26, 1978, Seattle.

67. E. D. Feit, "UV Curing: Science and Technology," S. P. Pappas, ed., Pub. Techn. Marketing (1978).

68. B. J. Lin, *J. Vac. Sci. Technol.* **12**, 1317-1320 (1975).

69. K. Jain, "Sources for UV Lithography," IBM San Jose Research Laboratory, Report RJ3697 (1982).

70. K. Jain, C. S. Willson, S. Rice, L. Pedersen and B. J. Lin, IBM Research Laboratory, Report RJ3737 (1982), and references therein, submitted to *Polym. Eng. and Sci.*

71. W. M. Moreau and P. R. Schmidt, 138th Electrochem. Soc. Meeting, Extended Abstracts, No. 178 (1970).

72. F. W. Dill, W. P. Hornberger, P. S. Hang and J. M. Shaw, *IEEE Trans. Electr. Dev.* **ED22**, 445 (1975).

73. a) E. Reichmanis and C. W. Wilkins, Jr., "Polymer Materials for Electronic Applications," ACS Symposium Series 184, Washington, DC (1982), pp. 29-43; b) E. Reichmanis and C. W. Walkins, Jr., Organic Coatings and Plastics Chem. Preprints, 180th Meeting Am. Chem. Soc. **43**, 243-251 (1980).

74. H. Hiraoka, W. L. Welsh, Jr. and J. Bargon, IBM Research Laboratory, Report RJ3942 (1982), and *J. Vac. Sci. Technol.* **B1**, 1062-1965 (1983).

75. a) J. R. Callum and C. K. Schoff, *Trans. Faraday Soc.* **67**, 2372-2388 (1971); b) R. B. Fox, Prog. Polym. Sci., A.-D. Jenkins, ed., Pergamon Press, 1, Chapter II (1967).

76. N. Grassie, Symposium on Macromol. Chem., Academiai Kiado, Budapest (1971), p. 725.

77. A. Levine, "Photopolymers, Principles, Process and Materials," Soc. Plastics Eng., Tech. Papers (1973), p. 106.

78. M. Tsuda, S. Oikawa, Y. Nakamura, H. Nagata, A. Yokata, H. Nakane, T. Tsumori, Y. Nakane and T. Mifune, *Photographic Science and Eng.* **23**, 290-296 (1979).

79. L. F. Thompson and M. J. Bowden, *J. Electrochem. Soc.* **120**, 1722-1726 (1973).

80. T. Bowmer and J. O'Donnell, *J. Polym. Sci., Polym. Chem. Ed.* **19**, 45 (1981).

81. T. Bowmer, J. O'Donnell and P. Wells, *Makromol Chem., Rapid Comm.* **1**, 1 (1980).

82. J. Pacansky, R. Kroeker, E. Gipstein and A. Gutierrez, IBM Research Laboratory, Report RJ3384 (1982).

83. E. Gipstein, A. C. Ouano and T. Tompkins, *J. Electrochem. Soc.* **129**, 201-204 (1982).

84. B. D. Grant, N. J. Clecak, R. J. Twieg and C. G. Willson, *IEEE Trans. Electr. Dev.* **ED22**, 1300-1305 (1981).

85. A. C. Ouano, *Polym. Eng. Sci.* **18**, 306 (1978).

86. a) T. Iwayanagi, T. Kohashi, S. Nonogaki, T. Matsuzawa, K. Douta and H. Yanazawa, *IEEE Trans. Electr. Dev.* **ED28**, 1306-1310 (1981); b) T. Matsuzawa and H. Tomjoka, *ibid.* **ED28**, 1285-1288 (1981).

87. M. Tsuda, S. Oikawa, Y. Nakamura, H. Nagara, A. Yakota, H. Nakane, T. Tsumori, Y. Nakane and T. Mifune, paper presented at Int. Symp. on Adv. in Photopolym. Sys., Washington, DC (1978), paper summaries, p. 77.

88. H. Hiraoka, and L. Welsh, Jr., IBM Research Laboratory, Report RJ3379 (1982).

89. H. Hiraoka and J. Pacansky, *J. Electrochem. Soc.* **128**, 2645-47 (1981).

90. H. Hiraoka and J. Pacansky, *J. Vac. Sci. Technol.* **19**, 1132 (1981).

91. H. Hiraoka and J. Pacansky, *IBM Techn. Discl. Bull.* **24**, 1468 (1981).

92. P. van Pelt, *Proc. Soc. Photo-Opt. Instr. Eng.* **275**, 150 (1981).

93. R. Allen, P. Foster and Y. T. Yen, *J. Electrochem. Soc.* **129**, 1378-81 (1982).

94. R. F. Leonard and W. F. Cordes, SPIE Opt. Microlith. Technology Symp., Santa Clara, California (1982).

95. R. F. M. Thornley and T. Sun, *J. Electrochem. Soc.* **112**, 1151 (1965).

96. D. O'Sullivan, P. B. Price, K. Kinoshita and C. G. Willson, *J. Electrochem. Soc.* **129**, 811-813 (1982).

97. I. Haller, M. Hatzakis and R. Srinivasan, *IBM J. Res. Dev.* **12**, 251 (1968).

98. H. Harada, O.Kogure and K. Murase, *IEEE Trans. Electr. Dev.* **ED29**, 518-524 (1982).

99. P. D. Davidse, *Vacuum* **17**, 139 (1967).

100. E. Kay, this volume, and references therein.

101. W. Kulcke, this volume and references therein.

102. a)B. Broyde, *J. Electrochem. Soc.* **117**, 1555-1556 (1970); b) J. M. Shaw and M. Hatzakis, *IEEE Trans. Electr. Dev.* **ED25**, 425-430 (1978); c) H. Hiraoka and A. R. Gutierrez, *J. Electrochem. Soc.* **126**, 860-865 (1979).

103. J. Pacansky and H. Coufal, *J. Am. Chem. Soc.* **102**, 410-412 (1980).

104. H. Itoh and C. G. Willson, IBM Research Report, submitted to ACS Symposium Series, "Polymers in Electronics."

105. R. Feder, I. Haller, M. Hatzakis, L. T. Romankiw and E. A. Spiller, IBM Research Center, Yorktown Heights, New York (1974), private communication.

106. L. Kevan and W. F. Libby, *Advances Photochem.* **2**, 183 (1964).

107. H. Hiraoka, *J. Am. Chem. Soc.* **95**, 1664 (1973).

108. A. D. McLean and M. Yoshimine, IBM Research Laboratory, San Jose, California, private communication.

109. a) J. Bargon, E. Gipstein and H. Hiraoka, U.S. Patent 3,964,908 (1976); b) H. Hiraoka, E. Gipstein, J. Bargon and L. W. Welsh, Jr., *J. Appl. Polym. Sci.* **22**, 3397 (1978).

110. J. Bargon, E. Gipstein and H. Hiraoka, *IBM Techn. Disc. Bull.* **18**, 2622 (1976).

111. a) E. D. Roberts, Applied Polym. Symp. No. 23, John Wiley and Sons, Inc., New York, (1974), pp. 87-97; b) E. D. Roberts, *A. C. S. Coatings and Plastics Preprints* **35**(2), 281-286 (1975); *ibid.* **43**, 231-241 (1980); c) E. D. Roberts, in "Polymer Materials for Electronic

27. A. I. Paramonov, Yu M. Prokhotskii and R. D. Erlikh, *Zh. Prikl. Spektrosk.* **27**, 473 (1977).

28. L. Wolff, *Justus Lebigs Ann. Chem.* **394**, 25 (1912).

29. J. Pacansky and D. Johnson, *J. Electrochem. Soc.* **124**, 862 (1977).

30. J. Pacansky and J. R. Lyerla, *IBM J. Res. Dev.* **23**, 42-55 (1979).

31. J. G. Calvert and J. N. Pitts, "Photochemistry," John Wiley and Sons, New York (1967).

32. a) J. Bargon, K. Tanaka and Y. Yoshimine, in "Computational Methods in Chemistry," J. Bargon, ed., Plenum, New York (1980), pp. 239-274; b) Y. Yoshimine and K. Tanaka, *J. Am. Chem. Soc.* (1980).

33. L. Wolff, *Liebigs Ann. Chem.* **325**, 129 (1902); *ibid.* **333**, 1 (1904).

34. G. Schroeder, *Chem. Ber.* **42**, 2336 (1909); *ibid.* **49**, 2697 (1916).

35. A. M. Trozzolo and S. R. Fahrenholtz, Abstracts of the 151st Meeting of the American Chemical Society (1966), K23.

36. A. Padwa and R. Layton, *Tet. Lett.* 2617 (1965).

37. K. P. Zeller, *J. Chem. Soc.* **D**, 317 (1975).

38. a) D. F. Jlten and R. J. Sutton, *J. Electrochem. Soc., Solid State Sci. and Technol.* **119**, 539 (1972); b) D. F. Jlten, *ibid.* **119**, 537 (1972).

39. a) J. M. Shaw and M. Hatzakis, *IEEE Trans. Elec. Dev.* **ED25**, 4 (1978).

40. a) S. Fujimori, *J. Appl. Phys.* **50**, 615 (1975); b) W. T. Tsang, *Appl. Opt.* **16**, 1918 (1977).

41. a) A. C. Livanos, A. Katzir, J. B. Shellan and A. Yariv, *Appl. Opt.* **16**, 1633 (1977); b) S. Austin and F. T. Stone, *ibid.* **15**, 1071 and 2126 (1976); c) R. A. Bartolini, *ibid.* **11**, 1275 (1972), and **13**, 129 (1974); d) M. S. Htoo, *Photogr. Sci. Eng.* **12**, 169 (1968); e) F. J. Loprest and F. A. Fitzgerald, *ibid.* **15**, 260 (1971).

123. T. Batchelder and C. Takemoto, Semiconductor Int'l (April 1982), pp. 213-220.

124. P. S. Burggraf, *ibid.* (June 1983), pp. 48-55.

125. J. Havas, *Electrochem. Soc., Extended Abstracts* **76-2**, 743-744 (1976).

126. M. Gazard, J. C. Dubois and C. Duchesne, *Appl. Polym. Symp.* **23**, 107-116 (1974).

127. E. D. Roberts, *Philips Techn. Rev.* **35**, 41-52 (1975).

128. The polyisilicon top layer by virtue of its e-beam and deep UV sensitivity can function as a dual resist as well as in a Same Level Mixed Lithography approach.

129. W. R. Goodner, Proc. of Kodak Microelectronic Seminar (1979), pp. 51-59.

130. a) G. N. Taylor and T. M. Wolf, *J. Electrochem. Soc.* **127**, 2665 (1980); b) G. N. Taylor, T. M. Wolf and M. R. Goldrick, *ibid.* **128**, 361-366 (1981).

131. M. S. Chang, *J. Vac. Sci. Technol.* **16**, 1973-1976 (1979).

132. K. L. Tai, W. R. Sinclair, R. G. Vadimsky and J. M. Moran, *ibid.* **16**, 1977-1979 (1979).

133. B. Griffing and P. West, Semiconductor Int. (June 1983), pp. 17-18.

134. G. L. Gaines, Jr., "Interfaces," Interscience, New York (1966).

135. K. P. Blodgett and I. Langmuir, *Phys. Rev.* **51**, 964 (1937).

136. A. Barand, C. LeGressus, H. Okusumi and A. Mogami, 8th Int'l. Congress on Electron Microscopy, Canberra, J. W. Sanders and D. J. Goodehild, eds., Part 1 (1974), p. 682.

137. a) A. Barand C. Rosilio and A. Ruaudel-Teixier, *Solid State Techol.* (August 1979), pp. 120-124; b) A. Barand, C. Rosilio and A. Ruaudel-Teixier, *J. Colloid. and Interface Sci.* **62**, 509 (1977).

138. C. A. Brown and T. Wu, IBM Research Laboratory, San Jose, California, private communication.

# REACTIVE ION ETCHING AND RELATED POLYMERIZATION PROCESSES

Eric Kay

IBM Research Laboratory
5600 Cottle Road
San Jose, California 95193

The major technological potential of "dry etch" processing arises from the demonstrated control that can be exercised over the etch features even at a submicron resolution level. In addition, high selectivity through judicial choice of chemical etchants and type of radiation, as well as the compatibility with well established vacuum technology practices, further serve to promote several of the evolving dry etching approaches. With the growing importance of "radiation" induced processing technologies, such as reactive ion plasma etching and laser induced chemical etching and deposition, it has become necessary to understand the mechanisms whereby ions, photons and electrons influence surface chemical reactions. This review will emphasize the progress that has been made in understanding the role which ion-surface interactions play in increasing the rates of surface etching processes and how this relates to control over etch feature geometry.

## ION SURFACE INTERACTIONS

The role of inert ion bombardment leading to physical momentum transfer sputtering of both elemental and multicomponent (alloys, compounds) target surfaces has been discussed widely in the literature and will not be dealt with in any detail in this review.

The role of energetic ion bombardment of a surface in a chemically reactive environment, is somewhat different than that associated solely with physical sputtering. In the chemically reactive environment the key role of ion bombardment can be the conversion of strongly bound (chemisorbed) layers at or near the surface into new chemical species which are then very weakly bound, and consequently desorb rather readily, especially in the presence of ion bombardment. This process we choose to refer to as "chemical sputtering." For a number of systems of technological interest, the rates of etching by chemical sputtering are very much higher than that achieved by pure momentum transfer physical sputtering. However, exceptions to this general rule do exist.

A typical prototype system which demonstrates the essential conceptual features of *chemical sputtering* is the reaction of fluorine atoms with a silicon surface in the presence of ion bombardment to form volatile silicon fluorides at room temperature.

## FLUORINE-SILICON SURFACE CHEMISTRY

The overall surface chemistry of fluorine atoms reacting with a silicon surface as a function of coverage and in the *absence* of ion bombardment is qualitatively understood. Fluorine atoms chemisorb with high sticking probability and penetrate the Si lattice up to about the equivalent of several monolayer coverage after which the sticking probability of F on such a fluorinated surface drops dramatically. The exact chemical composition of this layer is not yet clear but is most probably best described as an assembly of fluorinated silicon moeities such as SiF, $SiF_2$, $SiF_3$, and even trapped $SiF_4$. However, with further fluorine dosage this fluorinated "surface" layer continues to react, but very much more slowly, and converts to volatile silicon fluorides (85% to $SiF_4$ and the remainder to $SiF_2$ and SiF). In this spontaneous fluorine-silicon reaction, $SiF_3$ is hardly detectable as a volatile product. Winters *et al.*,[1] have recently developed in some detail the notion that the mechanism of this type of etching reaction has much in common with the much more thoroughly studied surface oxidation processes where the formation of negative surface ions followed by a Cabrerra-Mott-like field assisted anion-cation replacement process provides a major driving force for mass transport into the lattice and subsequent reaction. Recent ab initio molecular orbital calculations by Bagus *et al.*,[2,3] are quite consistent with the notion that negative ions are formed at the Si surface with a strong binding site for fluorine, not only on top of the silicon, but between the first and second silicon lattice layer with only a small activation barrier for lattice penetration. XPS data by Chuang[4] is also consistent with this fluorine surface penetration concept. More recent studies by Houle[6] on the effect of visible laser light on etching of silicon by fluorine have led to a better understanding of the effect of doping on the fluorine-silicon reaction. The data have been interpreted in terms of a model in which adsorption of electronegative fluorine causes band bending in the semiconductor, the direction of which depends on whether the material is doped n- or p-type. The light-generated charge carriers enhance the rate of fluorine diffusion below the surface and insertion into Si-Si bonds. By studies of relaxation after the light has been turned off, it has been found that the principal difference between n- and p-type silicon in the context of etching chemistry is that fluorine is retained more easily below the surface of n-type material. This latter result is consistent with the observations made by many workers that n-type Si etches faster than p-type.

## ION ENHANCED SURFACE CHEMISTRY

The superposition of energetic ion bombardment dramatically enhances the overall rate of reaction from that described above between fluorine and silicon. This result can be seen clearly from data presented in Fig. 1, where the rate of etching of fluorine at the clean Si surface suddenly increases by an order of magnitude as a 500 eV Ar ion beam is switched on. $XeF_2$ is used as a convenient way of delivering atomic-like F to the surface. It has been demonstrated that $XeF_2$ dissociates at the Si surface leaving F atoms free to bond to the surface and, at sufficient coverage, react to form volatile $SiF_x$ with no evidence of Xe being left on the surface. Comparison of data from this laboratory Tu *et al.*[7] for $XeF_2$ on Si with that of Flamm *et al.*,[8] on F atoms on Si shows that near room temperature the reaction probability of $XeF_2$ is much higher than that of F atoms incident on a fluorinated Si surface which leads to the conclusion that $XeF_2$ will lead to the formation of volatile $SiF_x$ more efficiently than F atoms. The degree to which ion bombardment can enhance the volatile product molecule formation will in part be related to the reaction probability of a particular etchant specie at various levels of coverage. Why the sticking probability varies from one incident specie to another on any surface is purely speculative at this time especially in the presence of ion bombardment. As the supply of "F" etchant is cut off (Fig. 1), only $Ar^+$ physical sputtering remains. Clearly the etching rate when both energetic ions and F atoms are involved at the surface far exceeds the sum of either individually. Yield measurements by Coburn *et al.*,[7] have demonstrated that one inert gas ion, say $Ar^+$, at a 1,000 eV, incident on a fluorinated Si surface can desorb the equivalent of about 30 Si containing species in the form of $SiF_x$. This event inevitably exposes a clean Si surface "patch" with high sticking probability for subsequent rapid fluoridation, thereby initiating the sequence of steps as a function of coverage as described above as long as a large enough supply of F containing etchant is available above the surface. We have shown that absolute cross section measurements for this chemical sputtering event give values of $10^{-14}/cm^2$ at fluorine coverages of a monolayer or greater.[9] This is an order of magnitude greater than the cross section associated with pure physical sputtering of $SiF_x$ at very low coverage (see Fig. 2).

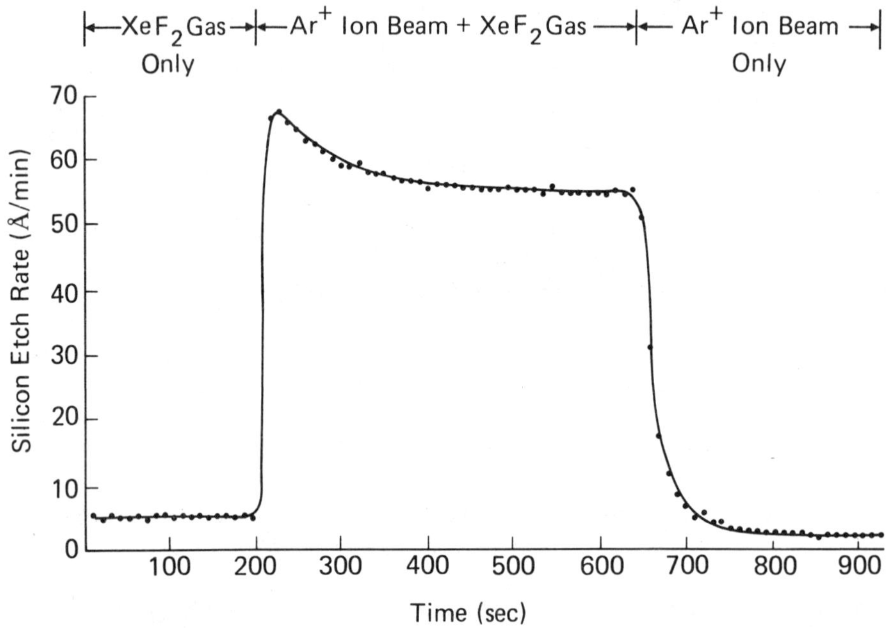


Figure 1. Si etching in $XeF_2$ with and without ion beam bombardment.

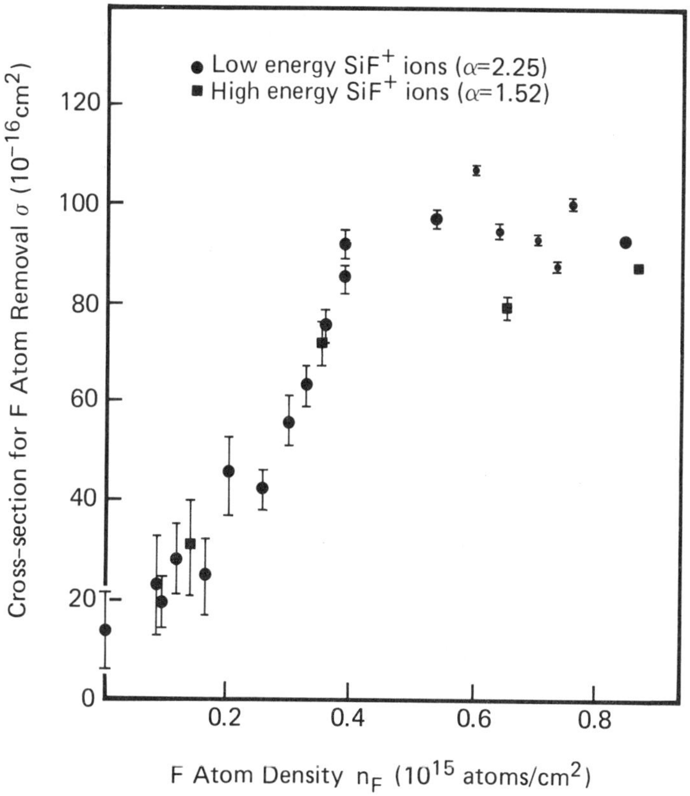


Figure 2. Cross sections for ion bombardment induced F atom removal as a function F coverage. The F removal is measured as emitted $SiF^+$ (SIMS) $SiF^+$ signal was independently calibrated as a f(F) coverage.

Ion bombardment is likely to disorder the receding Si surface so that the initial surface crystallographic orientation is likely to play a much less important role in the etching kinetics in the presence of ion bombardment than in the absence. The degree to which radiation damage, such as the production of vacancies and interstitials and their subsequent diffusion in the presence of ion bombardment influences the observed surface kinetics is not clear at this time. However, initial yields (*i.e.*, $SiF_x$ removed per incident ion) from a fluorinated Si surface which was exposed to a very low ion beam fluence ($10^{12}$ ions/$cm^2$, *i.e.*, 1 ion/10,000$Å^2$ of surface) were shown to be equivalent to that from a Si surface which had been subjected to prolonged high current density ion bombardment prior to simultaneous exposure to $XeF_2$ and energetic ions. Furthermore, the reaction rate on Si after prolonged ion bombardment dropped to its original rate as soon as the ion beam was shut off within only 1 or 2 additional monolayers being removed, even though the structural damage produced by the previous ion bombardment must have penetrated for several tens of monolayers. From these observations it would appear that *permanent* lattice damage resulting from ion bombardment, though undoubtedly present, cannot be the main driving force for the observed enhancement in the reaction rate between F and Si. Experimental evidence exists in this laboratory that this conclusion cannot be extrapolated to other chemical systems at random. Whereas *physical* sputtering is generally believed to arise from the very outermost one or two lattice layers, *chemical* sputtering is likely to be much less surface constrained and can involve the ion bombardment enhanced formation of volatile product species deeper in the lattice followed by subsequent ion enhanced diffusion to the surface. The degree to which ion bombardment facilitates the penetration of F into the lattice is not clear at this time but qualitatively this enhanced F penetration and compound formation is thought to be as key a factor with ion bombardment as it is with laser bombardment,[5,6] but mechanistically for different reasons. Electronic excitation is not thought to play a key role in the ion bombardment case.

## ETCHING DIRECTIONALITY

*It will become evident that etch feature geometry will be determined by the balance between neutral etchant versus energetic ion enhanced surface etchant chemistry together with selective blocking of the etch feature side-wall chemistry.*

The straightforward connections between ion enhanced etching rates and the concept of "directionality" can now be conceptualized using F and Cl interacting with Si as a prototype example.

Consider energetic ions only bombarding a Si surface *orthogonally* through a mask opening as shown in Fig. 3a. Such directional ion bombardment alone can be expected to result in physical sputtering giving rise to a highly directional etch feature as schematically illustrated in Fig. 3a. If instead of energetic ions one exposes the Si to neutral etchant species only, *e.g.*, F atoms, which arrive from all directions from above the mask opening, then Si would be expected to etch isotropically as illustrated qualitatively in Fig. 3b. The rate of this isotropic etching under the mask will be determined by the neutral etchant flux and the corresponding spontaneous etch rate at a given substrate temperature and orientation. This spontaneous F etch rate of Si is considerably higher than that obtained by pure momentum transfer physical sputtering alone in the respective F flux or ion flux and energy regimes commonly encountered in practical systems to be described later. If now orthogonally directed energetic ions *and* isotropically directed neutral F atoms arrive simultaneously at the Si surface through the mask opening, then an etch pattern is obtained as qualitatively represented in Fig. 3c. This etch pattern demonstrates the logical consequence of etching very much faster in the forward direction (in the direction of the incident ions) than the spontaneous isotropic etching. *The degree of control over the geometry of the etch feature depends first and foremost over the control we have over the spontaneous isotropic etch chemistry vis-à-vis the ion enhanced directional chemistry.* A quantitative way of demonstrating the spontaneous (isotropic undercut) reaction rate due to neutral etchant species vis-à-vis the reaction rate obtained by *simultaneous* ion bombardment, as a function of etchant flow rate is shown in Fig. 4 for the case of $XeF_2/Ar^+/Si$.

The overall set of circumstances described above is conceptually similar to that encountered in the more complicated plasma environments as commonly used. Ion bombardment with incident ion direction perpendicular to the plane of a Si wafer is a condition which exists in reactive ion etching (R.I.E.) in glow discharge configurations where the Si wafer is placed on an electrode surface which is at a negative potential with respect to the plasma potential. In the most commonly used capacitively coupled R.F. diode configurations, the powered negatively self-biased electrode is subject to high energy ion bombardment (usually several hundreds of electron volts) and the grounded electrode is subject to low energy bombardment (usually of the order of several tens of electron volts). The exact energy distribution at the respective electrodes is determined by many factors of which surface area of the powered electrode versus total grounded surface area exposed to the plasma and the plasma frequency and pressure are key factors. This perpendicular ion bombardment in a plasma configuration is a key source conducive to vertical etching. On the other hand, the isotropic etching which determines the degree of sideways undercut etching will be caused by all the

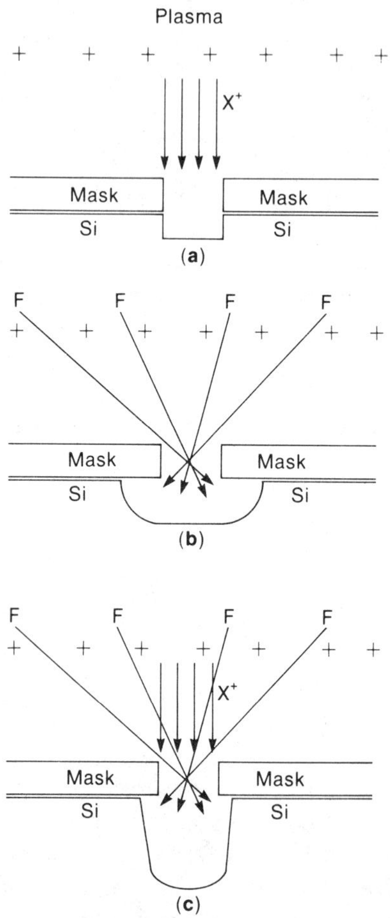


Figure 3. Schematic etch feature geometry for (a) normal incidence ion bombardment only; (b) random incidence F atom etching only; (c) combined ion and atom etching.

neutral plasma etchant species which inevitably arrive at the Si surface with random directionality from the plasma.

If the isotropic undercut chemistry could be completely eliminated, then the etch feature geometry would be determined primarily by the direction of the incident energetic ions. Several important situations will be described which tend toward this latter condition. The most obvious way to reduce or eliminate the isotropic (undercut) chemistry at the etch feature sidewalls is either to minimize the arrival of neutral etchant species or to block the side wall surfaces with a relatively unreactive barrier layer. (To a first order approximation sidewalls are assumed in this discussion not to be subject to ion bombardment.) This blocking situation can be approached, for example, using mixed halogen containing etchants. From earlier work by Pandey *et al.*[10] it can be deduced that chlorine atoms, whether arriving as such or formed by chemisorptive dissociation, will cover the Si surface to saturation after which little further reaction will take place. From potential energy interaction calculations by Bagus *et al.*,[3] it becomes clear that Cl atoms, in contrast to F atoms, have to overcome a very large activation barrier of ~10 eV before they can penetrate the Si lattice to form product molecules. Cl atoms will therefore tend to block the sidewalls and thereby *slow* down further reaction of Si with other etchant species. On the other hand, we have shown that wherever energetic ions hit such a chlorinated Si surface, the reaction will proceed to a volatile end product.[11] Chemical sputtering of a silicon halide will therefore result from the bottom surface of the Si etch feature. Whereas the mechanistic details of this type of ion enhanced reaction are not clear, the phenomenological results of F/Si versus Cl/Si were clearly demonstrated by beam experiments reported by Coburn *et al.*, (see Figs. 4 and 5). By comparing Fig. 4 with Fig. 5, it can readily be seen that even with ion bombardment, the $SiCl_x$ yields in the Cl/Si reaction are not as great as the spontaneous F/Si reaction. Although the ion enhanced reaction rate is much slower with Cl than with F, the directionality can be expected to be greater because of the sidewall blocking by Cl.

## FLUOROCARBON PLASMAS

In commonly used r.f. etching discharges, neither pure single nor pure mixed halide gases are ordinarily used, but instead halocarbons are often used, the simplest example of which is $CF_4$. In the ground state this neutral molecule is as inert as Ar and does not chemisorb to a measurable degree. On the other hand electron impact collisions in a $CF_4$ plasma fragment this molecule into several major species: F atoms, $CF_3^+$ ions, $CF_3$, $CF_2$, and CF radicals. All of these have been identified by the combination of mass

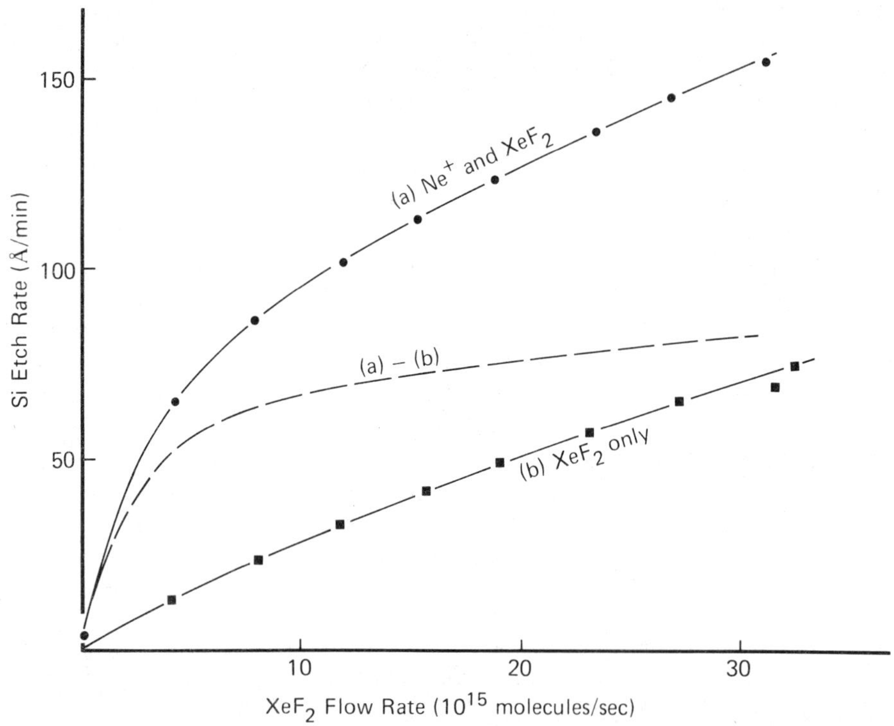


Figure 4. Si etch yield as a function of $XeF_2$ etchant flow rate with and without 1 keV Neon laser ion beam bombardment.

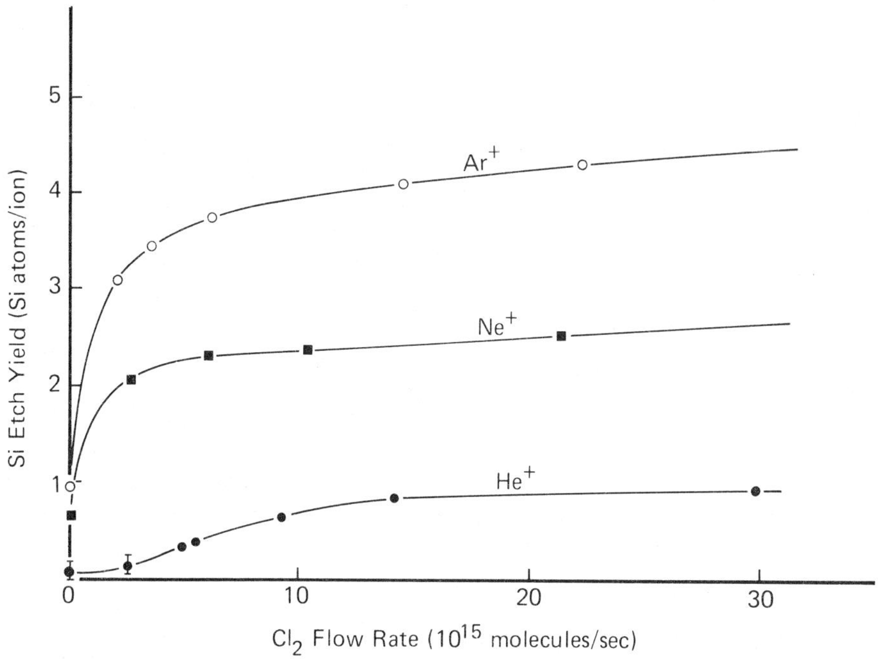


Figure 5. Si etch yield as a function of flow rate of $Cl_2$ with ion bombardment using helium, neon, and argon, 1 keV ion beams.

spectroscopy and emission spectroscopy and laser induced fluorescence for example, Refs. 12-15.

In a $CF_4$ plasma the major ion is $CF_3^+$. $CF_2^+$ is an order of magnitude less prevalent but may be more reactive. UHV beam experiments by Tu *et al.*,[7] have shown that the ability of energetic $CF_3^+$ ions to enhance the F/Si reaction is essentially equivalent to that previously found for $Ar^+$ at similar ion energies (see Fig. 5). Therefore, to a first order approximation the enhancement effect on the F/Si reaction is due to ion bombardment momentum transfer effects between the incident ion and the fluorinated Si surface and not primarily due to the chemical nature of the incident ion.

From Auger experiments by Coburn *et al.*,[16] involving U.H.V., $CF_3^+$ ion beam bombardment experiments of Si, it becomes clear that C is deposited on the Si surface in this bombardment process. The degree of carbon deposition will depend on the kinetic energy of the incident $CF_3^+$ since above the physical sputtering threshold two competitive processes, namely C deposition and simultaneous resputtering of previously deposited C will take place until a steady state C coverage will be reached. It has also been demonstrated that $CF_3$ radicals, which in a $CF_4$ plasma are expected to be more abundant than the short lived $CF_3^+$ will very effectively chemisorb and dissociate on a Si surface and thereby deposit C on the surface. The major difference between the $CF_3$ radicals and ions is that the radicals will not give rise to any physical or chemical sputtering as a competitive process to the C deposition and the radical collection efficiency at an electrostatically biased surface will be less than that for ions. This type of "carbonaceous" deposition process will serve as a blocking barrier on Si and will tend to slow down further F/Si reaction, especially on surfaces where energetic ion bombardment is expected to be low or absent, such as on the sidewalls of etch features. The more pronounced the carbonaceous deposition on these sidewalls, the more effectively the isotropic F/Si etching will be minimized, thereby leading to more pronounced vertical etch features.

$CF_3$ radicals and $CF_3^+$ ions are not the only possible source of a "carbonaceous" blocking layer. $CF_2$ and CF radicals are also present in a $CF_4$ discharge and these species have been shown to be precursors to oligomerization in the plasma gas phase and subsequent fluorocarbon "polymer" deposition.[13,17-19]

A commonly used plasma etchant gas mixture, where selective etching between $SiO_2$ and Si is required, is $CF_4$ and $H_2$. In this case $CF_2$ species become more and more abundant in the gas phase as the $H_2$ concentration is increased. H atoms react very rapidly with $CF_3$ radicals to convert them to $CF_2$+HF.[20,21] In addition, F etchant atoms are scavenged from the gas phase and pumped out of the system as stable HF molecules. The back reactions, *e.g.*, F with $CF_x$ radicals to ultimately form $C_nF_{2n+2}$, is therefore reduced. One can easily observe by optical emission that the F atom concentration goes down and that of $CF_2$ radicals goes up as $H_2$ is added to the $CF_4$ plasma. See, for example, data reported by Millard *et al.*, Fig. 6. $CF_3$ radicals are not monitored by emission spectroscopy. Furthermore, we have shown, directly by optical emission and indirectly by mass spectroscopy, that the fluorocarbon deposition rate is *proportional* to the concentration of $(CF_2)_n$ radicals in the plasma, see Figs. 7 and 8. Therefore, as the polymerization rate increases, so will its blocking action on the sidewalls of the etch features.

The effect of polymer deposition at the bottom of the etch feature, which is subject to ion bombardment, is more complicated and will be discussed in the next section. In order to understand the role of carbonaceous blocking layers not only in the context of minimizing isotropic etching (directionality) but also in the context of *selective* etching of two different materials such as, for example, $SiO_2$ *versus* Si, it is important to consider the consequences of $CF_3^+$ ion bombardment further. From Fig. 9, we can see that the relative effect of $CF_3^+$ *versus* $Ar^+$ ion bombardment on the etching rate of Si is very similar and not dramatically different on $SiO_2$. Far more significant is the fact that the absolute ion enhanced rate effect on $SiO_2$ per incident ion is approximately one order of magnitude lower than on Si. In fact, at room temperature in the absence of energetic ion bombardment $SiO_2$ will not etch *significantly* at the fluxes of etchant commonly used.[22,23]

From U.H.V. beam experiments as illustrated in Fig. 10, it becomes immediately evident that "C blocking" of the $SiO_2$ surface can be expected to be very much less than on the Si surface in the *presence* of energetic ion bombardment. The overall effect of this selective net deposition of carbonaceous material on Si versus $SiO_2$ in a typical $CF_4$+$H_2$ plasma is self-evident from Fig. 11. So although the ion enhanced etch rate of *clean* Si is very much greater than that of $SiO_2$, the more effective C blocking of Si *versus* $SiO_2$ in the $CF_4$+$H_2$ situation completely overshadows the etch rates of the pristine material.

Both from the U.H.V. $CF_3^+$/Si ion bombardment beam experiments and from mass spectrometric diagnostics of $CF_4$ plasma effluents, it can readily be shown that in the presence of energetic ion bombardment any source of carbonaceous deposit on $SiO_2$ is prone to conversion to volatile $CO_x$ and $CO_xF_y$ species; that is to say, ion bombardment helps to free O from the $SiO_2$ lattice, then tends to oxidize the C to volatile species. The mechanistic details of this process are unclear at this point. Injection of oxygen atoms into the gas phase of a fluorocarbon plasma from *any* source serves to remove unsaturates, *i.e.*, precursors to polymer film formation, thereby leaving F atoms

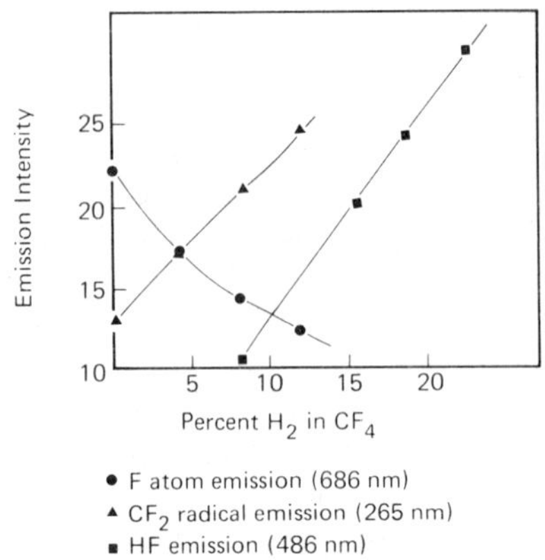


Figure 6. Emission intensities of F, $CF_2$, and HF as a function of hydrogen addition to a $CF_4$ plasma.

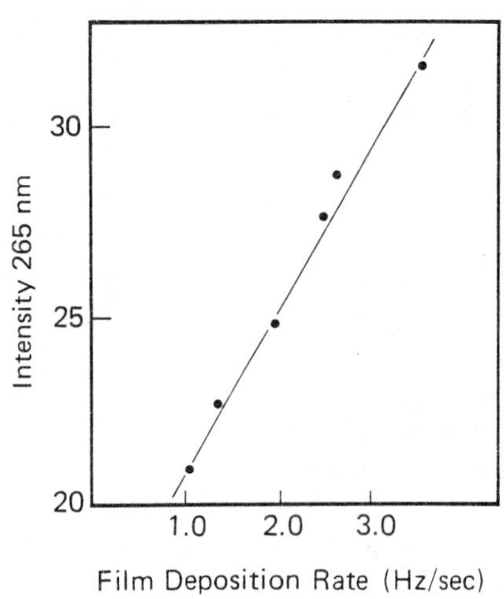


Figure 7. Deposition rate in a parallel plate reactor *versus* $CF_2$ band emission at 265 mm.

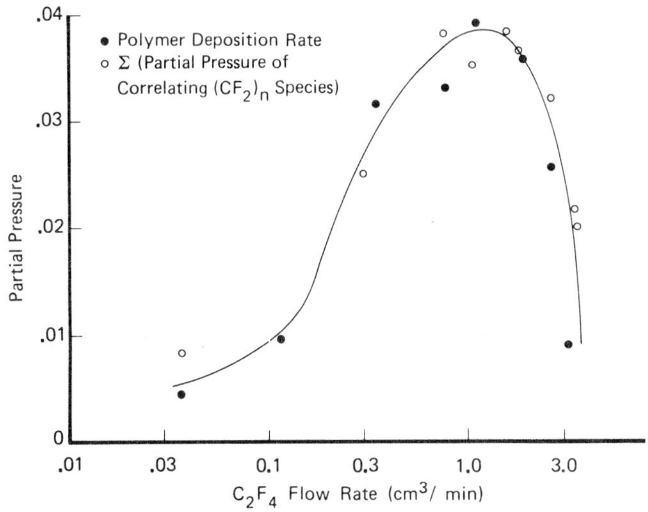


Figure 8. Deposition rate and summation of $(CF_2)_n$ partial pressures in a $C_2F_4$ plasmas as a function of flow rate of $C_2F_4$, demonstrating correlation and maximum as a function of residence time.

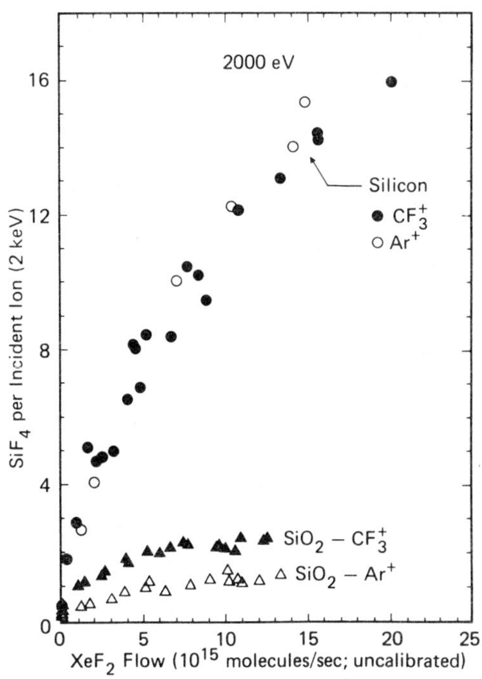


Figure 9. $SiF_4$ yield with 2 keV ion beams from Si and $SiO_2$ targets as a function of $XeF_2$ flow.

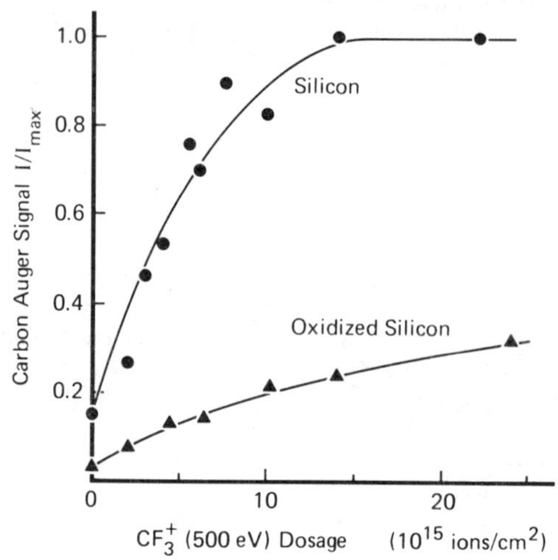


Figure 10. Surface carbon on Si *versus* $SiO_2$ as a function of $CF_3^+$ ion beam dosage.

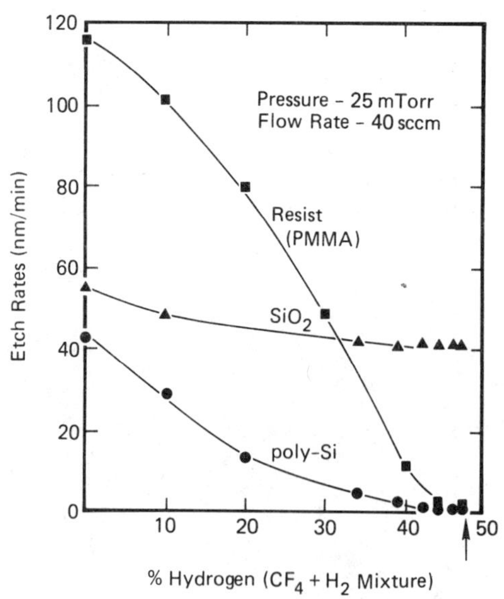


Figure 11. Etch rates of Si, $SiO_2$, and PMMA resist in a $CF_4$ plasma as a function of hydrogen injection into plasma after L. M. Ephrath *et al.*, J. Electrochem. Soc. 129, 2282, (1982).

free to etch with predictable consequences in etch rate and directionality. Furthermore $CF_3$ radicals are efficiently[20] converted to F atoms and $CF_2O$, thereby providing a further supply of F atom etchant species. Obviously, if *excessive* amounts of oxygen are added, oxidation of the surface to be etched will dominate.

In order to promote the etching of $SiO_2$ more rapidly than Si, it becomes necessary to deposit as much carbonaceous material on the surface to be etched such that etching of Si will be essentially eliminated by a blocking carbonaceous deposit whereas etching of $SiO_2$ will continue at reasonable rates because of removal of the blocking layers as $CO_x$ and $CO_xF_y$ volatiles in the presence of ion bombardment.

Any scheme, which deposits "carbonaceous" material on Si at a greater rate than on $SiO_2$ in the presence of energetic ion bombardment together with etchant species, will accomplish the same selective etching result in kind, if not in degree, as in the $CF_4$+$H_2$ work. So, for example, any scheme for increasing the polymerization precursors $(CF_2)_n$ in the gas phase by adding any other solid or gaseous F atom scavenger or by adding $C_2F_4$ to $CF_4$ instead of $H_2$ will decrease the etch rate of Si much more than that of $SiO_2$[12] (see Fig. 12). Recent results[29] have shown that monitoring the $CF_2$ radical concentration by laser induced fluorescence in a variety of etching fluorocarbon plasmas results in a direct measure of the subsequently found selective etch rates of Si *versus* $SiO_2$. These findings are further confirmation of the correlation of $CF_2$ and polymerization and its role in selectivity and directionality in R.I.E. as outlined above.

The detailed chemical composition of the "carbonaceous" deposits on various surfaces in an etching discharge is hard to predict and therefore the ability to predict its blocking capability from further etching is also speculative. It should not be considered surprising that any variable which leads to a change in the type and concentration of fluorocarbon species vis-à-vis the main halogen etchant in the plasma will influence both the degree to which an effective blocking layer is formed as well as its chemical structure. In addition, clear evidence exists that the degree of ion bombardment of a surface at which precursors to polymerization are present will greatly influence the *rate* of polymer formation as well as the chemical structure of the resultant carbonaceous deposit. The composition of the "carbonaceous" deposit will clearly relate to its etchability. Data by Kay *et al.*[24] as represented in Fig. 13 and 14 demonstrates an example of the effect of ion energy on rate of polymerization and Fig. 15 shows the effect of ion bombardment on the chemical structure of the "carbonaceous" deposits. In general, the rate of polymerization is enhanced with ion energy up to a maximum beyond which physical sputtering together with ion enhanced revolatilization of the polymer

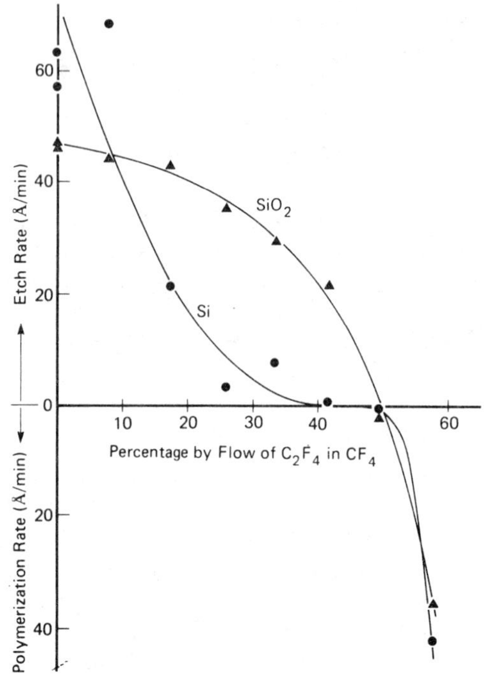


Figure 12. Etch rates of Si and $SiO_2$ in a $CF_4$ plasma as a function of $C_2F_4$ injection into plasma.

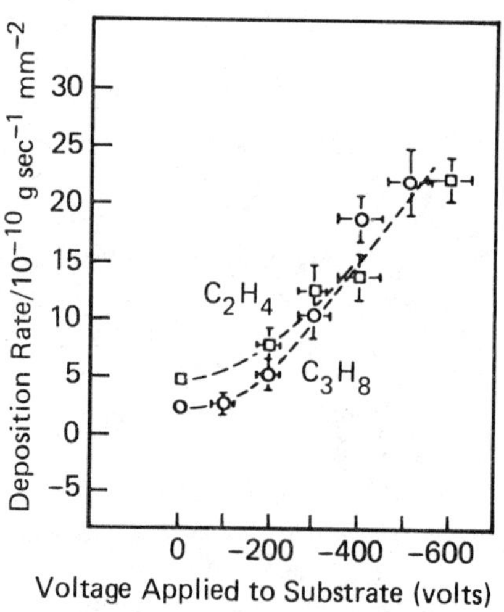


Figure 13. Polymer deposition rate in a $C_2H_4$ and $C_3H_8$ plasma as a function of ion bombardment energy during film growth.

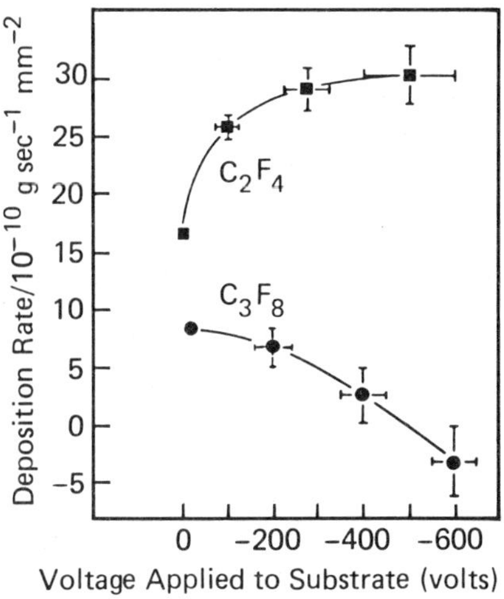


Figure 14. Polymer deposition rate in a $C_2F_4$ *versus* $C_3F_8$ plasma as a function of ion bombardment energy during film growth. In $C_3F_8$ etching dominates over deposition as ion energy increases.

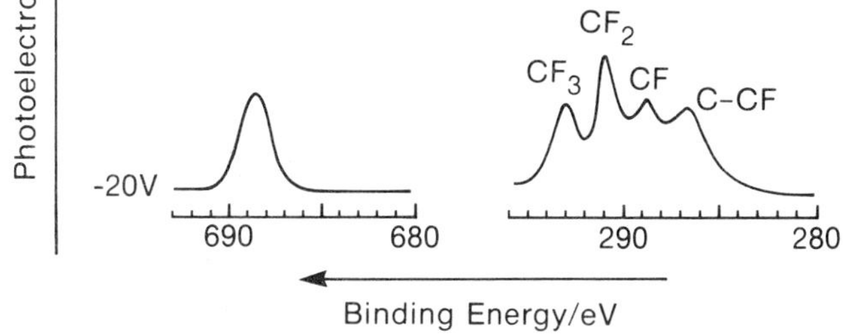


Figure 15. Effect of ion bombardment during film growth on structure of plasma polymerized fluorocarbon films at 20 and 400 eV substrate bias voltage, respectively.

deposit in the presence of F dominates over polymerization. The structure of the carbonaceous deposit becomes more and more C-like with increasing bombarding ion energy at the expense of Cf, $CF_2$, $CF_3$ component in the film deposit (see Fig. 15). However, not all the F atoms can be preferentially resputtered, thereby never reaching pure carbon films. In the case of hydrocarbon polymerization in the presence of enough energetic ion bombardment ultimately very hard "diamond-like" C deposits are formed.

## BLOCKING LAYERS ON SIDEWALLS *VERSUS* BOTTOM OF ETCH FEATURES

The mechanisms whereby these ion bombardment effects influence the formation of blocking layers at the bottom of an etch feature relative to the "vertical" sidewall where ion bombardment is assumed to be essentially absent, is only qualitatively understood. At the bottom of the Si etch feature a *competition* between the following four processes takes place, all involving energetic ion bombardment.

Enhanced etching of Si by F atoms in the presence of energetic ions,

$$Si_{(s)} + F_{(g)} + \text{ion bombardment} \rightarrow SiF_{x(g)} \uparrow . \qquad (1)$$

Enhanced deposition of polymeric material in the presence of CF, $CF_2$, *etc.*, and of energetic ions,

$$(CF_2)_{n(g)} + \text{ion bombardment} \rightarrow (CF_{2-x})_{n(s)} \downarrow . \qquad (2)$$

Enhanced etching of fluorine deficient polymeric deposit by F atoms and $CF_x$ radicals in the presence of energetic ions,

$$(CF_{2-x})_{n(s)} + F + \text{ion bombardment} \rightarrow C_nF_{2n+2(g)} \uparrow . \qquad (3)$$

Physical sputtering of surface material,

$$(CF_{2-x})_{n(s)} \text{ and/or } SiF_{(s)} + \text{ion bombardment} \rightarrow \text{sputt. species} \uparrow . \qquad (4)$$

The energy dependence of these four processes can be expected to be quite different as is their dependence on temperature of the surface. So, for example, the rate of polymerization by process (2) is far more sensitive to surface temperature than processes (1), (3). and (4). This is due to the observed large drop in sticking probability of the polymer precursor species with very small rise in surface temperature. This result will become quite clear in the following section.

Kay *et al.*, have recently published a number of papers in which advantage is taken of these competing etching and deposition processes to produce a new class of materials, *i.e.*, metal containing fluorocarbons. Injecting $C_3F_8$ monomer into a capacitively coupled r.f. diode system allows fluoridation to dominate at the electrode subject to high energy ion bombardment whereas polymerization dominates at the grounded electrode. If the powered electrode is made of Au, no fluoridation can take place and Au will be physically sputtered from the surface due to energetic positive ion impact from the plasma. If the powered electrode is made of material which forms a non-volatile fluoride, then again only physical sputtering can remove these fluorides from the surface and inject them into the plasma. If, on the other hand, volatile fluorides are formed at the powered electrode, then the rate of removal of electrode material as fluorides by chemical sputtering, as defined earlier, will dominate. In all three cases the material ejected from the powered electrode will traverse the plasma and in part be condensed into the polymer forming at the grounded electrode. Species etched from the powered electrode will thus be dispersed in the growing polymer matrix.

Detailed discussion of this film forming process is beyond the scope of this article but can be found in several recent publications.[18,19,23-27] Suffice it to say that the electrical characteristics of such metal containing polymers under some circumstances can be dramatically different from conventional dielectric plasma polymerized fluorocarbon polymer film deposits. As a result *all* the grounded surfaces being coated with such polymers can lead to electrical changes in the plasma, *e.g.*, the plasma potential, which in turn can impact the four processes, (1) through (4), described earlier, all of which depend on ion bombardment. To this degree these metal containing polymers do relate to plasma processing in general and plasma etching in particular.

Evidence of the lowering of the polymer formation rate as a function of increasing surface temperature has been measured in several ways. The most dramatic example we have encountered can be deduced from data presented in Fig. 16 and 17 on metal containing polymers at substrate temperatures differing only by several tens of degree centigrade.

In Fig. 17, the ratio of the polymer precursor $CF_2$ and Au atoms arriving at the substrate as measured by optical emission spectroscopy for various plasma conditions and the F and Au content subsequently found (ESCA) in the resultant polymer films are plotted at two substrate temperatures. The straight line behavior further illustrates the correlation between $CF_2$ arriving at the surface and the rate of polymerization.

Whereas the sticking probability of Au atoms is not expected to change dramatically between the substrate held at 15°C to that held at 60°C, the degree of polymer formation as measured by F content is very much less at

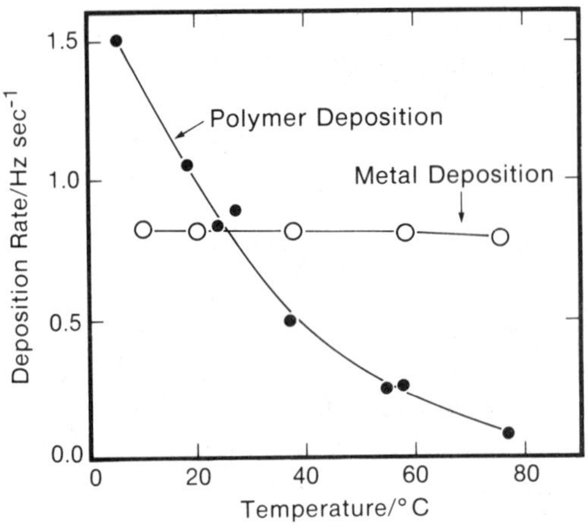


Figure 16. Deposition rate as function of substrate temperature in pure argon plasma (metal deposition only) and pure $C_2F_4$ plasma (polymer deposition only).

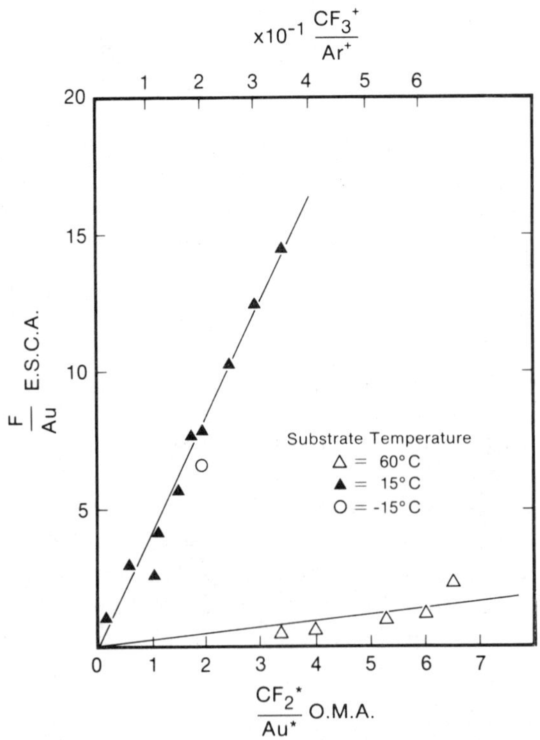


Figure 17. Key gas phase species, $CF_2$ and Au, arriving at substrate *versus* F and Au found in polymer films at two different substrate temperatures.

60°C than at 15°C. F content in the film can only be associated with polymer since fluorides of Au are absent.

The C content is not included in the ESCA data since C surface contamination in metal rich films confuses the issue and is not considered important to the theme of this article.

## ETCHING OF METALS

Whereas a detailed discussion of metal etching is beyond the scope of this review, a few general comments may be appropriate. Several classes of problems arise with metals of technological importance such as Cu and Al or their alloys. Low volatility of certain etch products at room temperature present a problem together with effective means of removing reactive etch products from the reaction zone without redeposition. Heating various surfaces, including the surface to be etched, is not considered a practical solution in many applications because of the deleterious effects of elevated temperatures on interfacial diffusion, alloy segregation, possible degradation of the photoresist material, differential thermal expansion and associated anisotropic film stress leading to interfacial adhesion problems - just to mention a few.

In addition, many metals of interest are susceptible to oxidation and most oxides are non-volatile which interferes with the volatilization of the metal, usually as a metal halide.

Energetic ion bombardment can help to remove the oxide initially but also requires that the partial pressure of oxygen contaminants in the system be kept low so that ion enhancement reoxidation does not interfere significantly with the volatilization of the metal halide. O and $H_2O$ scavengers can be added to the plasma to minimize this problem. In the case of aluminum, chlorine containing etchants such as $CCl_4$ and oxygen scavengers such as $BCl_3$ are commonly used. Chlorine is known to react spontaneously and very efficiently with *clean* aluminum. Ion bombardment of such clean Al surfaces does not enhance the etch rate quite in contrast to Si. Therefore, isotropic etching will result unless the sidewalls are blocked chemically, for example with polymeric material. Further problems are encountered, resulting from subsequent corrosion of the remaining metallurgy in a device, especially when a chlorine containing etchant is used. Various procedures to minimize this problem are being practiced usually involving a post etching treatment with an F containing atmosphere.

Numerous speculative mechanisms have been suggested for etching of metals and alloys, but to date far more insight is required in order to report any definitive mechanistic processes. This is especially true when chemical etching of alloys is considered in the presence of energetic ions in a plasma, since the role of ion induced segregation and the microscopic metallurgical character of the alloy can clearly be expected and does influence the chemical surface processes.

## REFERENCES


1. H. F. Winters, J. W. Coburn, and T. J. Chuang, J. Vac. Sci. Technol. B 1(2):469 (1983).
2. M. Seel and P. S. Bagus, Phys. Rev. B 23:5464 (1981).
3. M. Seel and P. S. Bagus, Phys. Rev. B 28:2023 (1983).
4. T. J. Chuang, J. Appl. Phys. 51:2614 (1980).
5. F. A. Houle, J. Chem. Phys. 79:4237 (1983).
6. F. A. Houle, J. Chem. Phys. (in press).
7. Y. Y. Tu, T. J. Chuang, and H. F. Winters, Phys. Rev. B 23(2):823 (1981).
8. D. L. Flamm, V. M. Donnelly, and J. A. Mucha, J. Appl. Phys. 52:3633 (1981).
9. E. A Knabbe, J. W. Coburn, and E. Kay, Surf. Sci. 123:427 (1982).
10. K. C. Pandey, T. Sakurai, and H. D. Hagstrum, Phys. Rev. B 16:3648 (1977).
11. U. Gerlach-Meyer, J. W. Coburn, and E. Kay, Surf. Sci. 103:177 (1981).
12. J. W. Coburn and E. Kay, IBM J. Res. Dev. 23:33 (1977).
13. M. M. Millard and E. Kay, J. Electrochem. Soc. 129:160 (1982).
14. P. J. Hargis and M. J. Kushner, Appl. Phys. Lett. 40:779 (1982).
15. A. Dilks and E. Kay, Macromolecules 14:855 (1981).
16. J. W. Coburn, H. F. Winters, and T. J. Chuang, J. Appl. Phys. 48:3532 (1977).
17. E. Kay, J. W. Coburn, and G. Kruppa, Le Vide 183:89 (1976).
19. E. Kay and M. Hecq, ISPC Montreal 2:490 1983, and J. Appl. Phys. in press, January 1984.
20. K. R. Ryan and I. C. Plumb, J. Phys. Chem. 86:4678 (1982).
21. I. C. Plumb and K. R. Ryan, ISPC-6 Montreal 2:326 (1983).
22. D. L. Flamm, C. Y. Mobag, E. R. Sklaver, J. Appl. Phys. 50:6211 (1977).
23. H. F. Winters and J. W. Coburn Appl. Phys. Lett 34:70 (1979).
24. E. Kay, A. Dilks, and U. Hetzler, J. Macromol. Sci.-Chem. A 12:1393 (1978).
25. E. Kay and A. Dilks, J. Vac. Sci. Technol. 16:428 (1978).
26. E. Kay, A. Dilks, and D. Seybold, J. Appl. Phys. 51:5678 (1981).
27. E. Kay and A. Dilks, Thin Solid Films 78:309 (1981).

28. M. Hecq, P. Ziemann, and E. Kay, J. Vac. Sci. Technol. 1:364 (1983).
29. P. J. Hargis, private communication.

# Organic Insulators

Markus A. Zuegel

International Business Machines Corporation

Sindelfingen, W. Germany

## Introduction

Insulators in semiconductors serve two purposes: electrical insulation between metallization levels and electrical, chemical and mechanical passivation of the total semiconductor structure.

In the first period of the semiconductor history insulation and passivation have been performed by inorganic insulators exclusively, an example is shown in fig. 1. Materials frequently used were frit glass, sputtered $SiO_2$ or $SiO_2$, $Si_3N_4$ etc. laid down in chemical vapor or plasma deposition processes.

In the early 70's the first experiments to implement organic materials for device insulation and passivation were reported. Today organic insulators are becoming the state of the art throughout the semiconductor industry. IBM typically uses polyimide for this function in it's high density memory products (see fig. 2).

The following paper "Organic Insulators" describes properties required for good insulators, composition and chemistry of polyimides, problems with organic insulators and finally the technology trend.

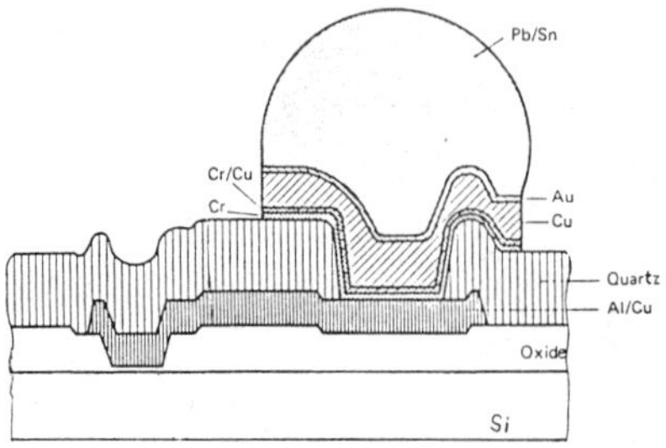


Fig. 1. Level Metal Structure of the 70's

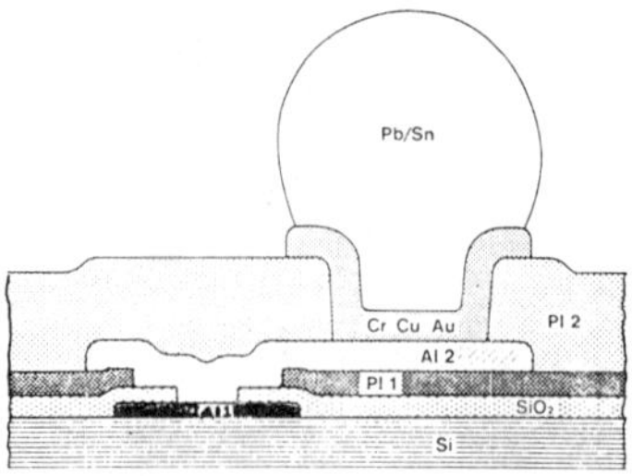


Fig. 2. Level Metal Structure in IBM S/C Memories

## Insulator Properties

In table 1 a qualitative properties comparison for the various insulator types is made, + denotes advantageous and - means disadvantageous.

Insulators have to withstand higher temperatures in subsequent fabrication steps or in real life. Thermal stability of inorganics is superior, but certain polymers like polyimides are stable enough (see fig. 3). Inorganics also show better chemical stability. On the other hand polyimide is attacked only by strong acids and alkalines, absent under normal circumstances.

Since inorganic insulators are purer and are deposited at elevated temperatures, adhesion to the underlying structures is good. Polymers require special effort to make them adhere to $SiO_2$, phosphosilicate glass, $Si_3N_4$ or metals.

Inorganic insulators provide an effective barrier against water permeation. Organic materials, however, can absorb significant amounts of water, as shown in fig. 4 for polyimide. Penetrated water in combination with contaminants will cause device parameter degradation and/or metal line corrosion.

Polymers are elastic and provide good mechanical protection, whereas inorganic insulators tend to be brittle and are susceptible to mechanical damage.

Table 1: Properties Required for Semiconductor Insulators

| PROPERTY | INORGANIC | ORGANIC |
|---|---|---|
| THERMAL STABILITY | + + | + |
| CHEMICAL STABILITY | + + | + |
| ADHESION | + | ± |
| RESISTANCE TO WATER PERMEATION | + + | - |
| CONTAMINATION | + | + |
| MECHANICAL PROTECTION | - | + + |
| PATTERN FORMATION | ± | + |
| LOW PROCESSING COSTS | - | + |
| PLANARIZATION | - | + + |
| DEFECT DENSITY | - | + + |
| EFFECT ON DEVICE PARAMETERS | - | ± |
| BREAKDOWN VOLTAGE | + + | + |

Tough competition in the semiconductor industry necessitates low costs for device insulation and passivation as well. The spin-on process used for polymers is simple and inexpensive in comparison to the sophisticated equipment and expensive techniques required to use inorganic materials. In general, also etching of polymers, especially of polyimides is easier.

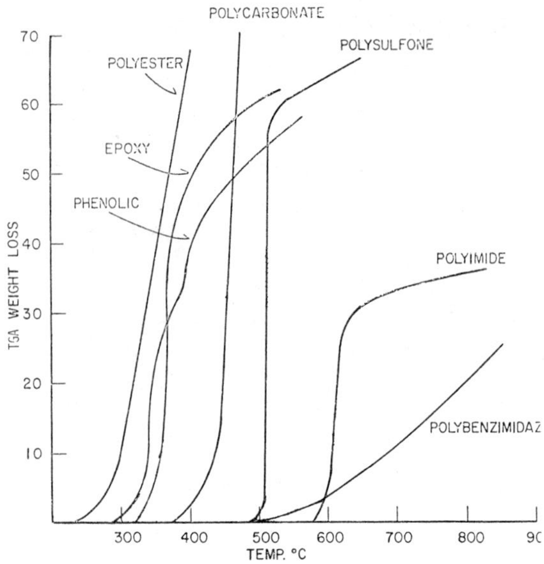


Fig. 3. Thermal Stability of Organic Insulators

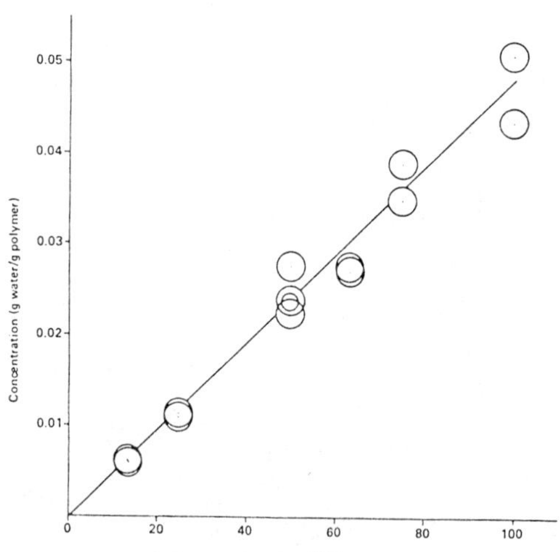


A Henry's law plot for 127-μm Kapton at 50°C.

Fig. 4. Water Absorption in Polymide Films

Good planarization is important to achieve dense structures in the metallization planes. Inorganic insulators provide conformal coatings with little planarization effect (see fig. 5). Very good planarization can be accomplished with spun-on polymers, as shown in fig. 6 for two polyimide types.

Good insulators must be defect-free. Fig. 7 shows the pinhole density in polyimide films as function of film thickness. Electrically measured defect levels with different insulation schemes between $1^{st}$ and $2^{nd}$ metal level are listed in table 2. The lowest defect levels are obtained with a dual insulator, since coincidence of defects in both, independently processed insulator components is highly unlikely.

Critical device parameters like threshold voltages in field-effect transistors and their stability can be affected by the insulator deposition process. Radiation damage or field enhanced contamination during deposition of sputtered $SiO_2$ in a radio frequency plasma is an example. Polyimide application and processing, in comparison, is a neutral process with no detrimental parameter effects.

To keep capacitive coupling between metallization levels low, insulator materials with low dielectric constants are preferred. The dielectric constant of polyimide is as low as that of $SiO_2$, approx. 3.5.

Finally, insulator breakdown voltage has to be high to support the field between metallization levels. A breakdown voltage comparison of sputtered $SiO_2$, of plasma deposited $Si_3N_4$ and of polyimide is shown in fig. 8.

The superiority of polymers especially of polyimide over inorganic insulators in important properties like good planarization (yield/product density), like good mechanical protection, low defect densities and negligible effects on critical device parameters (yield/reliability) and like ease of application and pattern formation (costs) is the reason for the wide-spread usage of organic insulators for semiconductors.

Fig. 5. Non - Planarization with Inorganic Insulators

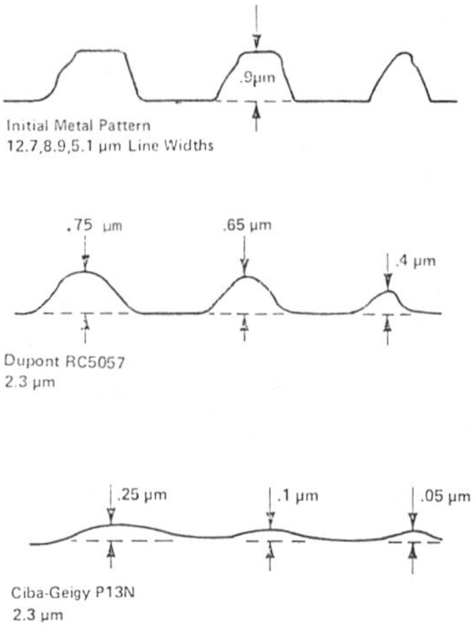


Fig. 6. Planarization with Polymides

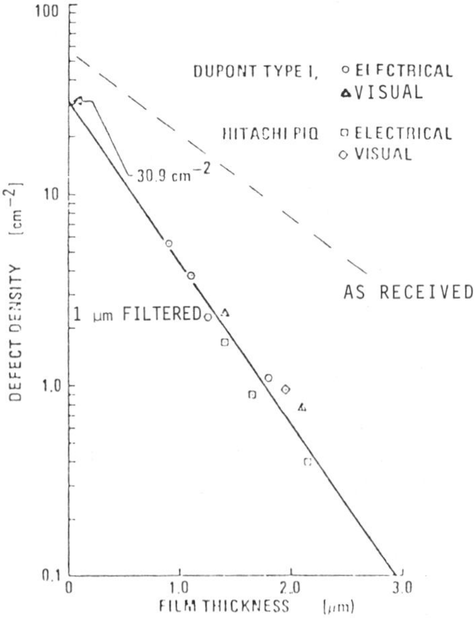


Fig. 7. Film Thickness Dependence of Polymide Pinholes

Table 2: Defect Densities M1 - M2 Shorts

| | |
|---|---|
| SPUTTERED $SIO_2$ 1.5 MICRON | 1 - 10 DEFECTS/$CM^2$ |
| POLYIMIDE 1.5 MICRON | 1 - 2 DEFECTS/$CM^2$ |
| DUAL INSULATOR SP. $SIO_2$ + PI | $<$.1 DEFECTS/$CM^2$ |

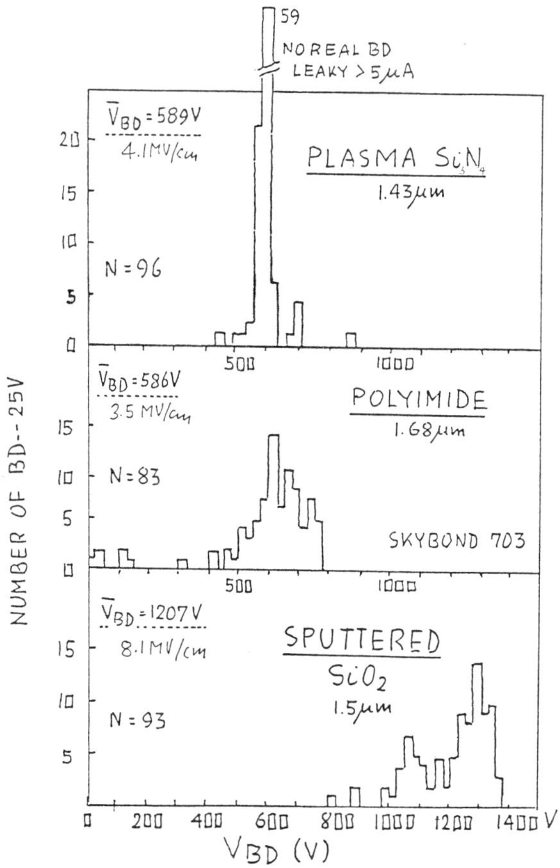


Fig. 8. Breakdown Voltage

## Polyimides - Properties, Composition, Chemistry

### Properties

Table 3 compares relevant physical properties of polyimide with other semiconductor materials. Thermal expansion of insulators should match with the substrate. Insulators should have a low modulus of elasticity, and their heat conductivity should be high. With the exception of heat conductivity, polyimide has very attractive insulator properties.

Table 3: Properties of Polymide and Inorganic S/C Materials

| Item / Material | Coefficient of thermal expansion (1/°C) | Young's modulus of elasticity ($gf/cm^2$) | Tensile strength ($gf/cm^2$) | Heat conductivity (cal/cm sec. °C) | Melting point (°C) | Specific inductive capacity | Dielectric break-down strength (V/cm) | Volume resistivity (Ω cm) |
|---|---|---|---|---|---|---|---|---|
| PIQ (HITACHI) | $2\text{-}7\times10^{-5}$ | $3\times10^{7}$ | $0.17\times10^{7}$ Limit of tensile strength | 0.0004 | 450-500 (heat decomposition) | 3.5-3.8 | $10^{6}$ | $>10^{16}$ |
| $SiO_2$ | $0.3\text{-}0.5\times10^{-6}$ | $0.738\times10^{9}$ | $0.14\times10^{7}$ | 0.005 | 1710 | 3.5-4.0 | $10^{6}\text{-}10^{7}$ | $>10^{6}$ |
| $Si_3N_4$ | $2.5\text{-}3\times10^{-6}$ | $1.55\times10^{9}$ | $0.64\times10^{7}$ | 0.028 | 1900 | 7-10 | $10^{6}\text{-}10^{7}$ | $10^{12}$ |
| $Al_2O_3$ | $9\times10^{-6}$ | $3.73\times10^{9}$ | $0.28\times10^{8}$ | 0.078 | 2050 | 7-9 | $10^{5}$ | $10^{14}$ |
| Si | $2.3\times10^{-6}$ | $1.62\times10^{9}$ | $0.12\times10^{9}$ | 0.072-0.34 | 1420 | 11.7-12 | $10^{5}$ | — |
| Al | $2.5\times10^{-5}$ | $7\times10^{8}$ | $0.7\text{-}1.4\times10^{6}$ | 0.57 | 660 | — | — | $2.5\times10^{-6}$ |

Composition

Commercial polyimide solutions contain the precursor polyamic acid dissolved preferably in N-methyl pyrollidone. Viscosity is a function of solids content and can be adjusted to specification. Thick layers require high viscosity material.

Chemical analysis of 2 polyimides is shown in table 4. For the same polyimide type vendors differ only in the solvent systems.

Table 5 describes precursors and solvent systems for commonly used polyimides.

Table 4: Chemical Analysis of Polymide

| Parameter | Du Pont RC 5878 | Beck Allotherm 610 – 16 |
|---|---|---|
| Solids content | 17.6 % | 17 % |
| Metallic Impurities | | |
| Na | 0.0010 | 0.0011 % |
| K | 0.0002 | 0.0001 % |
| Li | N. D. (0.0001) | N. D. (< 0.0001 %) |
| others total | 0.001 | 0.0030 % |
| Chloride | 0.001 | 0.0007 % |
| Solvent System | | |
| N-Methylpyrrolidone | 83.5 % | 99.9 % |
| Aliphatic Substituted Benzenes | Balance | None |
| Residue on Ignition | 0.1 % | 0.1 % |
| Acidity (total) | 37.0 mg KOH/g | 38.0 mg KOH/g |
| Density (25° C) | 1.059 g/cm$^3$ | 1.087 g/cm$^3$ |
| Viscosity (25° C) | Adjusted to spec | Adjusted to spec |

Polyimide Processing and Chemistry (see fig. 9)

As mentioned, polyimide requires adhesion promotion. The adhesion promoter is applied prior to polyimide application in a tank, on a spin-table or in the vapor phase.

Polyimide is simply spun-on like a photoresist. Film thickness depends on viscosity, spin-table acceleration and rotational speed. A bake prior to photoresist apply drives out the solvent. Polyamic acid is soluble in aqueous alkalines and is etched away by the photoresist developer together with the photoresist in exposed areas. To remove absorbed developer, neutralization with a weak acid is recommended.

After photoresist stripping the soluble precursor polyamic acid is cyclized at moderate temperatures forming insoluble polyimide. From the slope of the thermal gravimetric analysis (see fig. 10) a 2-step cyclization can be assumed (step 1 at > 125°C, step 2 at > 200°C). Below approx. 125°C solvent evaporation and above approx. 450°C polyimide decomposition take place.

Table 5: Commercially Available Polymeides

DU PONT 5057
5878
KAPTON

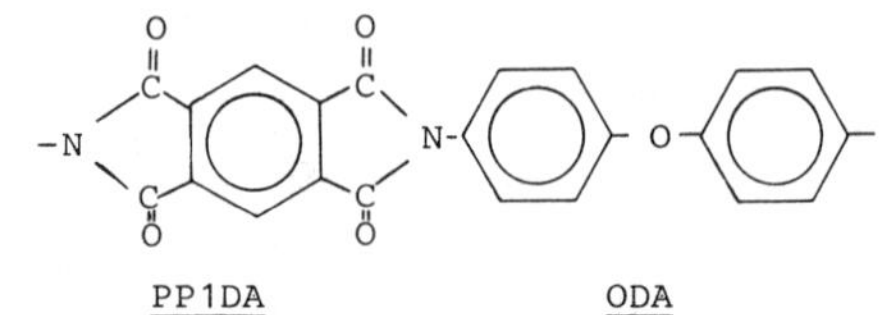


PMDA-ODA
NMP

MONSANTO
703

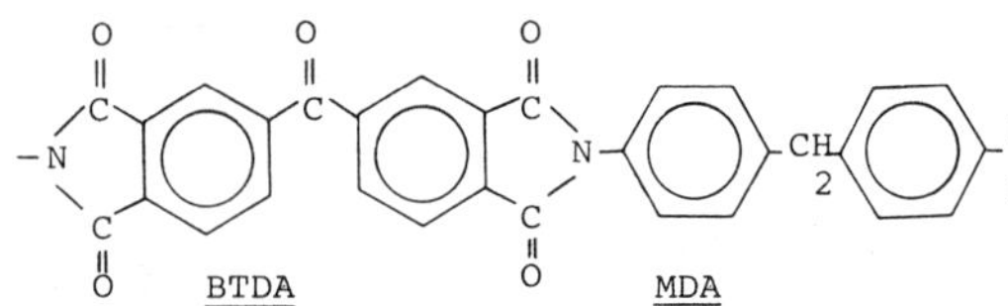


BTDA-MDA
NMP,ETOH,MIBK

CIBA-GEIGY
P13N

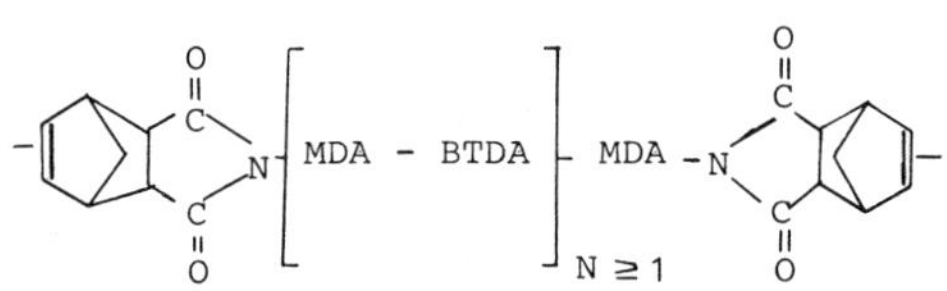


DMF

HITACHI
PIQ13

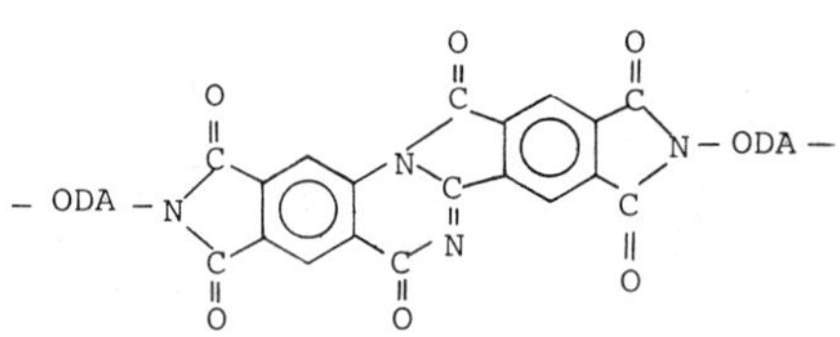


NMP

VENDOR/TYPE STRUCTURE CODE/SOLVENT

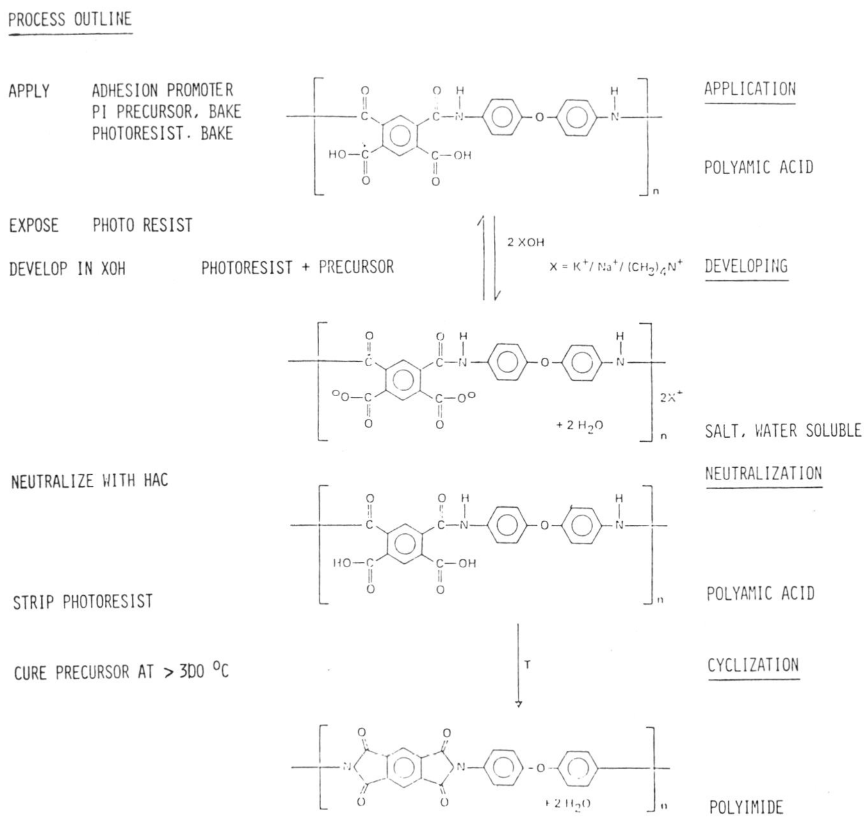


Fig. 9. Polymide Processing and Chemistry

Fig. 10. Thermal Gravimetric Analysis of Dupont PI2545

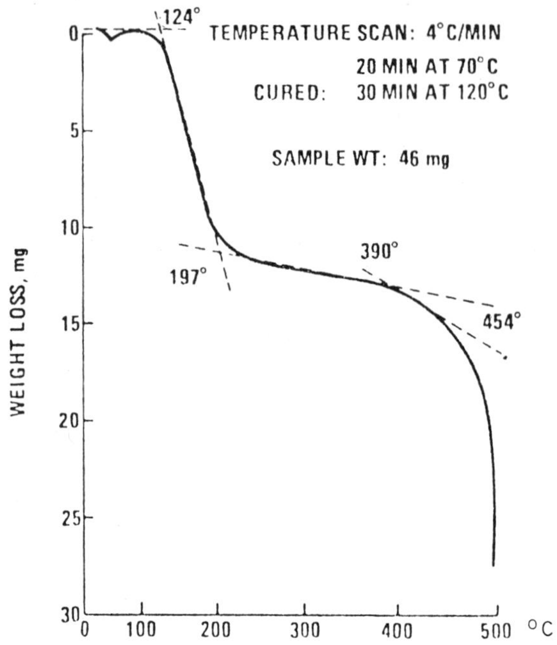


## Problems with Polyimide

During process development and manufacturing usage of polyimide engineers have been faced with several problems:

## Polyimide Adhesion

Marginal adhesion of polyimide to $SiO_2$ or phospo-silicate glass will result in temperature/humidity corrosion of metal lines, since water and contaminants are able to migrate along the weak interface.

One simple method to test polyimide adhesion on an oxidized silicon wafer is shown in fig. 11. Fig. 12 shows peel force traces of $SiO_2$ samples with and without adhesion promotion. The peel force is drasticly increased by the application of adhesion promoters.

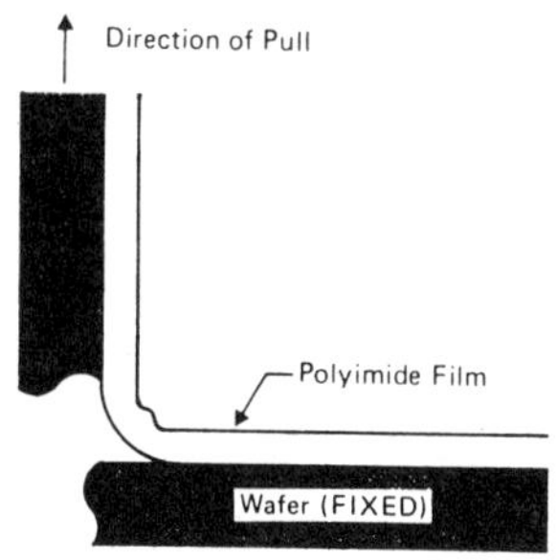


Fig. 11. Adhesion Testing

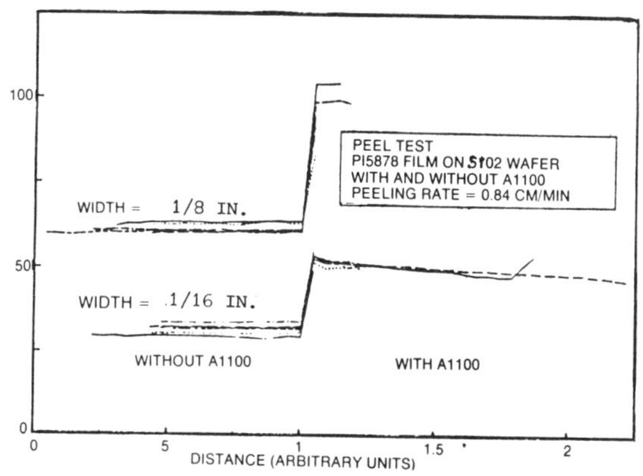


Fig. 12. Peeling Force

Adhesion promoters are organo-amino silicon compounds. The bonding mechanism between $SiO_2$ and polyimide by means of A1100 is illustrated in fig. 13.

A1100 can be applied simply by dipping the semiconductor wafer into a diluted aqueous A1100 solution. A1100 reacts readily with hydroxyl groups in the $SiO_2$ surface. After polyamic acid application the amino group of A1100 bonds with carboxylic groups of the precursor during polyimide bake or cure under release of water. Thus, the adhesion promoter produces a uniform transition from the inorganic surface ($SiO_2$) to the polymer (polyimide) and guarantees good and stable adhesion.

Since fully cured polyimide does not react with metals deposited on it, metal adhesion to polyimide can also become a problem. In such cases, incomplete polyimide curing prior to metal deposition with subsequent full curing will help.

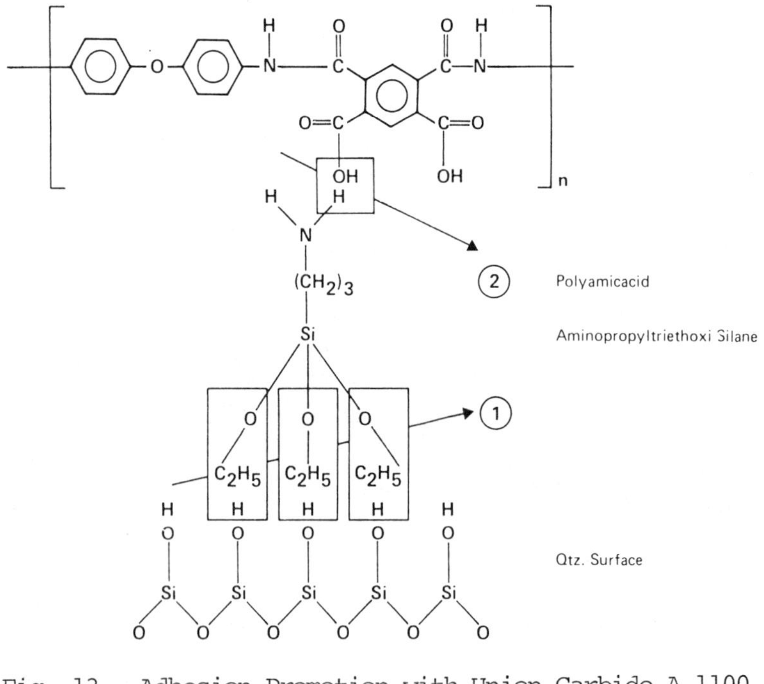


Fig. 13. Adhesion Promotion with Union Carbide A 1100

## Temperature/Humidity Corrosion

Poor polyimide adhesion leads to T+H corrosion, especially under electrical bias. Corroded metal lines are shown in fig. 14. Corrosion occurs in the presence of high humidity and contaminants (ions). Accumulation of sodium along negatively biased metal lines is shown in fig. 15. Therefore, contamination levels have to be kept low during application and processing of polymers.

Fig. 14. T+H Corrosion with Bias

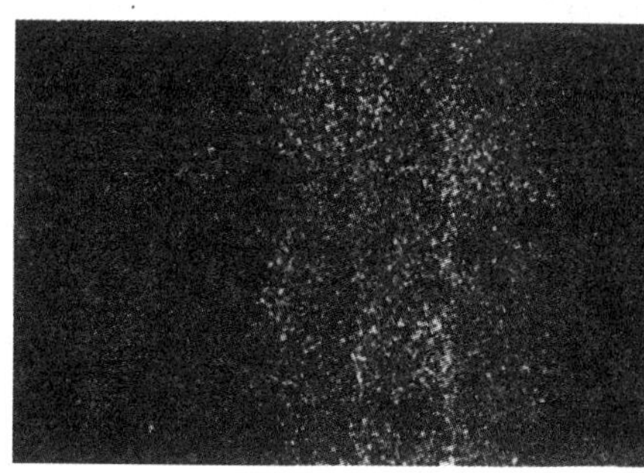

Fig. 15. Sodium Accumulation On-Bias

## Metal Opens over Topography Steps

Trapping of ions, e.g. from the photoresist developer ($Na^+$, $K^+$, $N(CH_3)_4^+$) results in incomplete curing causing formation of steep rims and overhanging ledges along the polyimide pattern. For this reason, neutralization of residual developer with acetic acid is part of the process.

Fig. 16 shows a typical wet etched polyimide pattern (via hole between 1st and 2nd metal). Notice the steep rims and ragged edges in fig. 17. Negative polyimide slopes and overhanging ledges (see fig. 18) lead to a narrowing of metal lines over the step and in extreme cases to line discontinuities (see fig. 19).

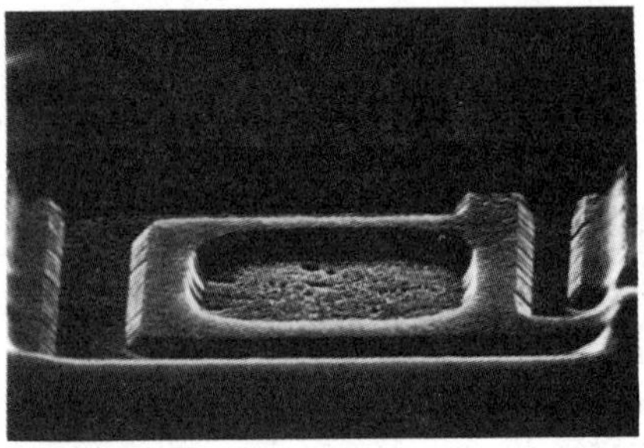

Fig. 16. Wet Etched Quartz and Polymide Vias

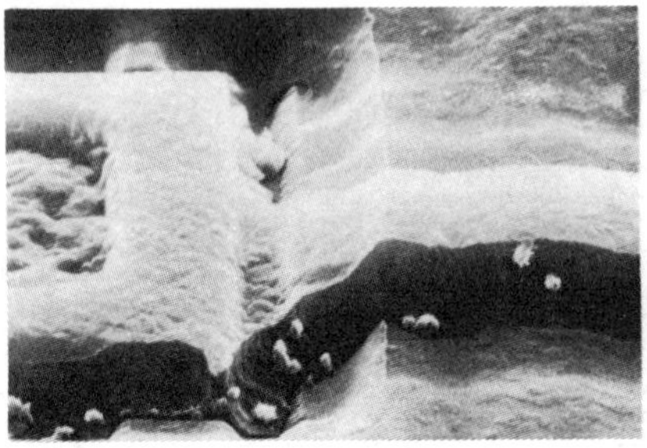

Fig. 17. M2 Edge Coverage

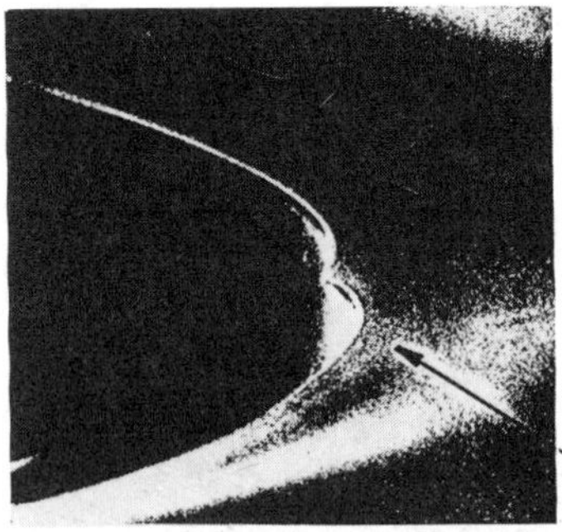


Fig. 18. Wet Etched Polymide Via

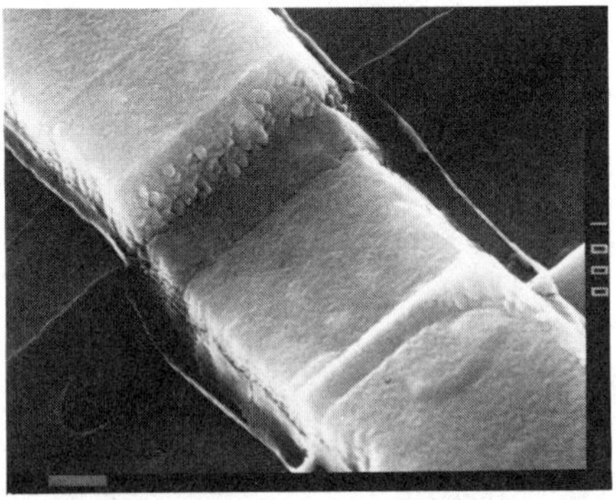

Fig. 19. M2 Open Over Polymide Ledge

Precursor Shelf Life

Fig. 20 shows the typical viscosity behaviour of polyamic solutions with time. The viscosity drops rapidly after formulation accompanied by a noticeable molecular weight decrease. Later on the viscosity of the solution increases again most likely due to starting polymerization. For good shelf life polyamic acid solutions should be stored in a refrigerator.

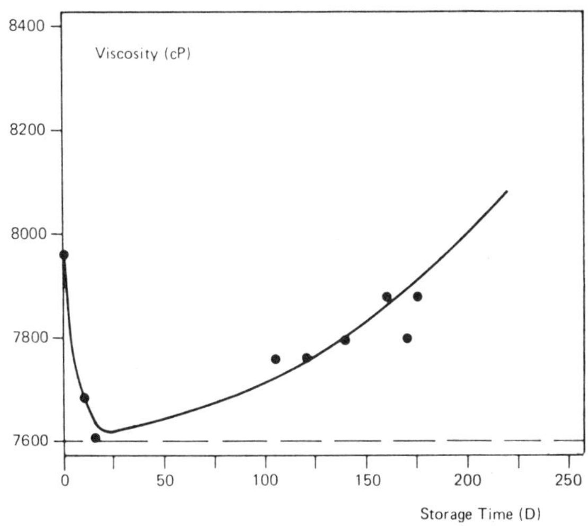


Fig. 20. Polymide Shelf Life

Organic Insulators Technology Trend

Technology trend goes in 3 directions:

Improved Materials

Polyimide, the prime polymer for device insulation and passivation was selected over a decade ago, at a time when little was known about organic insulation. With today's knowledge about requirements and material properties the search for better materials continues. The search is directed towards improved adhesion to semiconductor materials and resistance to water permeation.

Polymer Patterning With Improved Dimensional Control

Rapid memory bit density learning requires tighter groundrules in the metallization structure, which means smaller images in the insulator and tighter control. The high etch rate of polyamic acid in the photoresist developer leads to marginal dimensional control with little extendability to smaller dimensions. The other deficiencies of wet etched polyimide have been mentioned already - metal opens due to steep slopes and risk of T+H corrosion due to trapped ions.

Two potential ways to overcome these limitations are dry etched polyimide and usage of photosensitized polyimide.

Fig. 21 shows a polyimide via hole reactively ion etched in oxygen and fig. 22 the resulting good edge coverage with metal.

Fig. 21. 1-Step Dry Etched Polymide + $SI_3N_4$ Vias

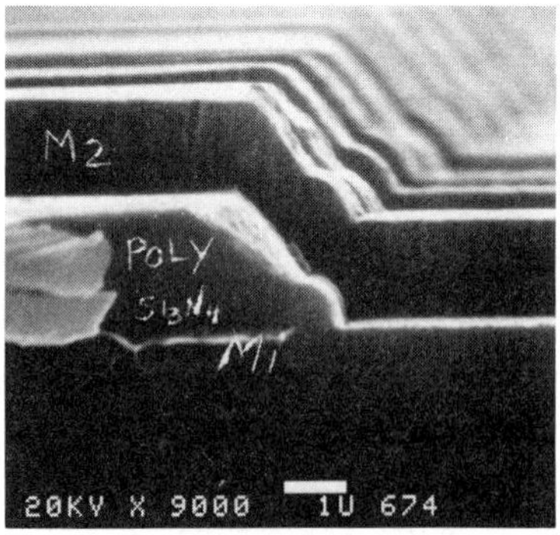


Fig. 22. M2 Coverage of Dry Etched Polymide Via

Photosensitive polyimide (see fig. 23) uses a photosensitive precursor, e.g. polyamic acid with the allyl group in a sidechain. The conjugated double bond can absorb light which by crosslinking makes the polymer insoluble in solvents. Subsequent curing forms polyimide under alcohol release.

Table 6 shows the outline for the in principle less complicated photosensitive polyimide process.

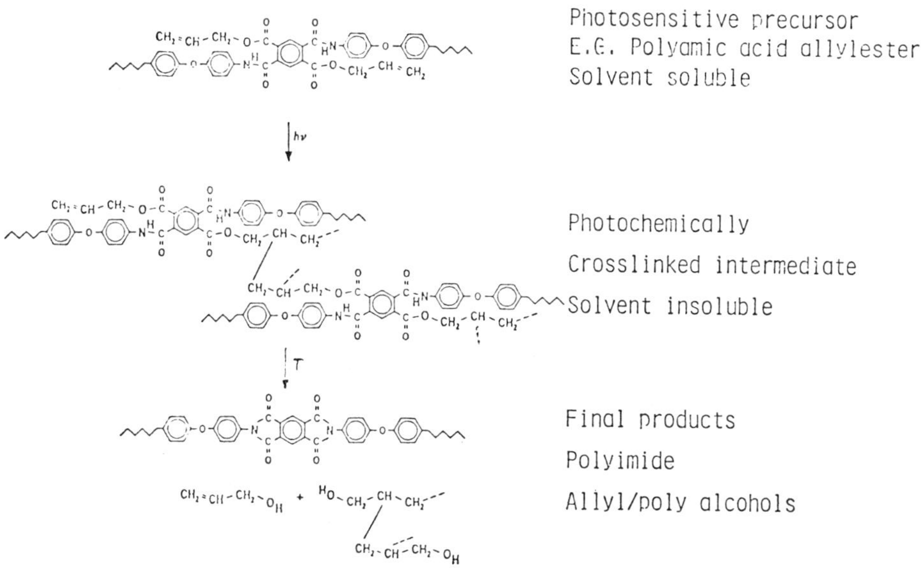


Fig. 23. Photosensitized Formation of Polyimide Pattern

Table 6: Process Outline

PHOTOSENSITIVE POLYIMIDE

| PROCESS STEPS | CONVENTIONAL | PHOTOSENSITIVE |
|---|---|---|
| ADHESION PROMOTER APPLY | X | X |
| PRECURSOR APPLY, BAKE | X | X |
| PHOTORESIST APPLY, BAKE | X | |
| EXPOSE | X | X |
| DEVELOP RESIST | X | |
| PRECURSOR | X | X |
| NEUTRALIZE | X | |
| STRIP PHOTORESIST | X | |
| CURE POLYIMDE | X | X |

Topography-free Metallization

A significant reduction of the metallization topography is a prerequisite to meet the tighter groundrule requirements of the near future. A topography-free metallization process can be developed by utilizing the good planarization properties of polyimide in combination with dry etched polyimide vias.
A schematic of this process is shown in fig. 24.

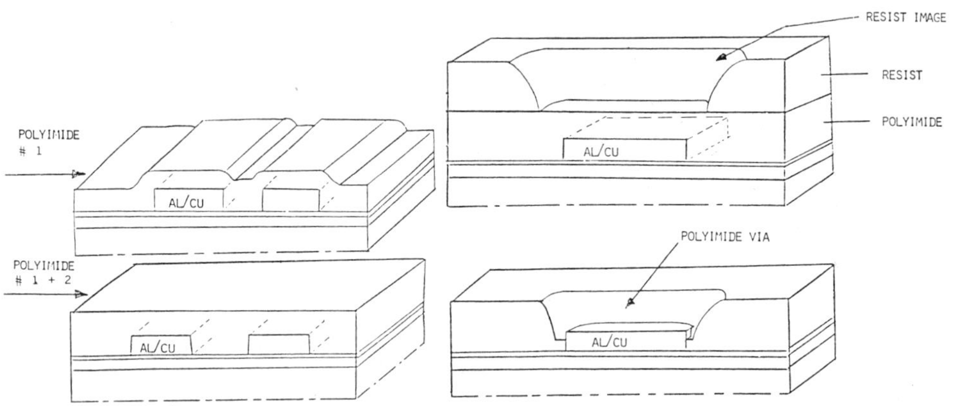


Fig. 24. Topography-Free Metalization Process

ACKNOWLEDGEMENTS

The paper "Organic Insulators" uses material from many experts inside and outside IBM, who all together made organic insulation and passivation a success in the modern semiconductor industry. The author is grateful for this information.

# MULTILAYER CERAMIC

Rainer O. Stahl

IBM Laboratories

Boeblingen, West Germany

ABSTRACT

The continuous advances in the semiconductor technologies involve smaller devices, faster circuits and increased circuit density, resulting in high performance large scale integrated (LSI) chips. The wiring between LSI chips requires a large number of high speed, high density interconnections. This paper describes the "Multilayer Ceramic (MLC)" process used in the IBM Corporation to fabricate multichip substrates with tight tolerances that meet these requirements by minimizing the interchip distances and providing multilayer wiring with dense vertical interconnection capability.

## INTRODUCTION

A decade ago the package for semiconductor chips had to accomplish the following tasks

- physical support for the semiconductor chip
- protection against environmental influences
- hold the chip temperature at the proper operating level
- supply of the proper voltage and current levels
- provision of a fast electrical path to the next level of package.

Today, in the age of LSI chips the package has also to be an integral part of the electrical design of the whole computer system. Significant enhancements of computer performance can no longer be achieved without equivalent advances in the packaging technologies. The benefits of fast LSI chips would be lost if the chips could not be packed closely together in order to minimize the signal delay between them.

This means a requirement for short, high density, impedance controlled interconnections. The fastest interconnections can be realized on multichip substrates. This type of substrate also gives relief to the next level of package by absorbing a large part of the interconnections between the chips. (Ref. 3, 4, 6, 7, 8).

Today's multichip substrates achieve a ten-fold increase in circuit density over the previous IBM package technology.

This paper describes the "Multilayer Ceramic (MLC)" process that was chosen to satisfy the requirements of modern high speed processing units (Fig. 1).

The basic material of MLC is an unfired ceramic sheet, the "greensheet". Holes are punched into the greensheet and filled simultaneously with the deposition of the desired metal pattern on the greensheet. Many metallized sheets are stacked and laminated together. The laminate is sintered. During sintering the substrate shrinks at a predefined rate.

## SUBSTRATE FABRICATION

### Selection of Materials

The material chosen for the ceramic is alumina ($Al_2O_3$). This is available in good quality and moderate cost, has high strength and good thermal conductivity. The conductor material is molybdenum. This metal is stable at the sintering temperature of alumina. It has good electrical conductivity, moderate cost and sufficient resistance to corrosion.

In order to ease sintering glass is added to the alumina.

### The Greensheet (Ref. 5)

The alumina and glass powders are mixed with an organic binder. This is done in a ball mill to assure good dispersion and coating of the powder particles. The re-

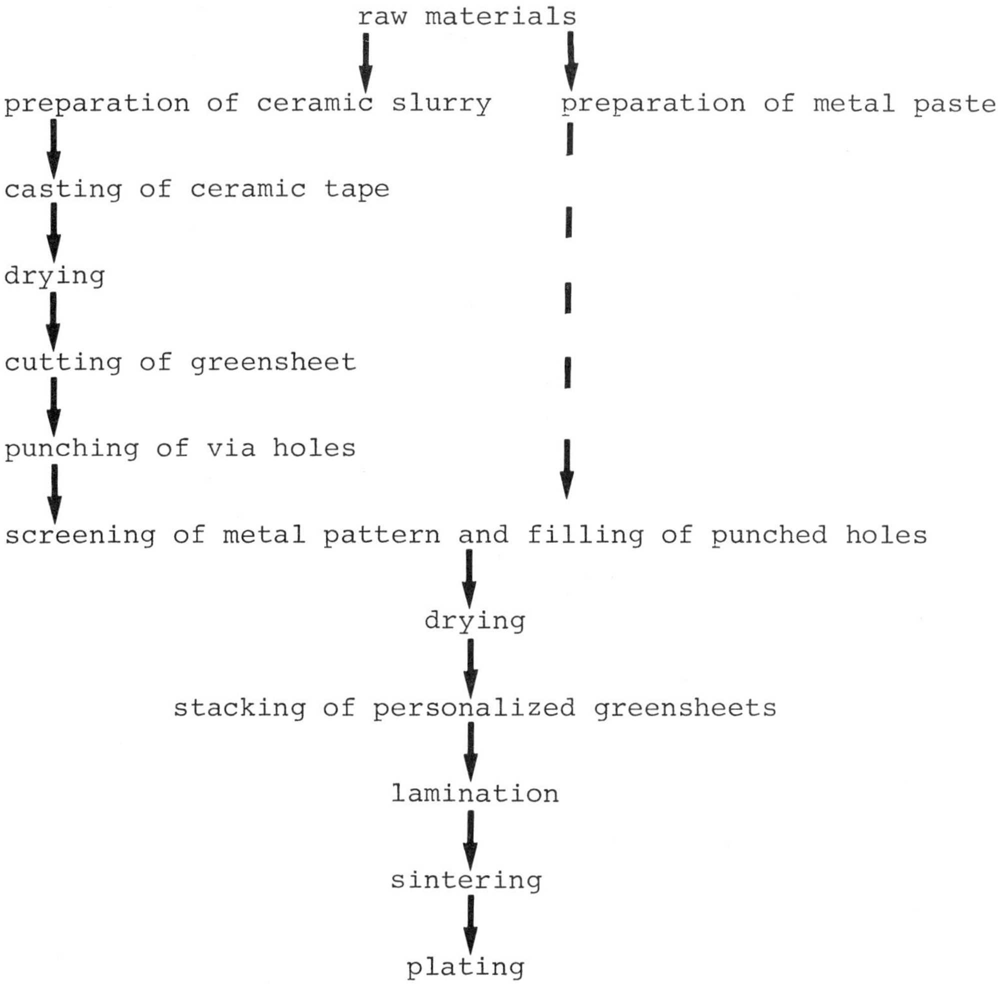


Fig. 1. Simplified process flow for fabrication of MLC substrates

sulting milk-like fluid is called the ceramic slurry. Very good dispersion of the ceramic particles is required to guarantee high quality greensheets, a prerequisite for good lamination and controlled shrinkage during sintering.

The greensheet is cast in the so-called doctor blade or tape casting process: The ceramic slurry flows from a reservoir under a blade onto a moving plastic tape. To achieve a void-free greensheet with constant thickness the slurry reservoir maintains a constant level and the tape moves with a constant speed depending on the viscosity of the ceramic slurry and the desired thickness of the greensheet.

The greensheet is subsequently dried, separated from the plastic tape and spooled. After inspection, the ceramic tape is cut into squares. Samples go into a quality test.

The thickness of the unfired greensheet is 0.20 mm or 0.28 mm, the width is 200 mm.

For the following process steps the greensheet must have good strength and dimensional stability. This was achieved by the proper selection of the organic binder system and a well controlled casting process. Nevertheless, the greensheet is very flexible to allow good lamination.

<u>Personalization of the Greensheet</u>

<u>Punching of Holes.</u> For interconnections in vertical direction holes are punched into the greensheet (Fig. 2). Each greensheet layer of the substrate has a different hole pattern. Therefore, a computer-controlled puncher was built. A multitude of individually movable punches finishes a greensheet with up to 36 000 holes in a few minutes. The punched greensheets are inspected after cleaning. Even after punching of that many holes, good strength and dimensional stability are the main characteristics of the greensheet. Deformation can not be tolerated.

<u>Metallization.</u> The material for the interconnections is a molybdenum paste. Similar to the fabrication of the ceramic slurry the metal powder is mixed with an organic solvent. The molybdenum paste is extruded through a nozzle and forced through a metal mask onto the greensheet as the paste head traverses the mask. The punched holes are filled simultaneously with the deposition of the conductor pattern (Fig. 3). The metal mask is placed in direct contact with the punched greensheet to prevent smearing of the printed lines.

The main difference to conventional "screen printing" techniques is the flexible surface of the greensheet. Although during drying of the greensheet a part of the binder system has evaporated, the remaining binder can interact with the solvent of the paste. To guarantee a good adhesion of fine paste patterns, such an interaction

Fig. 2. Comparison of holes in a greensheet with a punch card

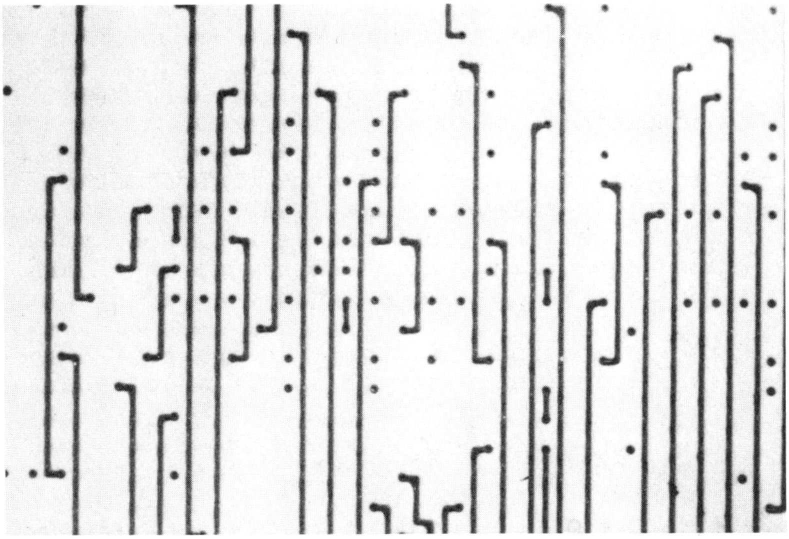

Fig. 3. Magnified section of a personalized greensheet

may even be desired, although every interaction is equivalent to distortion of the greensheet. For large areas of metallization an aggressive paste solvent is not permissible. That is why different paste compositions are used for the various metal patterns (Fig. 4).

After printing the metallized sheets of unfired ceramic are dried. The personalized greensheets are inspected with an optical scanner. Defects can be repaired. Repair capability at this stage of the process is a prerequisite for a good yield of the finished part, as the substrates consist of up to 36 layers, each with a different metallization.

## Stacking

The personalized greensheets are stacked in the correct sequence and orientation. Precision holes in the four corners and an orientation hole guarantee accurate positioning of the greensheets.

## Lamination

Lamination is done at high pressure and slightly elevated temperature. The purpose of lamination is to press the air out of the space between the layers, densify the material and achieve good adhesion between the layers. High quality of the laminate is obtained when the boundaries between the layers have disappeared. In this case separation of individual layers is no longer possible. If the lamination pressure is not high enough, the substrate will delaminate during firing. Under excessively high pressure the flexible greensheets will start to flow and later crack during firing.

During personalization of the greensheets, metal was deposited onto the layers. The lamination pressure produces higher densification around the metal lines, resulting in lower shrinkage during the following firing. This is a source for unequal shrinkage (warping) and distortion.

After lamination the substrate is still in the green state, as only a small part of the organic material has evaporated.

Lamination is one of the key steps in the MLC process. It determines, to a large degree, the behavior of the substrate during the next process step: sintering or firing.

Fig. 4. Various printed metal patterns

## Sintering

The goal of the sinter process is conversion of the porous, flexible green laminate into a hermetic, mechanically and thermally extremely stable piece of ceramic. Particle rearrangement, grain growth and material transfer, mainly surface diffusion cause the removal of the porosity. The linear shrinkage during these processes is approximately 17%. Sintering starts at 1250 $^{o}$C. The maximal used sintering temperature is 1560 $^{o}$C.

During the initial phase of the sinter process a part of the organic solvent and binder system evaporates through the interconnected pores of the material. The remaining organic materials start to decompose, causing a color change of the laminate from yellow to black, indicating pure carbon. Wet hydrogen oxidizes the carbon. At about 1200 $^{o}$C all of the organic material is burnt off. The laminate is now white, still porous and therefore weak.

During sintering the alumina particles as well as the molybdenum particles grow together. The substrate gains strength, porosity decreases, density increases and shrinkage happens. In order to accelerate sintering and still achieve maximal densification, the ceramic material contains about 10% glass. The glass allows sintering in the presence of a liquid phase (Ref. 9, pp. 498-501). As alumina has limited solubility in liquid glass, sintering can be done faster and at lower temperature. Glass is also advantageous for adhesion of the metal conductors to the ceramic and for reduction of grain boundary stresses during cooling, as alumina has anisotropic thermal expansion.

After a period of 33 hours in the sinter furnace, a hermetic material with extremely tight tolerances has been produced, although in the beginning of the temperature cycle high porosity was necessary to allow escape of the evaporating organic materials. Careful cooling is important to avoid stress in the ceramic/metal composite.

### Substrate Finishing

Features on the two surfaces and the exposed ends of the metal in the holes are electroplated with nickel and gold to prevent corrosion and to prepare the substrate for subsequent soldering of chips and brazing of pins. The substrate is electrically tested.

## OUTLOOK

The multilayer ceramic process is extendible to different material compositions. Future developments will tend to increase both electrical performance and reliability. Higher performance demands minimization of signal propagation delay. This can be achieved by further decrease of interchip distances and decrease of the dielectric constant of the ceramic material. Better reliability of the semiconductor to ceramic connections demands better matching of the coefficients of thermal expansion of semiconductor and carrier.

## SUMMARY

The multilayer ceramic process and material system represents a significant progress in the fabrication of ceramic carriers for semiconductor chips. A novel selection of materials and a complex series of process steps allow the integration of a high density, high performance three-dimensional interconnection system into a large ceramic body with extremely tight tolerances. This provides the basis for accommodation of a high number of LSI chips on a single substrate, thereby implementing a true LSI package (Fig. 5).

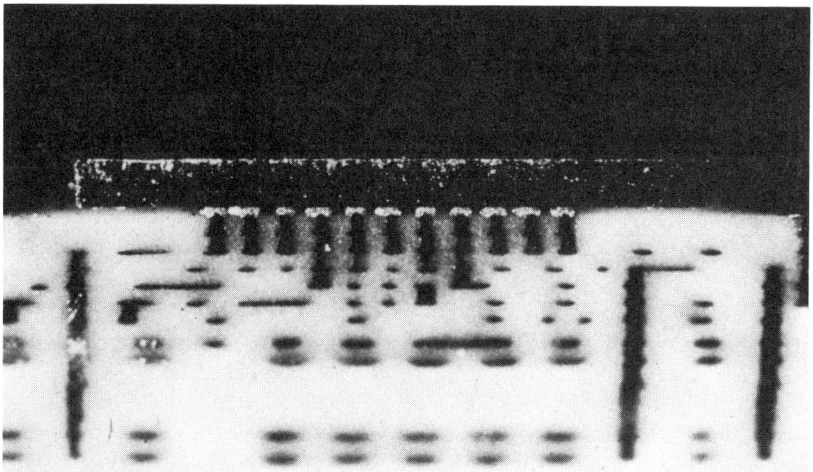

Fig. 5. Cross section of a multilayer ceramic substrate

Further development of the MLC technology will have to match the advances in the semiconductor technologies.

ACKNOWLEDGEMENT

It is not possible to acknowledge all contributions to the development of the MLC process. The earliest work was done in the 1960's in IBM groups in the United States, France and Germany (e.g. Ref. 1, 2). The author appreciates the opportunity to present a summary of this work.

REFERENCES

1. B. Schwartz and D.L. Wilcox, "Laminated Ceramics", Proceedings of the Electronic Components Conference, IEEE, New York, 1967, pp. 17-26.
2. H.D. Kaiser, F.J. Pakulski, and A.F. Schmeckenbecher, "A Fabrication Technique for Multilayer Ceramic Modules", Solid State Technol. 15, No. 5, pp. 35-40 (May 1972).
3. A.J. Blodgett, "A Multilayer Ceramic Multi-Chip Module", Proceedings of the Electronic Components Conference, IEEE, New York, 1980, pp. 283-285.
4. B.T. Clark and Y.M. Hill, "IBM Multichip Multilayer Ceramic Modules for LSI Chips - Design for Performance and Density", IEEE Trans. Components, Hybrids, Manuf. Technol. CHMT-3, pp. 89-93 (1980).

5. W.R. Swiss, "Fabricating Ceramic Sheets for Multilayer Substrates", Circuits Manufacturing, Vol. 19 No. 11, pp. 43-46 (1979).
6. M.S. Pittler, D.M. Powers and D.L. Schnabel, "System Development and Technology Aspects of the IBM 3081 Processor Complex", IBM Journal of Research and Development, Vol. 26, No. 1, pp. 2-11 (Jan. 1982).
7. A.J. Blodgett and D.R. Barbour, "Thermal Conduction Module: A High-Performance Multilayer Ceramic Package", IBM Journal of Research and Development, Vol. 26, No. 1, pp. 30-36 (Jan. 1982).
8. D.P. Seraphim and I. Feinberg, "Electronic Packaging Evolution in IBM", IBM Journal of Research and Development, Vol. 25, No. 5, pp. 617-629, (Sept. 1981).
9. W.D. Kingery, H.K. Bowen and D.R. Uhlmann, "Introduction to Ceramics", Second Edition, 1976, John Wiley and Sons, New York.

# HIGH PERFORMANCE COOLING AND LARGE SCALE INTEGRATION

Robert W. Keyes

IBM Thomas J. Watson Research Center

Yorktown Heights, NY 10598

Large scale integration has not invaded the highest performance regime of computation. The speed of circuits on LSI chips has not matched those available at lower levels of integration. The root of the difference is to be found in power; increases in number of components have not been accompanied by a proportionate increase in power per chip. The trade-off between power and speed means that the limited ability of technology to remove heat at high density from compact structures leads to a lower power dissipation per gate at higher levels of integrations and lower speed.

This paper first reviews methods of removing heat from large computing systems. A new method of removing heat from microelectronic chips that has recently been developed by Tuckerman and Pease at Stanford University offers a way to decrease the thermal resistance from chip to coolant by over an order of magnitude. Their method involves etching channels in the back of the chip and passing water through the channels. The method is described in detail and analyzed.

The new cooling technique should make the use of higher power circuits on integrated chips possible and increase the speed of large systems using large scale integration. The implications of the improved cooling for future large computers are assessed quantitatively in the context of a model recently developed by the author.

## INTRODUCTION

Modern solid state electronics has demonstrated a persisting capacity to provide increasing functional capability while decreasing the cost of components and the systems built from them. The technological themes that have made continuing advances possible are miniaturization and integration. Integration is the key to low cost because it permits many devices to be fabricated by handling and processing one object. Integration also leads to improvements in reliability because the interconnections formed by integrated semiconductor technology, in addition to being less costly, have turned out to be far more reliable than solder joints and pluggable connections.

Miniaturization, of course, is one of the ingredients that makes high levels of integration possible. In addition, however, it leads to increased speed of operation because electrical signals have shorter distances to travel and because capacitances are smaller. The lower capacitances also reduce the power dissipation of circuits. Thus, the benefits accruing from miniaturization and integration are to be found in cost, speed, power dissipation and reliability. It must be observed that low cost, low power dissipation and high reliability are the things that make it possible to construct and operate large systems containing a great many components.

A summary of advances in miniaturization and integration through the past two decades is presented in Fig. 1. The rapid growth of component count into the hundreds of thousands per chip has surpassed early expectations of the future of large scale integration and has come to be known as Very Large Scale Integration or VLSI.

VLSI has not, however, invaded the highest performance regime of computation. The speed of circuits on VLSI chips has not matched those available at lower levels of integration. The root of the difference is to be found in power; increases in number of components have not been accompanied by a proportionate increase in power per chip. Even though the reduction in device capacitances that accompanies miniaturization has made low power circuits possible there is still a trade-off between power and circuit speed. Limitations on the ability of technology to remove heat at high density from compact structures are directly reflected in the speed of circuitry.

Recently, however, Tuckerman and Pease at Stanford University have achieved chip-to-water thermal resistances of 0.1 degree/watt.(1) One kilowatt has been removed from a one centimeter square silicon

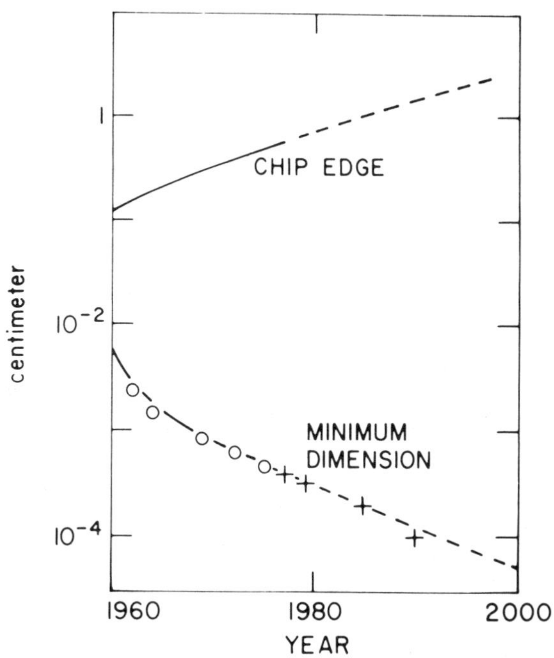


**Fig. 1** The advance of miniaturization and chip size. The combination of these two factors and ingenuity in device design has led to rapid increase in the number of components that can be produced on a chip.

substrate. The method involves etching channels in the chip to transfer heat directly from the chip to flowing water. Thus, an increase of up to two orders of magnitude in power dissipation per chip may make high performance VLSI systems possible in the not-too-distant future.

This paper will briefly review cooling techniques used in large, high performance computers. The novel cooling method developed by Tuckerman and Pease will be described. Their work will be referred to as "T and P". The new method represents a great advance as compared to conventional cooling of electronic computers and its possible application to computers will be described in some detail.

## COOLING LARGE SYSTEMS

The need for attention to the cooling of large systems has been recognized since the early introduction of computers. The heat dissipated must be transferred to some fluid that carries it out of the system and, eventually, out of the structure in which the system is housed. Thus, the physical specifications for installation of a large computer usually include an air conditioning requirement. In the machine itself, heat may be transferred to room air or to a liquid that carriers heat to a heat exchanger that subsequently transfers it to air.(2-8)

The basic problem encountered in the machine is that fast computation requires that all of the components of the system be close together, making the provision of large areas through which heat can be transferred from solids to fluids difficult. Liquid intermediaries are frequently used because solid-liquid heat transfer coefficients can be an order of magnitude larger than those to air and liquid cooling of computers was introduced before 1965.(2) However, compaction of systems is also limited by the space needed for wiring, interconnection, power supply lines, and filter capacitors and good results have also been obtained without liquids.(3,5)

For example, in the Amdahl 470 computers, heat is conducted through a metal stud from which it is transferred to a cooling air stream through an area ten to forty times the area of the chip.(3) In the IBM TCM, the heat flows from the chip to metal (or through helium gas to the metal) then through metal to a cold plate in which it is transferred to water.(6,7) Thermal resistances of about 10 deg/w are achieved in these systems. In both cases, the wiring is provided by a multilayer substrate on the opposite side of the chips from the cooling water.

The quantitative aspects of cooling are presented in Fig. 2. Here heat removal rates in large computers are plotted against the year of commercial introduction. Heat removal rates are expressed in watts per $cm^2$ of the first multi-chip substrate and per $cm^2$ of a chip. The drive toward denser packaging of chips has led to continual increase of the former quantity. For perspective, the rates of heat transfer in some more or less familiar situations are indicated. The power densities involved in electronics are very high in any comparison with common experience. In particular, the density of heat in the experiment of T. and P. is within an order of magnitude of the radiation intensity at the surface of the sun.

## THE FIN/CHANNEL COOLING METHOD

T. and P. observed that it was possible to scale more or less conventional liquid heat exchanger technology to dimensions of tens of microns and thereby apply it directly to silicon chips.[(1)] The basic idea in heat exchangers is the provision of a large area where a solid and a fluid are in contact and across which heat can flow from the solid to the fluid. The fluid, being mobile, carries the heat away. Fins attached to the solid are a convenient way to increase the interfacial area. Good performance requires that the fins have a high thermal conductivity; copper and aluminum are frequently used for the purpose. However, silicon also has a high thermal conductivity and can be used for the fins. Thus, T and P were able to fabricate the fins in intimate contact with a silicon chip by simply etching them into the back of the chip.[(1)] They thereby reduced the chip-to-coolant thermal resistance to around 0.1 deg/watt, well over an order of magnitude smaller than that found in contemporary electronic technology. The nature of the structure is shown in Fig. 3.

Some of the parameters of the experiment are given in Table 1. The design was based on known heat transfer principles.[(1)] The measured thermal resistances are given as a function of flow rate in Fig. 4. (Note that the structure was optimized for a pressure of 30 psi and is not optimum at the lower flow rates.) It is seen that the thermal resistance from chip to water was only 0.1 deg/watt at the highest flow rate. 800 watts was dissipated on the chip. A simplified analysis of the functioning of the cooling structure is presented in the Appendix.

T and P have considered and investigated various modifications of their original experiment.[(8,9)] Since the original design was already optimized, further improvement is obtained only with difficulty. For example, higher pressures drive the coolant through the channels faster and thereby decrease the temperature increase. However, the thermal

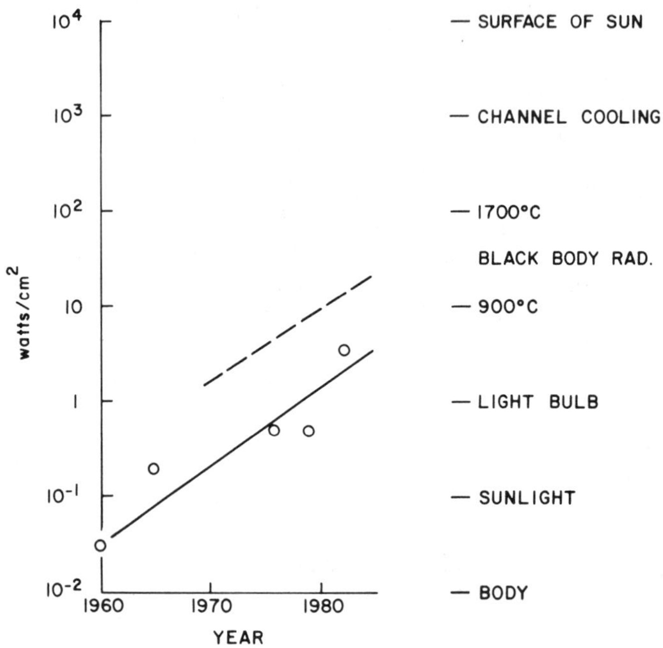


**Fig. 2** Trend of heat removal rates in large computers. The solid line refers to area of multi-chip substrates and the dotted line to chips themselves. The densities of heat flow encountered in certain other situations are indicated to provide perspective.

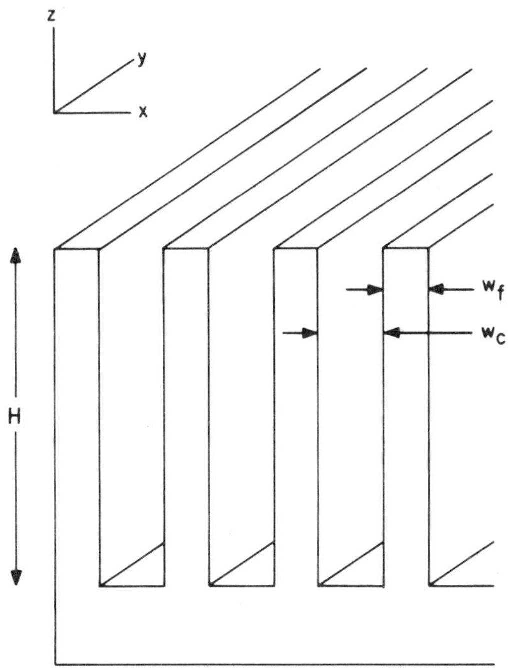


**Fig. 3** The silicon structure used by Tuckerman and Pease.[1] The dimensions are given in Table I and the coordinate system and notation used in the Appendix are indicated.

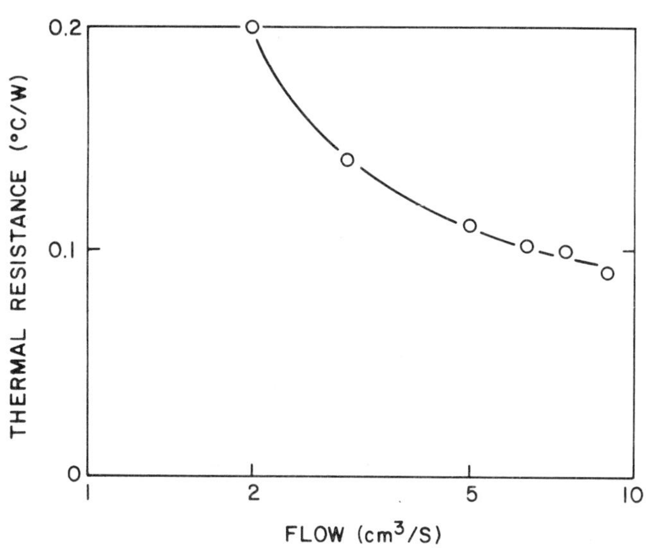


**Fig. 4** Measured thermal resistance vs. volumetric flow of water in the structure of Fig. 3.[1]

resistance depends on the pressure only as the 1/4 power. Other liquids have less favorable thermal properties than water.

Another significant aspect of the experiment is that the cooling is accomplished with a very compact structure. In addition to the silicon, a cover plate must be provided and sealed and water inlet and outlet manifolds provided. These can be fabricated in thin sheets of glass[1,9] and are of a size comparable to the silicon chip itself.

Table I. Parameters of the Tuckerman and Pease experiment[1]

| | |
|---|---|
| Chip size | 1 cm x 1 cm |
| Channel width | 50 $\mu$m |
| Fin width | 50 $\mu$m |
| Channel depth | 300 $\mu$m |
| Pressure | 31 psi |
| Water volume | 8.6 $cm^3/s$ |
| Thermal resistance | 0.09 deg/W |

## APPLICATION TO LSI

The significance of thermal resistance for integrated electronics is summed up in the following expression:

$$\Delta T = R_T N p \tag{1}$$

Here $\Delta T$ is the temperature increase of a chip, $R_T$ is the thermal resistance, N is the number of logic gates on the chip, and p is the power per gate. (The total power dissipated on the chip also includes a contribution from off-chip line-drivers that is smaller than the circuit power and is neglected here.) Chips are restricted to operate below some maximum $\Delta T$. The principal factor determining the maximum $\Delta T$ is reliability; atomic motions, such as those encountered in electromigration, are greatly accelerated by high temperatures and can destroy the functioning of the chip.

Equation (1) shows that a decreased $R_T$ can be used in various ways.[9] The number of gates on the chip can be increased, the power per gate can be increased, and the operating temperature can be decreased. These possibilities will be examined more closely. Various problems associated with the first two will appear.

The chief factor limiting N is yield. All of the devices and interconnections on a chip must operate and conform to certain specifications. The more gates on a chip, the higher the probability of a defect and the lower the yield. Improvements in technology through the past two decades have made it possible to produce chips with transistor counts approaching a million. One can look forward confidently to continuing improvements in processing and acceptable yields at ever-increasing N.

Another restriction on N, especially in large systems, arises from the need for chips to communicate with other chips and devices. More gates on a chip implies a need for more off-chip communication channels. Thus, the number of gates that can be usefully employed on a chip is limited by the availability of connections to the chip.[11] The number of connectors is limited by their size (they are large compared to features on the chip), and by reliability; off-chip connectors are one of the main sources of failure in electronic assemblies.[12]

A still further limit on the number of gates and the amount of off-chip communication arises from the simultaneous switching problem. Sending a large number of signals off of a chip simultaneously changes the current drawn by the chip from the power supply. The change in current through the inductance in the power supply leads perturbs the voltage and may cause erroneous operation.

Increasing p, the power per gate, by controlling more current allows capacitances to be charged faster; the circuit delays are decreased. The power that can be delivered to a device is limited by electromigration in the wires on the chip. The power that can be delivered to the chip is limited by the current-carrying capacity of the off-chip connections, which is, again, a question of electromigration.

Using the lower thermal resistance to decrease $\Delta T$ is the simplest way to apply new cooling technology and confers great advantages in reliability.[9] In fact, using part of the gain in cooling capability to operate the chip at a reduced temperature can be seen to be a key to relieving the difficulties associated with increasing the number of gates and increasing the power per gate. The limitations other than yield are rooted in reliability. For example, lowering the temperature decreases the rate of electromigration failure according to the formula

$$Rate \sim J^2 \exp\,(-Q/kT)$$

Here J is current density and Q is an activation energy, 0.5 eV for the aluminum conductors used on chips. Thus, for example, operating a chip at 40°C instead of 80°C reduces the rate of electromigration by a factor 8 or permits 3 times larger current at the same rate. The current carrying capacity of the chip connections would also be increased. Further, other mechanisms of failure of the connections would be decreased, permitting more connections to a chip. A large increase in the number of off-chip connections would not only satisfy the power delivery and signal channel requirements but could alleviate the simultaneous switching problem by permitting the transmission of both polarities of a signal off of the chip.

Finally, the ability to design logic chips with large numbers of gates is a practical limit on the use of very high levels of integration. A very large number of circuits can lead to great complexity in logic design, layout, and interconnection. Design becomes a major factor in the cost of a product. Therefore, memory chips with their simpler structure lead the advance of integration. The cost of the design of a specialized complex logic chip can only be supported when there is a large market for the chip. Several methods for making the advantages of VLSI available to small volume users have evolved. The microprocessor chip can be programmed to perform a variety of functions in the same way as a computer. Gate arrays or masterslices are chips that take advantage of volume production by manufacturing an array of standardized logic gates whose function is defined in the final stages of production by one or two layers of wiring. Computerized methods can be used to construct the wiring patterns that define the function of the gate array, and the steadily increasing power of computers can be expected to produce greater effectiveness of such design programs.

Thus, the problems associated with utilizing the high-intensity cooling on a chip are not insoluble. The goals of high reliability and low cost are served by increasing levels of integration and it seems safe to assume that progress in this direction will continue. Obviously, powerful single-chip processors will become possible. Perhaps the most important applications, however, will be in large systems, some of which already employ water for cooling.[2-8] This author has already projected the technology and performance of large systems through the next two decades.[13] These projections, then, constitute a convenient framework for exploring the implications of the new cooling technology for VLSI in large systems. Since it will probably take five years for the new technologies made possible by high-density cooling to reach the marketplace, it is most reasonable to apply the new cooling capability to the projections beyond 1987. The power per chip is taken to be 100 watts in 1987, increasing to 200 watts at the end of the century. The power is taken to be considerably less than the possible maximum to maintain a low chip temperature. The T and P method permits very dense packaging of chips and the module area per chip is only about twice the chip area. Such progress will place great pressure on the miniaturization of chip-module interconnections and module wiring.

The calculation was modified to determine the geometry of the unit cell by the wire-dominated chip model introduced by the author.(14,15) The size of the logic cell is then largely determined by the wiring requirements. The width of a wire channel in the model is taken to be five times the projected minimum lithographic dimension. The logic cells then turn out to be somewhat larger than those of the earlier projection.

The modified projection is presented in Figs. 5-8. Figure 5 shows the projected logic delay. It is several times smaller than the previous result(13) and should allow the construction of 1000 MIP processors by year 2000. The gap between logic speeds and main memory access is widened, further emphasizing the need for innovation in buffers and memory hierarchy.

The pressures on other aspects of the package resulting from the availability of high-density cooling are illustrated in Figs. 6 and 7. Figure 6 shows that the module area available per chip connection according to the model will be only of the order $10^{-3}$ $cm^2$. Denser module wiring will also be required; Fig. 7 shows the length of wire that must be placed in each $cm^2$ of module area. Highly miniaturized wire channels or larger numbers of wiring layers will be needed.

Figure 8 shows the projected utilization of silicon area. The area devoted to logic decreases less rapidly than in the earlier projection because of the use of the improved model of the logic cell. The area in memory is larger than in the previous case because higher projected processing rates call for larger main memories.

## CONCLUSION

The silicon fin/channel cooling method offers a way to increase the power dissipation of chips in large systems by at least an order of magnitude, offering an opportunity to introduce VLSI into such systems. Thus, T and P appropriately assert that "The dramatic improvement in

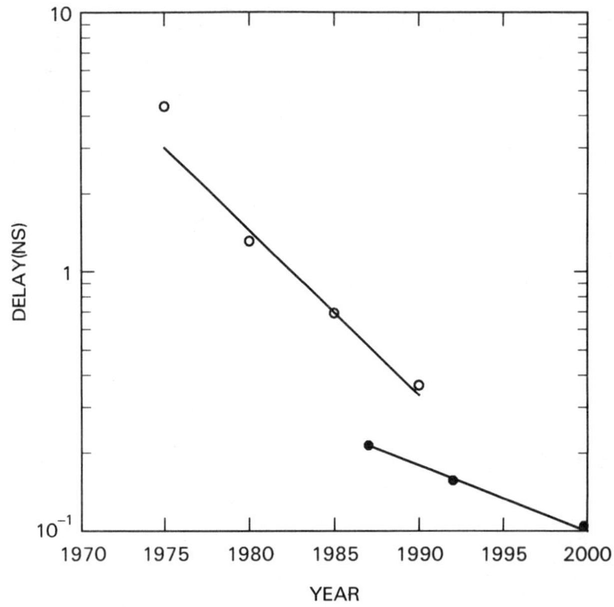


**Fig. 5** Projections of logic delay as in reference 13, modified to take account of the increased power dissipations made possible by the etched channel technique.

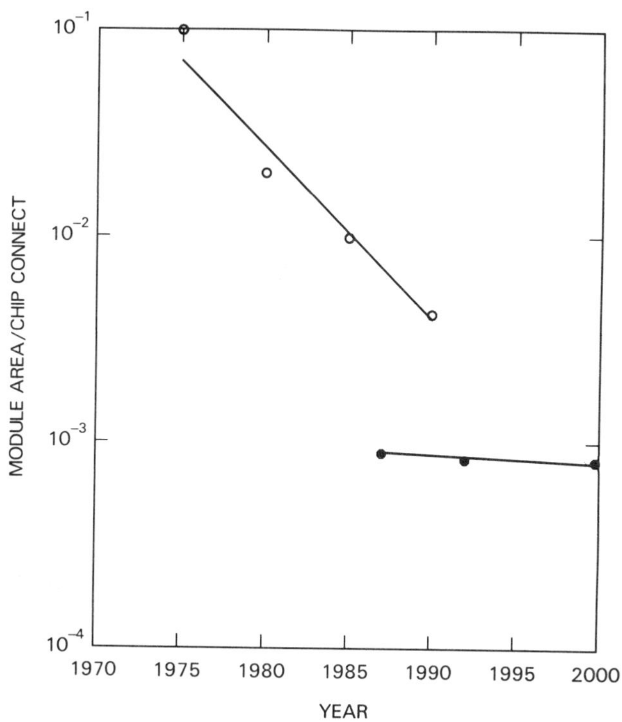


**Fig. 6** An illustration of the projected increased demand on packaging technology. Increased levels of integration and packaging chips closer together will decrease the area available per connection from a chip to the substrate on which it is mounted.

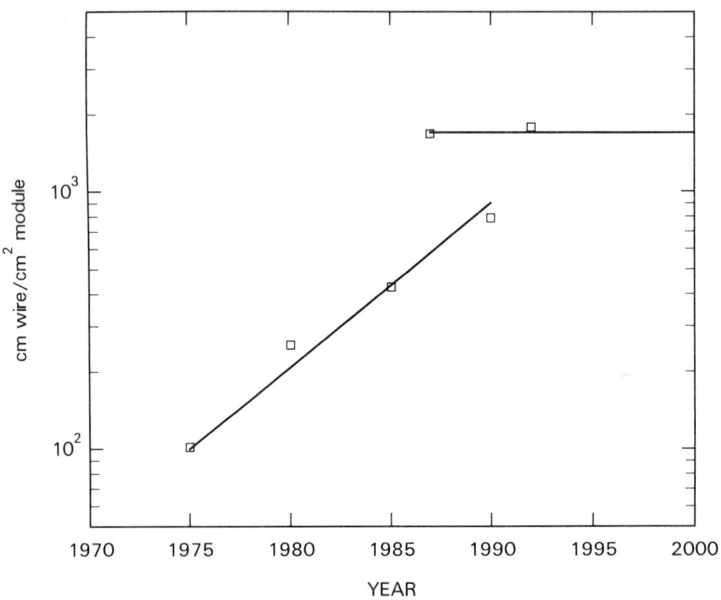


Projected density of package wiring

**Fig. 7** Another illustration of the increased pressure on packaging technology. A greater length of wire per unit area of substrate will be needed.

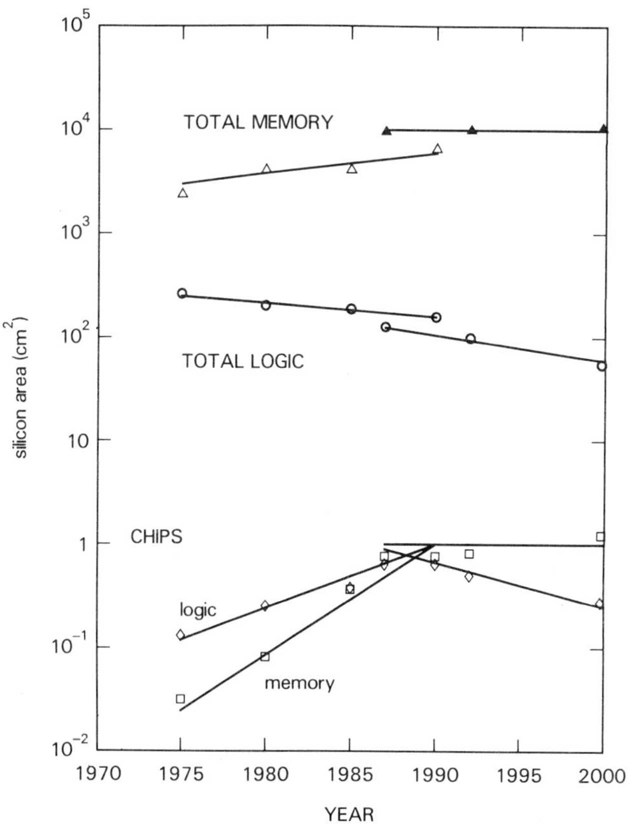


Fig. 8 Utilization of silicon area in the modified projection.

practical, compact IC heat-sinking capability presented....offers a new degree of freedom for the system designer."[1]

## APPENDIX: Analysis of Fin/Channel Heat Transfer

T and P used known engineering practice to design and optimize the heat transfer structure.[1] In order to make the physics of the fin/channel cooling structure transparent it will be analyzed here in terms of a highly oversimplified model that is susceptible to analytical description. The notation is explained in Fig. 3 and in Table II.

The heat flows from the chip to the fins, from the fins to the fluid, and is carried out of the chip by the fluid. It is assumed that heat is generated uniformly across the chip and enters the fins at the same rate throughout the chip. The heat current entering the fins is independent of y. The average temperature of the water thus increases linearly in the y direction. All of the heat generated on the chip flows into the fins at z=0. Also, the temperature gradient in the fins is independent of y. Since no heat flows through the ends of the fins, the z deriviative of the temperature vanishes at z=H. These boundary conditions are satisfied by the following representation of the temperature:

$$T = \alpha y - \frac{J_o}{K_f w_f}\left(z - \frac{z^2}{2H}\right) + \gamma \tag{1}$$

Here $\alpha$ and $\gamma$ are constants to be determined and the heat current per unit length at the base of the fins has been introduced:

$$J_o = \frac{Q}{WL}(w_c + w_f) \tag{2}$$

The current in a fin is

$$J = J_o\left(1 - \frac{z}{H}\right) \tag{3}$$

The divergence of the current in the fins must be accounted for by the heat flowing into the fluid. The fluid flow is laminar in all of the experiments.[1] The rate of heat flow into the channel must be calculated by taking into account the variation of velocity across the channel and its effect on the temperature profile. The rate as a function of wall temperature and velocity-weighted average fluid temperature is

$$j = \frac{140}{17} K_c \frac{T - \Theta}{2w_c} \tag{4}$$

The quantity (140/17) = 8.235 is known as 'Nu', the Nusselt number for a rectangular channel without end effects. In practice, the viscous drag from the ends of the channel will slow the passage of fluid and the effective Nu will be somewhat less. Thus the divergence of the heat current is

$$\frac{dJ}{dz} = -\frac{J_o}{H} = -2j = -Nu\,K_c\,\frac{T-\Theta}{w_c} \tag{5}$$

Further, the heat entering the fluid must be carried away by the flow

$$C\bar{v}\,w_c\,\frac{\partial\Theta}{\partial y} = 2j \tag{6}$$

Solving Eq. (5) for $\Theta$ and using Eq. (6) allows $\alpha$ in Eq. (1) to be determined:

$$\alpha = J_o/C\,w_c\,\bar{v}\,H \tag{7}$$

So far the approximate solution satisfies the equations defining the problem. However, the fluid enters the channels at $y = 0$ at the zero of temperature, and this condition can only be satisfied in an average way. The constant $\gamma$ will be determined by requiring that the temperature of the fluid averaged over z at $y = 0$ be zero. The expression for T then becomes

$$T = J_o\left[\frac{y}{C\,w_c\bar{v}H} + \frac{w_c}{NuK_cH} - \frac{1}{K_fw_f}\left(z - \frac{z^2}{2H} - \frac{H}{3}\right)\right] \tag{8}$$

The three terms in Eq. (8) have obvious interpretations. The first is the temperature rise due to heating of the fluid, the second represents the resistance to flow from the fin into the fluid, and the third is the temperature profile in a fin. The maximum temperature of the chip is found at $y = L$, $z = 0$:

$$T_m = \frac{Q}{WL}(w_c + w_f)\left[\frac{12\mu L^2}{C\,w_c^3PH} + \frac{w_c}{Nu\,K_cH} + \frac{H}{3K_fw_f}\right] \tag{9}$$

The velocity has been expressed in terms of the viscosity and pressure in Eq. (9).

Since $w_c$ appears in the denominator of the first term and the numerator of the second, there is a value of $w_c$ that minimizes $T_m$; the value is a compromise between the reduction of the flow of water caused by narrowing the channel and the addition of thermal resistance to heat flow into the channel that arises from widening it. There is also an optimum value of $w_f$ because of the factor involving the number of fins, $W/w_c + w_f$); making the fins too thick decreases their number, making them too thin increases their thermal resistance.

It is convenient in finding the optimal values to introduce a length parameter b defined as

$$b^4 = 12 Nu\, \mu\, K_c\, L^2/CP \tag{10}$$

and to write the expression for $T_m$ in the form

$$T_m = (Q/WL)(NuK_c)^{-1} \left[ f(w_c) + \frac{NuK_c H^2}{3K_f w_f} \right] \frac{w_c + w_f}{H} \tag{11}$$

Here f(w) is

$$f(w) \equiv w + (b^4/w^3) \tag{12}$$

Note that b is a characteristic parameter of the fluid. By taking the partial derivatives w.r.t. $w_c$ and $w_f$ it is found that the extremum of $T_m$ is at value of $w_c$ given by the solution of

$$d(w_c\, f(w_c))/d\, w_c = 0 \tag{13}$$

Furthermore, the optimum fin thickness is

$$w_f = H\left( \frac{NuK_c w_c}{3K_f f(w_c)} \right)^{1/2} \tag{14}$$

The extremum of the thermal resistance is

$$T_m = (Q/WL)(NuK_c)^{-1} (w_c + w_f)^2\, f(w_c)/w_c H \tag{15}$$

With the form of f given by Eq. (11), Eq. (13) yields

$$w_c = b \tag{16}$$

It will be observed that, since $w_f$ in Eq. (15) is proportional to H, there is a minimum of $T_m$ at the following value of H:

$$H_m = b\left( \frac{6}{Nu} \frac{K_f}{K_c} \right)^{1/2} \tag{17}$$

The minimum as a function of H is an artifact of our approximation; however, formulae similar to Eq. (17) are used in heat transfer practice as a criterion of the point at which diminishing returns are obtained from increases in fin height. T and P choose a fin height

$$H_o = b\,(K_f/NuK_c)^{1/2} \tag{18}$$

The optimal fin thickness when $H = H_o$ is $3^{-1/2}b$ and the temperature increase is

$$T_m = 2(1 + 3^{-1/2})^2\, (Q/WL) b\, (NuK_c K_f)^{-1/2} \tag{19}$$

The above optimization was carried out at constant pressure. A more appropriate optimization might be one carried out with constant power input driving the coolant.[9] The driving power is

$$\begin{aligned} U &= P \bar{v} w_c HW/(w_c + w_f) \\ &= P^2 w_c^3 HW/12\mu L (w_c + w_f) \end{aligned} \tag{20}$$

The value of P determined by Eq. (20) can be substituted into Eq. (9) and the values of $w_c$ and $w_f$ that minimize $T_m$ in the resulting equation for $T_m$ determined. There is, in fact, an optimum value of H in this case. A characteristic length parameter, a, is defined by

$$a^5 = 12\mu (NuK_c)^2 L^3 W/C^2 U \tag{21}$$

The best values of $w_c$ and $w_f$ turn out to be

$$w_c = w_f = (3K_f/32NuK_c)^{1/10} a \tag{22}$$

The optimal H is

$$H = (3K_f/NuK_c)^{3/5} a \tag{23}$$

With these values of the geometrical parameters the temperature increase is

$$T_m = 5(Q/WL)(3K_f)^{-2/5} (NuK_c)^{-3/5} a \tag{24}$$

The appearance of exactitude in the formulas of this section should not deceive the reader. Many approximations are involved, in particular, "fully developed flow," the velocity and temperature profiles that would be found far from the inlet in a long channel, as been assumed, and the properties of the fluid have been taken to be temperature-independent. The dependence of Nu on the aspect ratio of the channel has been neglected in differentiation. The approximations involved in writing the simple equation (1) have already been pointed out.

In addition, thermal resistances in the silicon, before the heat current enters the fins, must be added to the thermal resistance of the fin structure. A rough idea of these may be obtained: the bulk thermal resistance of a chip 8 mm x 8 mm and 0.25 mm thick is 0.025 deg/w. The spreading resistance into silicon from a 2$\mu$m diameter spot on the surface is about 1600 deg/watt (only the power generated in a single device encounters this thermal resistance.)

**TABLE II**

Notation employed in analysis of the fin/channel cooling structure

| | | |
|---|---|---|
| $T$ | = | temperature in a fin |
| $\Theta$ | = | average temperature in the fluid |
| $w_f$ | = | width of fin |
| $w_c$ | = | width of channel |
| $K_f$ | = | thermal conductivity of fin |
| $K_c$ | = | thermal conductivity of fluid |
| $J$ | = | heat current in fin (in z direction) |
| $j$ | = | heat current into a channel (in x direction) |
| $Q$ | = | total rate of heat production on chip |
| $L$ | = | length of chip and channels |
| $W$ | = | width of chip |
| $\mu$ | = | viscosity of fluid |
| $P$ | = | pressure drop in channel |
| $v$ | = | $P\,w^2/12\mu L$ = average velocity of fluid flow |
| $H$ | = | height of fin |
| $U$ | = | Power input to coolant |
| $C$ | = | specific heat per unit volume of fluid |
| $a$ | = | characteristic dimension, Eq. (21) |
| $b$ | = | characteristic dimension, Eq. (10) |

## REFERENCES

1. D.B. Tuckerman and R.F.W. Pease, "High-performance heat sinking for VLSI," IEEE Electron Device Letters, Vol. 2, pp. 126-129, 1981.

2. M.I. Donegan and B.R. Pagnani, "Thermal design of the Saturn V Launch Vehicle digital computer," National Electronic Packaging and Production Conference (NEPCON) Proceedings, 1965, pp. 1-20.

3. R.J. Beall, "Packaging for a Supercomputer," 1974 INTERCON Technical Papers (IEEE, New York, 1974), Paper 18/3.

4. E.A. Wilson, "True Liquid Cooling of Computers," 1977 National Computer Conference Proceedings, Montvale, N.J.: AFIPS Press, 1977, pp. 341-348.

5. D.P. Seraphim and I. Feinberg, "Electronic Packaging Evolution in IBM," IBM J. Res. Dev., Vol. 25, pp. 617-629, 1981.

6. R.C. Chu, U.P. Hwang and R.E. Simons, "Conduction Cooling for an LSI Package," IBM J. Res. Dev., Vol. 26, pp. 45-54, 1982.

7. S. Oktay and H.C. Kammerer, "A Conduction-Cooled Module for High-Performance LSI Devices," IBM J. Res. Dev., Vol. 26, pp. 55-66, 1982.

8. R.M. Russell, "The Cray-I Computer System," Comm. ACM, Vol. 21, pp. 63-72, Jan. 1978.

9. D.B. Tuckerman and R.F.W. Pease, "Ultrahigh Thermal Conductance Microstructures for Cooling Integrated Circuits," Electronic Components Conference, San Diego, 1982.

10. D.B. Tuckerman and R.F.W. Pease, "Optimized Convective Cooling Using Micromachined Structures," Electrochem. Soc., Montreal, 1982.

11. B.S. Landman and R.L. Russo, "On a Pin Versus Block Relationship for Partitions of Logic Graphs," IEEE Trans. on Computers, Vol. C-20, pp. 1469-1479, 1971.

12. A.J. Blodgett and D.R. Barbour, "Thermal Conduction Module," IBM J. Res. Dev., Vol. 26, pp. 30-36, 1982.

13. R.W. Keyes, "The Evolution of Digital Electronics Towards VLSI," IEEE Trans. Electr. Dev., Vol. ED-26, pp. 271-279, April 1979.

14. R.W. Keyes, "GaAs High-Speed Logic," in Digital Technology, Status and Trends, (ed. H. Painke), R. Oldenbourg, Munich, 1981; pp. 253-273.

15. R.W. Keyes, "The Wire-Limited Logic Chip," IEEE J. Sol. St. Circuits, to be published.

# LIMITATIONS OF DIGITAL ELECTRONICS

Otto G. Folberth

IBM Laboratory

Boeblingen, Fed. Rep. of Germany

## 1. INTRODUCTION

In the last two to three decades the evolution of Digital Electronics was absolutely spectacular. During this period, most of the significant properties of this technology were improved several orders of magnitude. This holds especially for the integration density, the power dissipation, the performance, the reliability, and - last not least - the cost per function.

These improvements were mainly the result of the continual miniaturisation and integration of more and more functional units on small Silicon chips. All this - however - came not for free: at each new miniaturisation level physical and technological phenomena posed barriers for future progress. Such phenomena were either unexpected and new, or they were already known, but insignificant at the larger scale of previous miniaturisation levels. Substantial R & D-effort was required to overcome these barriers and to develop viable solutions at continually higher integration densities[1].

There is no doubt that this trend will continue. In future, such barriers will probably become higher and higher and harder and harder to overcome. The increasing complexities and ever raising development costs of VLSI-chips will pose additional problems for further progress. All these obstacles will certainly reduce the rate of progress, but will probably not slow it down completely.

Contrary to the above mentioned "soft" barriers, there are - however - also "hard" physical limits, which cannot be surpassed even with unlimited means and which will bring to an end the evolution of the Silicon-Chip-Technology, eventually. If everything else is sorted out, there remain essentially three limitations of this kind:

- geometrical limitations
- thermal limitations
- signal propagation limitations.

In the remainder of this paper the significance and interdependence of these limitations[2] will be discussed, mainly for the digital Silicon Technology at (elevated) room temperature.

## 2. GEOMETRICAL LIMITATIONS

Future sophisticated fabrication methods will allow the production of elements with minimal linear dimensions of less than 0.1μm[3,4]. In this case, minimal transistor structures will not be limited essentially by the fabrication process but rather by such critical dimensions, which are not inherently affected by miniaturisation, e.g., the thickness of depletion layers. The area requirement[5] for such minimal transistors on silicon chips is in the order of a few $\mu m^2$. For reasons of simplicity the following calculations will be performed with one value for the minimal transistor size - namely, 2 $\mu m^2$ - keeping in mind that this figure is only a rough estimate and that it might be somewhat smaller for ultimate MOSFET's and somewhat larger for ultimate bipolar transistors.

An elementary logic gate with a fan-out of about 4 requires the area of approximately two transistors, provided that area-saving, "super-integrated" structures are used (e.g., "stacked CMOS", $I^2L$, polysilicon resistors above active elements, etc.). With these assumptions the maximal geometrical integration density[6] is $D_m = 2.5 \cdot 10^7$ gates/$cm^2$.

It is further assumed that a multitude of wiring layers on the chip surface will be available, so that wiring constraints can be neglected. It is understood that this is a "best case" assumption and that in all practical cases some fraction of the circuits will remain nonwirable.

## 3. THERMAL LIMITATIONS

With forced air-cooling about 2W/$cm^2$ thermal power can be removed from planar chips, without exceeding the maximal critical temperature of about 100°C. With boiling liquids, about one order of magnitude more heat, namely 20 W/$cm^2$ can be removed. Recently, a new scheme was proposed and experimentally proven using forced liquid-cooling with a liquid streaming along micro-fins etched directly into the backside of chips[7,8]. This scheme will allow to remove up to 1000 W/$cm^2$, a substantial improvement over the present state-of-the-art, which could boost the performance of future computers significantly.

## 4. INTERDEPENDENCE BETWEEN INTEGRATION-DENSITY, POWER, AND HEAT-REMOVAL RATE

For a specific circuit-family and specific geometrical groundrules usually some trade-offs between performance and power dissipation are possible: In a certain range the switching delay can be decreased with increased power (and vice-versa). This behaviour can be characterised conveniently with a figure of merit, the so-called power-delay-product (synonomous with "speed-power-product"). For state-of-the-art circuits this figure is in the order of $10^{-12}$ Wsec, while for ultimately miniaturised circuits it may become of the order of $5 \cdot 10^{-16}$ Wsec[9,10].

While this power-delay-product is a rather useful figure of merit for individual circuits, it is - unfortunately - less useful for the characterisation of logic networks comprising a multitude of circuits. With increasing integration levels (number of circuits per chip) the properties of the wiring determine more and more the performance of the network, while the switching properties of the individual circuits become of lesser and lesser importance[2,15].

For moderate integration-levels it has become customary to modify the power-delay-product by adding the wiring capacities to the device capacities, resulting in an overall increased (less favorable) power-delay-figure[11].

An example for a rather conservative state-of-the-art set of assumptions should illustrate this approach: Let us assume an integration-density of $D = 5 \cdot 10^3$ gates/cm², a power dissipation of $Q = 1$ Watt/cm² and an uncorrected power-delay-product of $K = 5 \cdot 10^{-13}$ Wsec. Let us further assume that the added wiring capacities will degrade the power-delay-product by a factor of 4, resulting in a corrected power-delay-product of $K_c = 2 \cdot 10^{-12}$ Wsec. Inserting these numbers, we find for the corresponding loaded delay:

$$t_d = \frac{K_c \cdot D}{Q} = 10 \text{ nsec} \qquad (1)$$

which is a reasonable result.

Such a procedure will produce good approximation as long as the power-dissipation (due to the switching action of the digital network) can be mainly treated as the charging and discharging of capacitancies through resistive lines. In other words, as long as a "short line"-RC-approach is applicable. Such an approach is appropriate as long as the pulse length is large compared with the wire length, this means large compared with the chip dimensions. The pulse length, on the other hand, depends on the internal transistor switching time $t_i$. In general, the shortest possible pulse requires several time elements $t_i$, say 5. Consequently, an RC line has the length of such a pulse if its time-constant $t_{RC} = 2.3 \cdot RC$ equals $5t_i$. Assuming state-of-the-art groundrules and switching times, the pulse lengths - according to the above given approximation - are in the order of 10 cm.

This means, present day chips are small compared with the pulse lengths and the usual RC calculations are, therefore, meaningful and will yield useful approximations.

5. THE "LONG LINE"-APPROACH

This situation will change, however, if the miniaturisation and the integration density progresses: Let us consider an "ultimate technology" with ultra-miniaturised fast devices and ultra-dense chips. For 0.1 μm groundrules and $t_i = 100$ psec the same approximation as before leads to pulse lengths in the order of 0.03 cm. This is considerably smaller than the chip dimensions (which will certainly even increase over the presently usual ones), and, therefore a "short-line"(RC)-approach yields grossly misleading and unacceptable (too pessimistic) results[2]. This holds especially for the power-delay-product - founded essentially on RC-models - which will loose its usefulness as a figure of merit for the characterisation of such large and ultra-dense logic networks.

## 6. THE TRANSMISSION-LINE MODEL

To overcome this problem, it is advisable to restart with approximations "from the other end", so to say. This means, to consider the wiring as the essential part of the network and to add the switching elements as small corrections later on. In this approach, the wiring is considered as being composed of "long lines" and the signal propagation on these long lines is treated with the transmission-line theory[12,13]. Let us assume that on hypothetical ultra-dense chips all long signal-lines are materialized in upper metallization layers with nearly ideal transmission-line properties, while only the short lines, especially the intra-circuit-connections, are realized within lower-level, especially first level metal-lines. (Please remember that we assume that a multitude of wiring layers is available on the chip surface, see paragraph 2).

The signal-propagation-time on the short-lines is relatively short, inspite of the fact that the signal-propagation-velocity is low ($10^9$ cm/sec), due to the so-called "slow-wave-mode" operation[12]. The signal-propagation-time on the long signal-lines is longer, since these lines are much longer, inspite of the fact that the signals travel in the fast TEM-mode with about $10^{10}$ cm/sec.

We will assume a synchronous operation of the network, for which the so-called "Wallmark-synchronisation-condition"[13] applies, i.e. all pulses at every fan-in point of the network must be in phase within one tenth of the cycle time $t_c$ ($t_c = 1/f_c$, with $f_c$ being the clock frequency of the system). Lastly, we will assume an effective chipsize of 1 cm², with a longest signal-line of 2 cm, e.g. from the lower-left corner to the upper-right corner of the chip. With these assumptions we can approximate[12] the upper limit for a synchronous operation to be $f_{max} = 5 \cdot 10^8$ Hz.

In order to support such an operation the internal switching times of the transistors should not be longer than 1/5 of $1/f_{max}$. (This is equivalent to the assumption made earlier for the approximation of the pulse lengths). Thus $t_i$ should be shorter than 0.4 nsec, which is certainly in the realm of the ultra-miniaturised transistors assumed for this study. Therefore, in this approach and with these assumptions it is justified to neglect the switching properties of the devices all together, since the performance of the network will be determined essentially by the wiring and not by the devices.

## 7. THE POWER-DELAY RELATION FOR "ULTIMATE" CHIPS

It was shown earlier for the "short-line"(RC)-model that the performance increases with increased "driving", i.e. increased power dissipation, see paragraph 4. It is reasonable to assume that - up to $f_{max}$ - such a relation exists also in the transmission-line model. As already stated, however, the power-delay-product as described in paragraph 4 is no longer applicable. A "global approach" derived by R.W. Keyes[3], seems to be more appropriate for this case of ultimately miniaturised gates on ultimately dense chips. In this approach the maximum power per gate P, consistent with a maximum power density Q, is given by

$$P = (\Lambda V^2)^2 \cdot \frac{f_c^2}{Q} \qquad (2)$$

where $\Lambda = 10^{-10}$ F/cm is an entity related to the dielectric constant of the free space, V is the voltage-swing and $f_c$ the clock frequency. Unfortunately, the power per gate according to (2) is only a rough estimate due to a fairly large uncertainty of $\Lambda$ and a strong dependency from V (4th power) so that small deviations in the assumption of the minimum voltage-swing have great effects. While R.W. Keyes'[3] assumption for $\Lambda$ seems reasonable, his value for V = 1V is rather high. A more appropriate value should be in the order of V = 0.4 V[9,14,15].

Note, that - given $1/f_c = t_c$ - this relation can be transformed in one similar to the power-delay-product:

$$P \cdot t_c^2 = \frac{(\Lambda V^2)^2}{Q} \qquad (3)$$

but now with $P \cdot t_c^2$ being constant and $\frac{(\Lambda V^2)^2}{Q}$ being the "figure of merit".

Power-delay-trade offs can be accomplished by varying the impedance of the circuits and networks in a similar manner as customary with the "constant-power-delay-approach". The performance, however, increases not linearly with increasing power - as in the case of the validity of a constant power-delay-product - but slower, namely with $f_c$ proportional to $\sqrt{p}$.

Notwithstanding R.W. Keyes' derivation of this relationship[3], there is some plausibility for such a less than linear increase of the performance as a function of power: In order to drive harder one needs lower impedencies, these are only achievable with larger cross-sections and therefore lower integration densities, this means larger capacities, which - in turn - means even more current (power) for charging and discharging. This extra power does not contribute to the performance improvement. Overall a $f_c$ proportional to $\sqrt{P}$ relations looks plausible. Basically responsible for this less favorable tradeoff is the "wrong" scaling of the resistivities, i.e. the fact that the overall resistivities increases with decreasing dimensions[12,16].

## 8. INTERDEPENDENCY OF THE VARIOUS LIMITATIONS

Figure 1 shows a diagram similar to the usual power-delay-representation with the following details:

- The performance is marked on the horizontal axis, with increasing clock frequency, i.e. decreasing switching delay towards the left.
- The power dissipation is marked on the vertical axis, low power in the lower part of the diagram and high power in the upper part.
- State-of-the-art power-performance figures are in the right-upper corner of the diagram with improvements pointing towards the left-lower corner.
- Assuming a maximal power-dissipation-density of - say - $Q = 20\ \mathrm{W/cm^2}$ (as discussed in paragraph 3) the maximal - thermally limited - integration density is

$$D_{th} = Q/P \quad [\mathrm{Gates/cm^2}]$$

if P is the power-dissipation per gate, as delineated on the vertical axis. Therefore, the thermally limited maximal integration density $D_{th}$ can be marked on the same axis, with higher densities pointing down and lower densities pointing up. The product of two figures

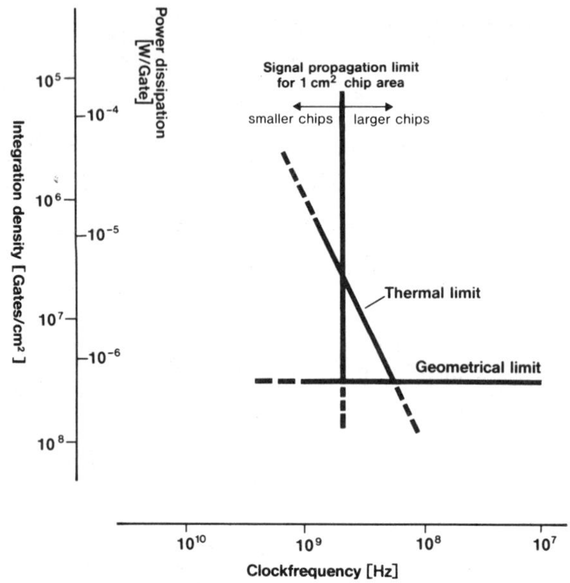


Fig. 1. Power-Delay-Density Diagram for $Q = 20\ W/cm^2$

(each one from the two different scales, but taken from the same point of the axis) is then always Q. (In the case chosen for Figure 1 : $Q = 20\ W/cm^2$).

- By lowering P (the power dissipation per gate), the thermally limited integration density increases up to $D_{max} = 2.5 \cdot 10^7$ Gates/cm² the geometrical limited integration density, which is a solid limit as described in paragraph 2. This limit is independent of the performance and therefore represented by a horizontal straight line in the power-delay-density-diagram of Figure 1.
- If we assume, as we did in paragraph 6, that the performance-limit is essentially determined by the signal-propagation-time, then this limit is independent of the power-dissipation and the integration density, but dependent on the length of the longest

line, therefore the chip size. In the power-delay-density diagram this limit is a straight vertical line, at $5 \cdot 10^8$ Hz for an 1 cm² chip (as described in paragraph 6). This line moves to the left (higher frequencies) for smaller chips and to the right (lower frequencies) for larger chips.

- Assuming again a maximum power density of Q = 20 W/cm², then the thermally limited power-delay-relation, as given by equation (3), which is of the form $P_G/f_c^2$ = const., is represented in the double logarithmic power-delay-density diagram of Figure 1 by a straight line with an inclination of 116.5°. (A constant power-delay-product would be represented by a 135°-line).

Consequently, a 1 cm² chip with 20 W power-dissipation is limited in various regions by different effects. Let us start this discussion with low integration densities going towards higher densities:

- At low integration densities and high power-dissipation per gate the performance is limited by signal-propagation delays. (This holds, however, not for very low integration densities, where the standard RC-models are more appropriate and the limitations are best described by the standard power-delay-product.
- In a medium density and medium power-dissipation per gate range the power-performance limit is governed by equation (2), with a $P_G/f_c^2$ = const. relation.
- At $2.5 \cdot 10^7$ gates/chip the geometrical density limit is reached forming a solid limit regardless of other parameters.

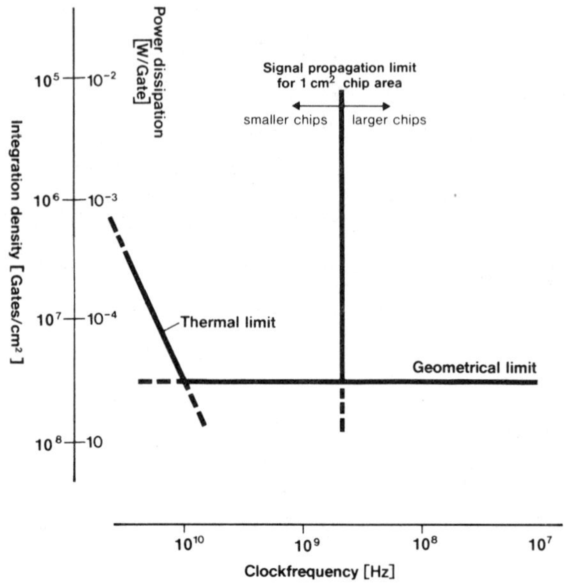


Fig. 2. Power-Delay-Density for Q = 1000 $W/cm^2$

Further progress in cooling techniques (namely forced liquid cooling, as described in paragraph 3) may shift the maximal power-dissipation-density to - say - 1000 $W/cm^2$. Thereby the thermal limit shifts to the left in the diagram and - for the same integration densities - the maximum power-dissipation per gate is higher, this means that the two scales of the vertical axis have to be shifted with regard to each other, accordingly (Figure 2). Therefore, geometrical limitations and - predominantly - propagation delay limitations will become even more stringent for future VLSI logic.

## 9. SYSTEM ASPECTS

As discussed above, the signal-propagation-time will become the most prominent limiting factor for ultra-miniaturised digital VLSI-chips. Therefore, a clever partitioning of digital systems will become even more significant than today: The network should be devided into units ("islands") as small as possible, with many of those located on large chips. Synchronous operations should be restricted to intraisland traffic, while a loose coupling should govern the interisland traffic, which should be spread as much as possible time- and space-wise (multiplexing). As smaller the "islands" are, as higher clock frequencies are feasible since all arguments in the previous paragraphs pertaining to "chip-size" are equally valid for "island-size" if the "system-chip" is partitioned and operated as described. Ordered structures with short wires should be favored over the "spagetti" approach of across-the-chip (island) wiring. This holds especially for memory functions of the one or the other kind, which will be incorporated into the chips. These memory functions have to be matched performance-wise with the logic. Extensive use of the hierarchy-techniques is required to achieve such a match, otherwise the overall system-performance will suffer. Furthermore, the various digital functions should be arranged into groups, which can be switched off whenever their operation is not required, thereby reducing the overall power dissipation. Also, off-chip driving should be reduced as much as possible since drivers dissipate more power than regular circuits, aggravating thereby the cooling problem.

## 10. FINAL REMARKS

The "signal-propagation-time"-limit is a very universal one, being not only a fundamental limit for the room-temperature Silicon

Technology, but for any other technology using electrical and optical signals. Replacing Silicon with other materials and reducing the temperature will not help to move or to revome this limitation. This does not mean, however, that faster devices than those one which can be made out of Silicon are useless. It only means that for the here described application in ultra-dense digital VLSI not much can be gained by replacing Silicon. On the other hand in cases where thermal limitations are predominant low temperature operations offer quite some relief. The non-linearities of the I-V-characteristics become more pronounced with lower temperatures. Therefore the operating voltages, and - in turn - the power dissipation can be reduced. This holds especially and prominently for the Josephson-Technology.

REFERENCES

1. O.G. Folberth, "Die Physik der Großintegration," Phys. Bl. 37 pp. 83-90, (1981).

2. O.G. Folberth, "The Interdependence of Geometrical, Thermal and Electrical Limitations for VLSI Logic," IEEE Journ. Solid-State Circ., vol. SC-16, pp. 51-53, (1981).

3. R.W. Keyes, "Physical limits in semiconductor electronics," Science, vol. 195, pp. 1230-1235, (Mar. 1977).

4. J.T. Wallmark, "A statistical model for determining the minimum size in integrated circuits," IEEE Trans. Electron Devices, vol. ED-26, pp. 135-142, (Feb. 1979).

5. B. Hoeneisen and C.A. Mead, "Fundamental limitations in microelectronics," Solid-State Electron., vol. 15, pp. 819-829, 891-897, (1972).

6. O.G. Folberth and J.H. Bleher, "The fundamental limitations of digital semiconductor technology," Microelectron. J., vol. 9, no. 4, pp. 33-41, (1979).

7. D.B. Tuckerman and R.F.W. Pease, "High-Performance Heat Sinking for VLSI," IEEE El. Dev. Letters, vol. EDL-2, pp. 126-129, (1981).

8. R.W. Keyes, "High Performance Cooling and Large Scale Integration," this volume.

9. K.N. Ratnakumar, J.D. Meindl, and D.J. Bartelink, "Performance limits of E/D NMOS VLSI," in Dig. IEEE Int. Solid-State Circuits Conf., vol. 23, pp. 72-73, (1980).

10. O.G. Folberth, "The ultimate speed-power product of logic transistor circuits," Proc. IEEE, vol. 54, p. 1125, (1966).

11. P.W. Cook, D.L. Critchlow, and L.M. Terman, "Comparison of MOSFET logic circuits," IEEE Journ. Solid-State Circ., vol. SC-8, pp. 348-355, (Oct. 1973).

12. O.G. Folberth, "Signalfortpflanzung in integrierte Schaltungen," Int. Elektr. Rundschau, vol. 28, pp. 9-12, 29-31, (1974).

13. J.T. Wallmark, "Fundamental physical limitations in integrated electronic circuits," Inst. Phys. Conf. Ser., vol. 25, pp. 133-167, (1975).

14. F. Faggin, "How VLSI impacts computer architecture," IEEE Spectrum, pp. 28-31, (May 1978).

15. J.D. Meindl, K.N. Ratnakumar, L. Gerzberg, and K.C. Sarawat, "Circuit Scaling Limits for Ultra-Large-Scale Integration," in Dig. IEEE Int. Solid-State Circuits Conf., vol. 24, pp. 36-37, (1981).

16. R.W. Keyes, "Physical problems and limits in computer logic," IEEE Spectrum, pp. 36-45, (May 1969).

# ELECTRON BEAM MICROCIRCUIT INSPECTION TECHNIQUES

Erich Kubalek and Eberhard Menzel

Werkstoffe der Elektrotechnik
Universität Duisburg
D-4100 Duisburg, West-Germany

## INTRODUCTION

The progressive miniaturization of integrated circuit (IC) components more and more demands micron scaled, highly sensitive and possibly nondestructive "material-microcharacterization techniques" for the recognition of the cause for the malfunction of microcircuit devices, thus improving the device fabrication process. Furthermore, the growing package density and complexity of ICs causing more difficulties in function control and failure analysis of large scale integration (LSI) and very large scale integration (VLSI) microcircuits calls for an improvement of existing and additionally the development of new testing techniques, especially enabling the direct investigation of individual IC parts and with that a localization of a possible malfunction.

To fulfil these demands a new generation of instruments and a new array of techniques is required in order to guarantee the quality and reliability of microelectronic devices, a task which becomes more important with increasing intelligence and responsibility of these devices.

For testing and microcharacterization of microcircuits the most successful techniques are those based on the interaction of an electron beam (e-beam) probe with the IC. These e-beam inspection techniques in a unique combination offer the possibility of both testing the functioning and localizing a possible malfunction by different e-beam testing techniques and additionally in many cases of investigating the cause of this malfunctioning by evaluating a great variety of material parameters by means of different e-beam microcharacterization techniques.

Topographic Structures :

- Scanning Electron Microscopy (SEM)
- Transmission Electron Microscopy (TEM)
- Scanning Transmission Electron Microscopy (STEM)

Chemical Structure :

- X-Ray Microanalysis
- Auger Microanalysis
- Cathodoluminescence (CL)
- Electron Loss Spectroscopy (ELS)

Cristalline Structure :

- Transmission High Energy Electron Diffraction (THEED)
- Reflection High Energy Electron Diffraction (RHEED)
- Low Energy Electron Diffraction (LEED)
- Electron Channeling Pattern (ECP)

Electronic Structure :

- Electron Beam Induced Current (EBIC)
- Electron Beam Induced Voltage (EBIV)
- Cathodoluminescence (CL)

Magnetic Structure :

- Magnetic Contrast
- Lorentz-Microscopy

Elastic Structure :

- Electron Acoustic Microscopy (EAM)

Fig. 1. Survey of the most important e-beam microcharacterization techniques

As the greater part of the e-beam microcharacterization techniques - a survey of the most important techniques is listed in Fig. 1 - are well-known conventional techniques this paper is restricted to the modern e-beam testing techniques [1 - 22], which have received a tremendous impetus in the past few years, because in combination with the classical testing methods they enable new test strategies for electronic devices.

The e-beam testing techniques use the attractive properties of an e-beam probe lying in its capability of being used as a finely focussed, easily aligneable contactless probe, which can be operated in a "passive" mode - the so called Voltage Measuring Mode - as well as in an "active" mode, the Current Impressing Mode.

In the "Voltage Measuring Mode" the e-beam acts as a nondestructive, nonloading, easily aligneable micron scaled "contact" enabling the evaluation of time dependent voltages on free selectable IC nodes under real IC operation conditions in the sampling mode. Thus the e-beam represents the only alternative to the mechanical probe used so far for electrical signal measurements within medium scale integration (MSI) and LSI-microcircuits by completely avoiding its disadvantages. Moreover the e-beam can be scanned over an IC area and in the stroboscopic mode enables the evaluation of voltage distributions at this IC area for a fixed IC operating state.

Besides the Voltage Measuring Mode, where the e-beam is required to have as little influence on the device operation as possible, the e-beam can also take over an "active" role in IC-testing. In the "Current Impressing Mode" the e-beam serves as a micron scaled current source taking advantage of absorbed currents flowing into the irradiated area [19, 23, 24], or of carrier multiplication in thin insulating films [25, 26] or in pn-junctions [27] providing volatile switching operations. By means of specially designed e-beam sensors on the chip logical states within a device can be changed allowing e.g. the IC operation mode (normal-selftesting) to be e-beam controlled [19, 23, 24] or IC subsystems to be connected to a common data bus [27].

The full capability of e-beam testing is obtained when the e-beam is operated in a combination of these two modes and when special test aids on the chip are provided [19, 23, 24]. This offers the possibility for on wafer tests without the necessity for applying wafer probes [19, 23, 24] or would allow waferscale integrated systems on which defective subsystems could be easily localized and the remaining subsystems be interconnected by the "Restructuring Mode" of the e-beam probe [27]. In this latter operation mode the e-beam can be used to open or close switches in the subsystems in a nonvalatile manner by utilizing e-beam induced irradiation damage in MOS devices or e-beam induced charges in floating gate field effect transistors [27].

The already existing importance of e-beam testing will increase further, because its field of application will enlarge by the additional use in other parts of microelectronics and by the use of other e-beam modes. Thus e-beam testing has been already sucessfully used for contactless testing of three dimensional conductor networks imbedded in a dielectric matrix as employed in complex multichip packages [28, 29] and for example the "Cathodoluminescence Mode" is discussed for testing of integrated optics devices.

In the following the capabilities and the limitations of the most important e-beam testing techniques are treated. The fundamentals and the apparative details are discussed elsewhere [14, 15, 17, 18, 21, 22].

## VOLTAGE MEASURING MODE

The different currently known electron beam test techniques using the electron probe as a microcontact can be divided into two main types, the qualitative and the quantitative techniques (Fig.2). Some of these different techniques use a continuous e-beam during measuring, whereas for time resolved measurements a chopped e-beam is necessary.

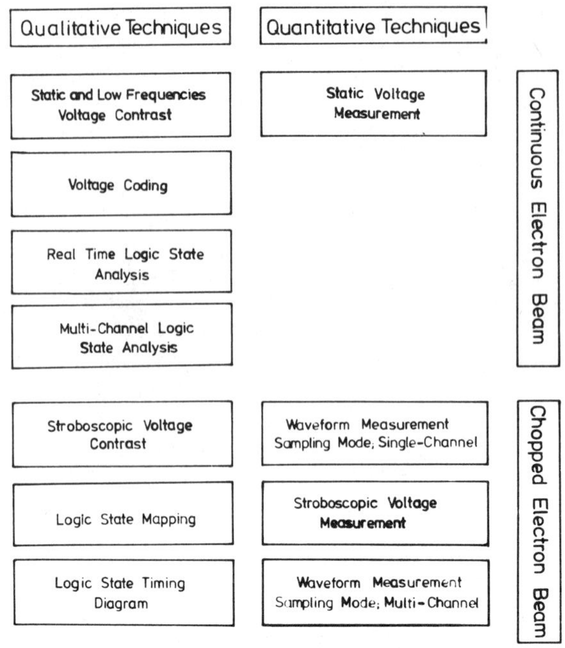


Fig. 2. Survey of the different currently known electron beam test techniques based on the "Voltage Measuring Mode"

The first type techniques are only usable for a coarse evaluation of the applied voltages. But this on the one hand already enables an IC-internal logic state analysis and in this way a localization of soft- and hardware failures [6, 17, 20, 21, 30-33]. The e-beam logic state analysis techniques like Voltage Coding (VC), Logic State Mapping (LSM), Logic State Timing Diagram (LSTD) and Multichannel Logic State Analysis (MLSA) require repetitive signals, e.g. short program loops whereas signals with low repetition rate or nonrecurring signals are only detected by the Real Time Logic State Analysis (RTLSA)-Technique.

On the other hand the coarse evaluation of the IC internal voltages enables only a coarse function control and failure analysis by aid of the Static and Low Frequencies and the Stroboscopic Voltage Contrast [7, 11, 16, 34-42].

The second type techniques allow exact quantitative voltage measurements in ICs. The voltage information then is displayed in different ways as

(i) a one or two dimensional static voltage distribution (Static Voltage Measurement) [8]

(ii) a two-dimensional voltage distribution (micrograph) in the stroboscopic mode showing an IC area at a fixed operating state (Stroboscopic Voltage Measurement) [8]

(iii) a waveform in the sampling mode (Waveform Measurement; Sampling Mode; Single and Multi-Channel). Operating in this quantitative voltage measuring mode the e-beam provides informations on rise and fall times, amplitudes and delays of IC-internal signals [7, 9, 11-15, 18, 36, 43] and in this way allows the control of IC-internal signal timing, failure localization and classification [13] and even the control of computer simulation program [44, 45].

In the following some of these techniques mostly used in practice are discussed in detail.

## Qualitative Techniques

All qualitative techniques are based on the phenomenon known as voltage contrast. This contrast appears superimposed to the topography and material contrast in a Scanning Electron Microscope (SEM) and is caused by the local electric fields above the operating ICs. Those parts of the device to which a positive voltage is applied like for example in regions at logic level "1" show up dark and grounded or negatively biased parts - for example regions at logic level "0" - show up bright in a voltage contrast micrograph. The reason for this contrast is the existence of retarding electrical fields above positively biased parts of the IC, which influence the amount of the collected secondary electrons making the SE-signal highly sensitive to the electrical fields on the surface of an operating microcircuit [22]. The voltage contrast is unsuitable for the exact quantitative measurements of voltages because the correlation between the SE-signal and the electrical field on the IC-surface is non definite (the local retarding fields depend on the geometry of the IC surface and on the geometry of the SEM specimen chamber) and is additionally affected by surface topography and by different materials. Nevertheless a coarse function control and failure analysis of ICs and IC-internal Logic State Analysis are already possible.

### Static, Low Frequencies and Stroboscopic Voltage Contrast

The Static and Low Frequencies Contrast Techniques are the simplest techniques and can be carried out in any SEM if a socket for the IC, electrical leads in the vacuum chamber, electrical vacuum feedthroughs and an IC drive unit are provided and the e-beam is scanned over the IC. The IC drive unit consists of an IC-power supply in the Static Voltage Contrast Technique and addition-

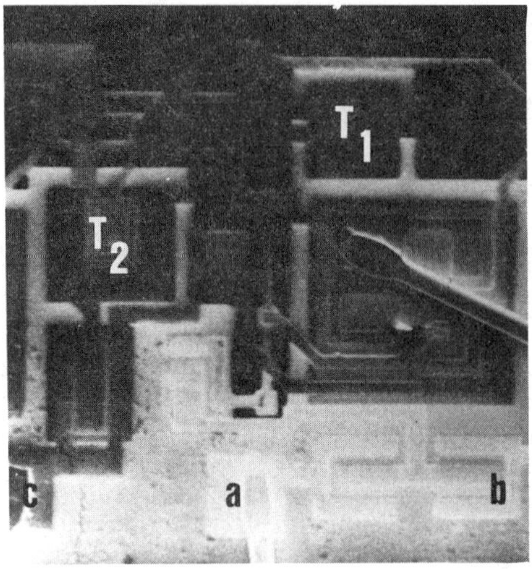


Fig. 3a. Stroboscopic micrograph of a TTL-NAND gate (delay time 20 ns); passivated

Fig. 3b. TTL-NAND gate (delay time 50 ns); passivated

ally of a clock or function generator in the Low Frequencies Voltage Contrast Technique. These techniques give informations about the distribution of IC-voltages (logical states of digital devices; voltage drops at resistors in analog devices; shorts and opens in both kinds of devices) and the operating of ICs at very low frequencies respectively. As the highest change of states frequency the human eye can follow is only a few Hertz, this Low Frequencies Voltage Contrast Technique is restricted to those devices that are functioning at these low frequencies. Testing ICs at high frequencies necessitates the use of a Stroboscopic Voltage Contrast technique employing a chopped e-beam. The principle of this stroboscopic method is to generate e-beam pulses that impinge on the IC with the same repetition rate as the periodical signal on the IC and at a certain fixed phase position of the IC-signal. Variable delay times between the e-beam pulse and the IC signal allow in this way the visualization of different IC states, thus analyzing the slew rates of signals and possible propagation delays. As an example Fig. 3a and 3b show two operational states of a NAND-gate (TTL; 5 Volts swing) operating at 350 kHz. To demonstrate the propagation of the slew rate through the circuit it was necessary to use electron beam pulses of 10 ns duration. The time delay of the beam pulse with respect to the input signal was changed from 20 ns (Fig. 3a) to 50 ns (Fig. 3b). Notice the input (a, b) and output (c) connections and the two transistors $T_1$ and $T_2$. At 20 ns delay transistor $T_1$ is nearly conducting and the other blocking. At 50 ns delay or 30 ns later the operation is reversed. As the generation of e-beam pulses shorter than 1 ps is possible [39] even microwave devices such as gunn diodes may be investigated besides the more commonly used bipolar and MOS devices.

These techniques may be applied universally to testing bipolar and MOS devices which show voltage swings of several volts and are nondestructive and non loading if the e-beam is operated with low electron irradiation doses and low e-beam energies (see Fig. 8; energy of about 700 eV for this aluminum material)[46]. Both non passivated and passivated bipolar devices can be investigated, the latter because the primary electrons cause an e-beam induced conductivity in the energy dissipation volume [25] forming a conducting channel that shorts the conducting line with the irradiated surface [46, 47]. When this procedure is used with MOS ICs irradiation damage occurs. Therefore at present a nondestructive testing of unpassivated MOS devices is possible only by reducing the primary electron energy such that the primary electrons cannot reach the sensitive gate oxides [48]. These test techniques can be carried out at passivated ICs, too [49] using the so called capacitive coupling voltage contrast [50].

The obtainable spatial resolution is dependent on the chosen e-beam current or in the stroboscopic technique on the peak beam current in the pulse, on the chosen primary electron energy, on the lens aberrations and on the chopping degradation [22]. Therefore a nondestructive and non loading device test requiring low primary electron energies and currents is only possible with a consequent loss in spatial resolution, which under these conditions is in the range of 1 to 3 microns. Because of this spatial resolution loss an optimization of the electron optics for low primary electron energies is necessary if VLSI-ICs are to be investigated.

The voltage resolution is strongly dependent on the field distribution at the IC and between the IC and the detectors, but usually is in the range of 1 to 2 Volts (in several cases voltage differences of down to 50 mV have been observed). Concerning ICs with low voltage swings of less than 1 Volt such as ECL (Emitter Coupled Logic) or $I^2L$ (Integrated Injection Logic) these techniques are unsuitable, because the local fields on these devices are not high enough to cause a visible voltage contrast.

Electron Beam Logic State Analysis Techniques [21]

These techniques are based on the qualitative voltage contrast, too, and allow the display of IC-internal logical states in unpassivated and passivated ICs. All the different techniques except the Real Time Logic State Analysis require repetitive signals and do not allow real time analysis, because the e-beam probe is either scanned across the IC and/or is chopped, so that interference pulses, missing or additional bits are only detected by chance.

Fig. 4. Voltage Coding Micrograph of a MOS circuit showing different bit patterns of three neighbouring lines

Signals with low repetition rates or non recurring signals can only be detected by the RTLSA-Technique [20].

In the Voltage Coding Technique [6] a continuous electron beam scans the surface of an IC which is operated under real conditions from an IC drive unit. This drive unit may consist of DC power supplies, function and word generators, complete microprocessor programming units, or a test computer. The qualitative voltage contrast on the IC nodes modulates the brightness of a monitor. If the time varying IC-signal, e.g. the clock signal or a short program loop, is synchronized to the frame or line frequency of the monitor the time dependence of logic states is transformed into a spatial distribution of states on the monitor screen. Horizontal or vertical steady state stripes are generated when the scan generator triggers a short program loop on the IC or triggers the IC clock signal via a frequency multiplier. This technique allows the observation of logical states on the IC, where the logical "1" appears as dark and "0" as bright contrast, enabling the test of program loops on IC internal data lines. Fig. 4 shows an example. Different logic signals are applied to three neighbouring metal lines, in this case 16 bit words, the logic "1" appearing dark, equivalent to a plus 10 Volts signal, and the logic "0" appearing bright (digital MOS devices; metal line width 8 $\mu$m; 1.2 keV primary electron energy; $10^{-8}$A electron beam current). The performance data and requirements of the VC technique are listet in comparison to the

| | voltage resolution | $I_{PE}$ | time resolution | IC frequency | information | e-beam deflection | e-beam state | periodic events required | data acquisition time | data evaluation |
|---|---|---|---|---|---|---|---|---|---|---|
| Voltage Coding | ≈1 V | $10^{-8}$A | 500 ns | < 2 MHz | micrograph | frame scan | conti-nuous | short program loop | 50 ms ... 100 s | operator |
| Logic State Mapping | ≈0.4 V | $10^{-11}$A | ns .... ps is depending on e-beam blanking systems | MHz ... GHz | micrograph | line scan | pulsed, delay shifted with frame | short program loop | 50 clock cycles : typ. 60 s | operator |
| Logic State Timing Diagram | ≈0.4 V | $10^{-11}$A | ns .... ps is depending on e-beam blanking systems | MHz ... GHz | electrical signal | vector scan | pulsed, delay shifted | short program loop | 10 channels, 50 clock cycles: ≈1 s | logic state analyzer, computer |
| Real Time Logic State Analysis | 0.7 V<br>2.5 V | $10^{-7}$A | 120 ns<br>50 ns | 4 MHz<br>9 MHz | electrical signal | vector scan | continuous | no | clock rate 4 MHz, 4096 clock cycles : ≲1 ms | logic state analyzer, computer |
| Multichannel Logic State Analysis | 0.7 V<br>2.5 V | $10^{-7}$A | 120 ns<br>50 ns | 4 MHz<br>9 MHz | electrical signal | vector scan | continuous | repetiti-ve pro-gram | 8 channels, 4 MHz, 4096 clock cycles : ≲12 ms | logic state analyzer, computer |

Fig. 5. Performance data and requirements of IC-internal logic state analysis techniques

other techniques in Fig. 5. The voltage resolution, which in the logic state analysis technique is the ability to distinguish between "0" and "1" of an IC signal, is about 1 V at a primary electron (PE) beam current of $10^{-8}$A and a time resolution of about 500 ns. The maximum reasonable signal repetition rate is determined by the spatial resolution of the monitor in the TV mode and is 2 MHz at best, corresponding to about 100 stripes per line. With decreasing line and frame frequencies the maximum observable signal frequency reduces, too [6]. The necessary periodic events may be short program loops or a multiple of the IC clock. The data acquisition time is the time for recording a micrograph and is typically 50 ms ... 100 s. The data interpretation is done by the operator.

The performance of the VC-technique was greatly improved in the LSM-technique by combining the VC-technique with the Sampling Technique [30, 33], where the e-beam-scanning across a line perpendicular to the IC conductor lines - is pulsed with the same repetition rate as the repetitive signal on the operating IC and the phase between e-beam pulses and the program loops on the IC is varied synchronously to the frame scan. A further reduction of the total data acquisition time was then achieved by application of a vector scan and a logic state analyzer in the LSTD technique, where the e-beam now jumps from one conductor line to another instead of scanning across a line as in the LSM-technique. At last the data acquisition time was reduced further in the MLSA-technique which is based on the RTLSA-technique.

Comparing the performance data of the different e-beam logic state analysis techniques requiring repetitive signals show that the LSTD-techniques and the MLSA-technique are the most efficient. Whereas the MLSA-technique allows an extremely fast detection of logic states on a number of freely programmable IC internal modes, the LSTD-technique uses a very low primary beam current and thus a better spatial resolution is achieved.

As on many ICs, especially microprocessors, a large number of signals are nonrecurring and program loops cannot be kept very short always, a Real Time Logic State Analysis (RTLSA) technique was developed [20, 21], which is able to measure nonrecurring signals, does not require repetitive signals, is independent of the program length and which allows short data acquisition times. In this technique a continuous e-beam probe is positioned to the node of interest, where it detects the time dependent IC signal in a real time manner. This signal is fed to a logic state analyzer providing an additional channel. In contrast to the other e-beam logic state analysis techniques the secondary electrons are analyzed in this technique in a retarding field spectrometer, allowing a proper control of the secondary electrons and thus rendering the voltage contrast definite. Measurements, which were in agreement with calculation [51] showed that with this technique an internal logic

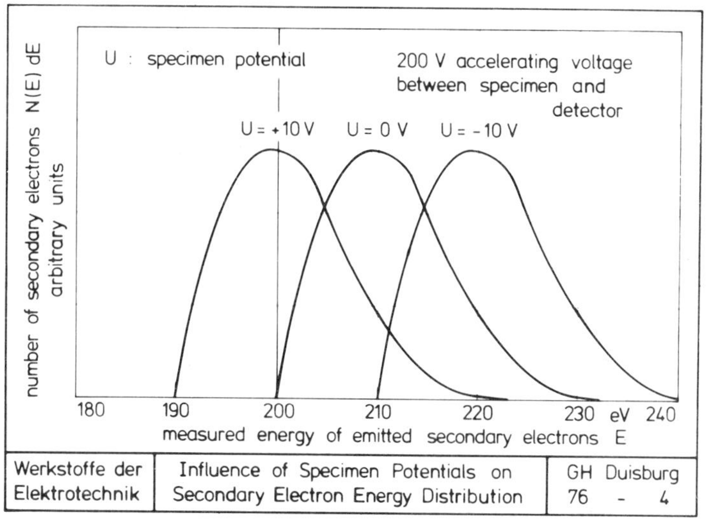


Fig. 6. Energy shift of secondary electron spectrum with applied specimen voltage (schematically)

analysis is possible with 0.7 Volts voltage resolution ($I^2L$, ECL) up to frequencies of 4 MHz and 2.5 Volts voltage resolution (TTL, MOS) up to frequencies of 9 MHz both with a spatial resolution of 3 µm ($10^{-7}$A primary beam current; 700 eV primary beam energy).

## Quantitative techniques

The basis of all quantitative techniques is the fact that all secondary electrons of the secondary electron energy spectrum emitted from an electrode at a certain potential gain the same kinetic energy towards the detector. In this way the whole energy spectrum is linearly shifted with applied specimen voltage (Fig.6). The linear shift of the maximum of the spectrum is a direct measure of the applied specimen voltage. Obviously this can only be true if the shape of the SE spectrum is in no way influenced as for example is the case if retarding fields as mentioned above are present on an IC surface. From these considerations the two main preconditions for a quantitative measurement are:

(i) Suppression of local fields above the IC surface

(ii) Energy analysis of the SE.

If the voltage measurements have to be carried out on different materials or if quantitative voltage distribution images of the IC surface are to be made, additionally the change of the SE-spectrum due to topography and material has to be taken into account and then the third precondition has to be:

(iii) Elimination of topography, material contrast and surface condition (variations of work function).

Numerous experimental set ups have been constructed to fulfil these preconditions and to make the voltage contrast technique a quantitative and reliable measuring method (see ref. in 18). Some of these fulfil all above mentioned preconditions and result in quantitative techniques for imaging IC areas like Static and Stroboscopic Voltage Measurement Technique. In these technique the voltage information is isolated from the total signal by a substraction of the secondary electron signal bearing voltage information from that of the same measuring point with no voltage information. This subtraction is realized by different techniques (see Ref. in 18 or 22) requiring highly sophisticated setups. A really complete elimination of these influences, however, has not been achieved so far. In practice quantitative voltage spot measurements measuring the voltages for fixed local points on the IC as a function of time are much more important than quantitative voltage micrographs. Therefore this drawback is not a real problem as for spot measurements a complete voltage isolation is not necessary. This is true, because the spot measurements are carried out on known material, such as IC aluminum lines, so that the material contrast can be neglected. Additionally the influence of the topography contrast due to elevations and voids on the aluminum surface can be drastically reduced by scanning along a line during the measurement and thus eliminating the topography effect to a negligible extent. Finally, the influence of different surface conditions can be reduced by using oil free vacuum systems [52] and by measuring on cleaned surfaces [18]. In this way the necessary equipment can be reduced to a minimum still allowing quantitative voltage spot measurements to be performed on a single spot (Waveform Measurement, Sampling Mode, Single Channel) or on several points in the IC (Waveform Measurement, Sampling Mode, Multi-Channel).

Waveform Measurement - Sampling Mode (Single Channel)

With this technique which is the most important quantitative practical method, the electron beam measures signal waveforms on metal lines in ICs. The electron probe is easily positioned to the interesting node on the IC and the voltage measurement is done with a setup, which is working similar to a sampling oscilloscope. In this method the periodical signal is sampled at certain phase positions of the signal on the IC. Integrating over a number of samples and subsequent linear change of the phase position yields the waveform.

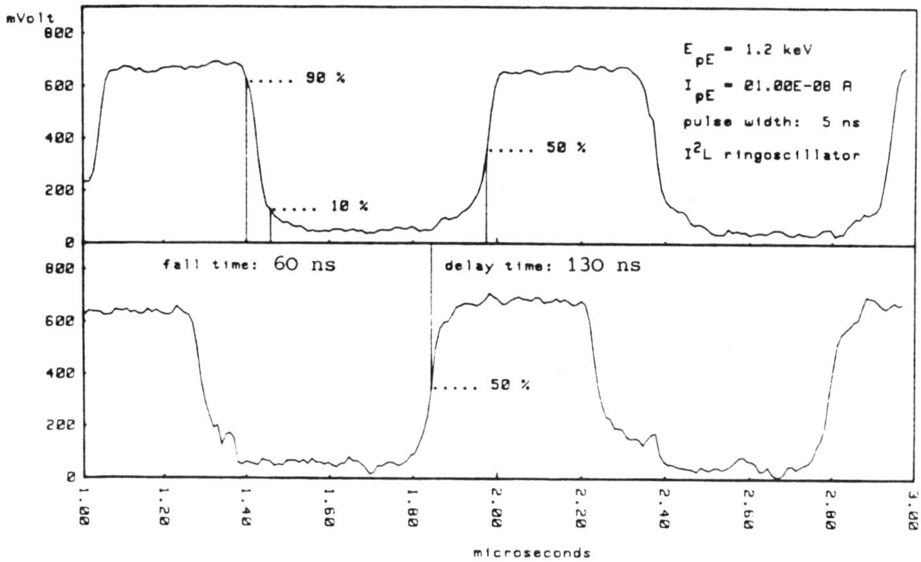


Fig. 7. Quantitative voltage measurements on two nodes of an $I^2L$-ringoscillator. The voltage variation superimposed on the square wave signal is noise. The common time reference in this case was established by using the output signal of the ringoscillator as a trigger signal for the generation of e-beam pulses.

As an example for the performance of this technique the measurement of two waveforms on an $I^2L$ ringoscillator are presented in Fig. 7. The fall time has been measured to about 60 ns and the delay time between the signals to about 130 ns. The voltage swing of this device is about 600 mV.

With this technique all internal nodes of an unpassivated IC are accessable and waveforms my be traced. Thus signal amplitudes, rise and fall times, and propagation delays of signals may be analyzed and computer simulation programs may be controlled. The attained resolutions of this technique are nowaday about 20 ps, 1 $\mu$m and 1 mV simultaneously at 1.2 keV and $10^{-8}$A and fulfill the demand of VLSI devices, at least the time and voltage resolution.

For measurements on passivated ICs the same statements already given are true.

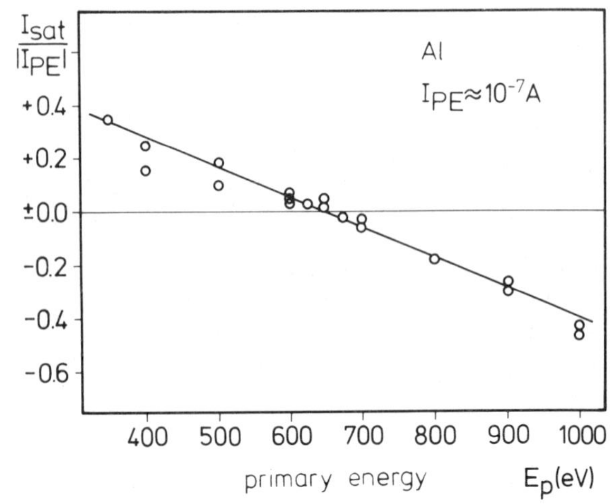


Fig. 8. Dependence of impressed current $I_{sat}$ on primary electron energy

CURRENT IMPRESSING MODE

Currents impressed to the IC nodes by the e-beam may alter the IC operation momentarily if the beam current is of the same order as the IC switching currents. This effect my be taken advantage of, thus converting the role of the e-beam from a passive to an active one. The effect of impressed currents offers the possibility of using the e-beam as a contactless probe that applies input signals to the device or subsystem under test. Due to the dependence of the secondary electron emission coefficient on the primary electron energy the current flowing into the irradiated spot also is energy dependent. The behaviour of this current ($I_{sat}$) is shown in Fig. 8 for an aluminum specimen. Positive (left), negative (right) or zero currents (centre) may be impressed by proper choice of the PE energy. A primary electron energy of 700 eV appears to be the best choice for having a nonloading measuring probe on this material. If PE energies of larger than 1 keV are chosen, about half the primary beam current is impressed. By a special design of the IC circuitry it is possible to use the beam as a noncontacting "switch". This has been realized by Mucha and coworkers and our group [23, 24] by initiating a built-in selftest on a 1 kbit ROM device. The ROM was designed to be functionally convertable from the normal mode (Fig. 9 left) to a testing mode

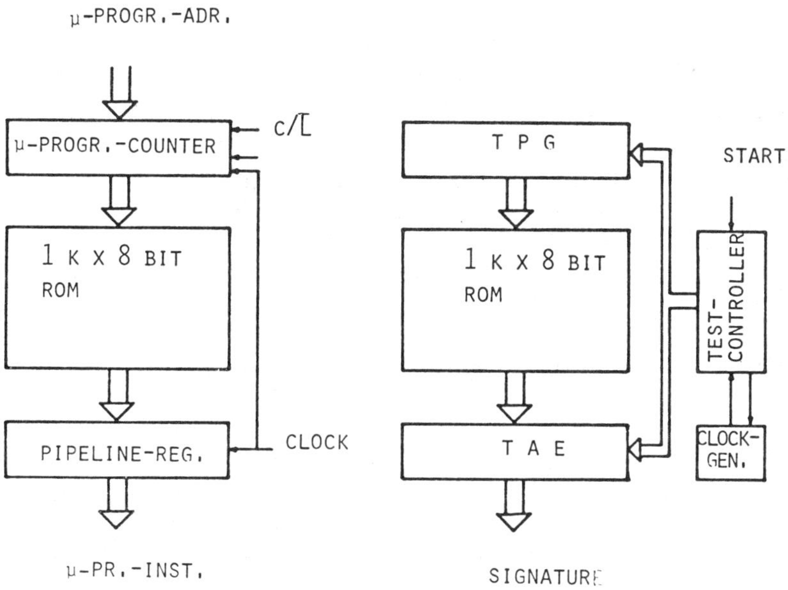


Fig. 9. Selftesting ROM device [23, 24]

(right), where the microprogram counter is converted to a test pattern generator and the pipeline register to a test answer evaluator. The test controller is initialized by an electron beam pulse of some 100/us compensating a built-in current source of 50 nA. With the 1024 pulses of the built-in clock the self-test is carried out. Fig. 10 shows the two different e-beam operation modes: on the left the impressing mode where the beam current compensates the internal current source. Then the beam is deflected toward the clock circuitry where it measures the clock pulses. The two top curves show the x- and y-beam deflection signals; the x-deflection being zero and the y-deflection was about 300/um. After the 1024 pulses of the clock the compressed test data may be read out at the signature register by the beam in the Voltage Measuring Mode. As a consequence of using the electron as a contact as well as a current source there exists the possibility of testing each chip on the wafer without wafer probers provided that a wafer interconnection assures that each chip is connected to the supply voltage. This is already realized [23, 24].

## CONCLUSIONS

Microcircuit inspection of LSI and VLSI devices would be unthinkable without the e-beam inspection techniques. This is because these techniques on the one hand enable testing of microcircuits

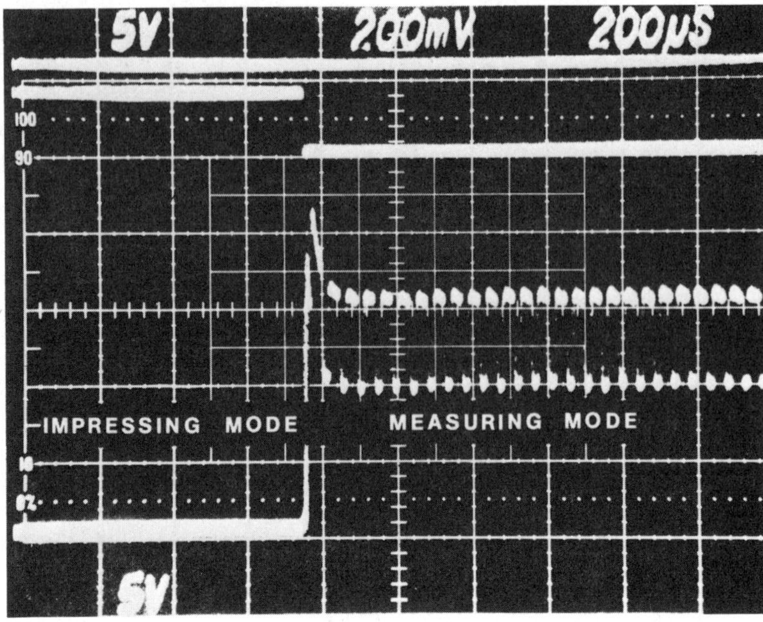


Fig. 10. Electron beam operating modes on the selftesting ROM device

in a nondestructive manner and thus localizing malfunctions by means of different e-beam testing techniques and on the other hand evaluating a great variety of material parameters thus investigating the cause of this malfunction by means of various e-beam micro-characterization techniques. This paper presents in an overview the conventional, well known e-beam microcharacterization techniques and discusses the capabilities and limitations of the modern e-beam testing techniques in detail.

ACKNOWLEDGEMENT

Thanks are due to the federal ministry of research and technology (BMFT) of Germany and to the ministry of science and research (MWF) of North-Rhine-Westfalia for financial support of parts of the work presented here.

REFERENCES

1. O.C. Wells, C.G. Bremer: Voltage Measurement in the Scanning Electron Microscope, J. Phys. E.: Sci. Instr. Series 2, 1: 902 (1968)

2. J.P. Fleming, E.W. Ward: A Technique for Accurate Measurement and Display of Applied Potent Distributions Using the SEM, Scanning Electron Microsc.: 465 (1970)
3. G.S. Plows: Stroboscopic Scanning Electron Microscopy and the Observation of Microcircuit Surface Voltage, Ph.D. Thesis Cambridge University Library, Cambridge, Great Britain (1969)
4. A. Gopinath, C.C. Sanger: A Technique for the Linearization of Voltage Contrast in The Scanning Electron Microscope, J.Phys. E.: Sci. Instr., 4:334 (1971)
5. J.M. Hannah: Scanning Electron Microscope Applications to Integrated Circuits Testing, Ph.D. Thesis, University of Edinburgh Library, Edinburgh, Scotland (1974)
6. G.V. Lukianoff, T.R. Touw: Voltage Coding : Temporal Versus Spatial Frequencies, Scanning Electron Microsc.: 465 (1975)
7. A. J. Gonzales, M.W. Powell: Internal Waveform Measurements of the MOS Three Transistor, Dynamic RAM Using SEM Stroboscopic Techniques, Technical Digest of IEDM, IEEE, New York, USA: 119 (1975)
8. L.J. Balk, H.P. Feuerbaum, E. Kubalek, E. Menzel: Quantitative Voltage Contrast at High Frequencies in the SEM, Scanning Electron Microsc.:615 (1976)
9. P.R. Thomas, K.G. Gopinathan, A. Gopinath, A.R. Owens: The Observation of Fast Voltage Waveforms in the SEM Using Sampling Techniques, Scanning Electron Microsc., I:609 (1976)
10. V. Dyukov, M. Kolemeytsev, S. Nepijko: A Simple Technique of Potential Distribution Mapping in a Scanning Electron Microscope Micr. Acta 80, 5:376 (1978)
11. H. Fujioka, T. Hosokawa, Y. Kanda, K. Ura: Submicron Electron Beam Probe to Measure Signal Waveform at Arbitarily Specified Positions on MHzIC, Scanning Electron Microsc.: 755 (1978)
12. H.P. Feuerbaum, D. Kantz, E. Kubalek: Quantitative Measurement with High Time Resolution of Internal Waveforms on MOS-RAMs Using a Modified Scanning Electron Microscope, IEEE, SC-13: 319 (1978)
13. P. Fazekas, H. Lindner, R. Lindner, J. Otto, E. Wolfgang: On-Wafer Defect Classification of LSI-Circuits Using a Modified SEM, Scanning Electron Microsc. I:801 (1978)
14. H.P. Feuerbaum: VLSI Testing Using the Electron Probe, Scanning Electron Microsc. I:285 (1979)
15. E. Wolfgang, R. Lindner, P. Fazekas, H.P. Feuerbaum: Electron-Beam Testing of VLSI Circuits, IEEE J. Sol. State Circuits, SC-14, (2):471 (1979)
16. L. Kotorman: Non-Charging Electron Beam Pulse Prober on FET Wafers, Scanning Electron Microsc. IV:77 (1980)
17. G. Crichton, P. Fazekas, E. Wolfgang: Electron Beam Testing of Microprocessors, IEEE Test Conference: 444 (1980)
18. E. Menzel, E. Kubalek: Electron Beam Test Techniques for Integrated Circuits, Scanning Electron Microsc., I:305 (1981)

19. E. Kubalek, E. Menzel: Microcircuit Inspection Techniques Using an Electron Beam Probe, Microcircuit Engineering "Proceedings of the International Conference on Microlithography", Lausanne: 395(1981)
20. M. Ostrow, E. Menzel, E. Kubalek: Real Time Logic State Analysis of IC-Internal Signals with an Electron Beam-Probe-Logic State Analyzer Combination, Microcircuit Engineering "Proceedings of the International Conference on Microlithography", Lausanne: 514 (1981)
21. M.Ostrow. E. Menzel, E. Postulka, S. Görlich, E. Kubalek: IC-Internal Electron Beam Logic State Analysis, Scanning Electron Microsc.: to be published (1982)
22. E. Menzel, E. Kubalek: Fundamentals of Electron Beam Testing of Integrated Circuits, Scanning, Int. J.of SEM and Related Methods: to be published (1982)
23. B. Könemann, J. Mucha, G. Zwiehoff: Built-In Test for Complex Digital Integrated Circuits, IEEE J. Solid-State Circuits, SC-15, (3):315 (1980)
24. U. Theus, H. Leutiger: Gleichzeitiger Selbsttest an der unzersägten Si-Scheibe, NTG Fachberichte "Großintegration" 77: 89 (1981)
25. L. Pensak: Conductivity Induced by Electron Bombardment in Thin Insulating Films, Phys. Rev. 75:472 (1949)
26. D.M. Taylor: The Effect of Passivation on the Observation of Voltage Contrast in the Scanning Electron Microscope, J. Phys. D.Appl., 11:2443 (1978)
27. D.C. Shaver: Electron Beam Testing and Restructuring of Integrated Circuits, Proceedings OF THE Second Caltech Conference on VLSI, Pasadene, Ca., USA:(Jan. 1981)
28. T.H.P. Chang, F.J. Hohn, P.J. Coane, D.P. Kern: Electron Beam Testing of Packaging Modules for VLSI Chip Arrays, 16th Symp. on Electron Ion and Photon Beam Technology, Dallas, Texas: (May 1981)
29. H.C. Pfeiffer, G.O. Langner, W. Stickel, R.A. Scinpson: Contactless Electrical Testing of Large Area Specimens Using Electron Beams, 16th Symp. on Electron Ion an Photon Beam Technology, Dallas Texas:(May 1981)
30. H.P. Feuerbaum: Funktionsprüfung von hochintegrierten MOS-Schaltungen mit der Elektronensonde, Ph.D. Thesis, University, Duisburg : (1980)
31. G. Crichton, P. Fazekas, J. Otto, E. Wolfang: Interne Prüfung von Mikroprozessoren mit der Elektronensonde, NTG Fachberichte "Großintegration", 77:99 (1981)
32. E. Wolfgang, P. Fazekas, J. Otto, G. Crichton: Electron Beam Testing of Microprocessors in "Hardware and Software Concepts in VLSI",North-Holland: to be published (1982)
33. P. Fazekas, H.P. Feuerbaum, E. Wolfgang: Scanning Electron Beam Probes VLSI Chips, Electronics 14:105 (1981

34. G.Y. Robinson: Stroboscopic Scanning Electron Microscopy at Gigahertz Frequencies, Rev. Sci. Instr., 42 (2):251 (1971)
35. A. Gopinath, M.S. Hill: A Technique for the Study of Gunn Devices at 9.1 GHz Using a Scanning Electron Microscope, IEEE Trans. Electron. Dev. ED-20:610 (1973)
36. A. Gopinath, M.S. Hill: Some Aspects of the Stroboscopic Mode: A Review, Scanning Electron Microsc.:235 (1974)
37. A. Gopinath, M.S. Hill: Deflection Beam-Chopping in the SEM, J. Phys. E.: Sci. Instr. 10:229 (1977)
38. T. Hosokawa, H. Fujioka, K. Ura: Two-Dimensional Observation of Gunn Domains at 1 GHz by Picosecond Pulse Stroboscopic SEM, Appl. Phys. Let. 31:340 (1977)
39. T. Hosokawa, H. Fujioka, K. Ura: Generation and Measurement of Subpicosecond Electron Beam Pulses, Rev. Sci. Instr., 49: 624 (1978)
40. T. Hosokawa, J. Fujioka, K. Ura: Gigahertz Stroboscopy with the Scanning Electron Microscope, Rev. Sci. Instr. 49:1293 (1978)
41 H.P. Feuerbaum, J. Otto: Beam Chopper for Subnanosecond Pulses in Scanning Electron Microscopy, J.Phys. E.: Sci. Instr. 11: 529 (1978)
42. E. Menzel, E. Kubalek: Electron Beam Chopping Systems in the SEM, Scanning Electron Microsc. I:305 (1979)
43. E. Menzel: Elektronenstrahltestsystem für die Funktionskontrolle und Fehleranalyse höchstintegrierter Schaltkreise, Ph.D. Thesis, University Duisburg, Duisburg:(1981)
44. H.P. Feuerbaum,K. Hernaut:Application of Electron Beam Measuring Techniques for Verification of Computer Simulations for Large-Scale Integrated Circuits, Scanning Electron Microsc.I: 795 (1978)
45. B. Höfflinger, H. Sibbert, G. Zimmer, E. Kubalek, E. Menzel: Model and Performance of Hot-Electron MOS Transistors for High-Speed Low Power LSI, IEEE/IEDM Technical Digest:463 (1978)
46. E. Wolfgang: Electron Beam Testing : Problems in Practice, Scanning, Int. J. of SEM and Related Methods : to be published (1982)
47. J. Fujioka, K. Nakamae, K. Ura: Function Testing of Bipolar ICs and LSIs with the Stroboscopic Scanning Electron Microscope, IEEE J. Solid State Cir. 15:177 (1980)
48. K. Nakamae, H. Fujioka, K. Ura: Measurement of Deep Penetration of Low-Energy Electrons into Metal-Oxide-Semiconductor Structure, Journ. Appl. Phys. 52 (3):1306 (1981)
49. S. Görlich, E. Menzel, E. Kubalek: Capacitive Coupling Voltage Contrast, Beitr. Elektronenmikroskop.Direktabb.Oberfl.15:133 (1982)
50. D.L. Crostwait, F.W. Ivy: Voltage Contrast Methods for Semiconductor Divice Failure Analysis, Scanning Electron Microsc. IV:935 (1979)

51. A. Gopinath: Estimate of Minimum Measurable Voltage in the SEM, J. Phys. E.: Sci. Inst., Vol. 10:911 (1977)
52. E. Menzel, E. Kubalek: Electron Beam Test System for VLSI Circuit Inspection, Scanning Electron Microsc., I:297 (1979

# INDEX